212
Advances in Polymer Science

Editorial Board:
A. Abe · A.-C. Albertsson · R. Duncan · K. Dušek · W. H. de Jeu
H.-H. Kausch · S. Kobayashi · K.-S. Lee · L. Leibler · T. E. Long
I. Manners · M. Möller · O. Nuyken · E. M. Terentjev
B. Voit · G. Wegner · U. Wiesner

Advances in Polymer Science
Recently Published and Forthcoming Volumes

Self-Assembled Nanomaterials II
Volume Editor: Shimizu, T.
Vol. 220, 2008

Self-Assembled Nanomaterials I
Volume Editor: Shimizu, T.
Vol. 219, 2008

Interfacial Processes and Molecular Aggregation of Surfactants
Volume Editor: Narayanan, R.
Vol. 218, 2008

New Frontiers in Polymer Synthesis
Volume Editor: Kobayashi, S.
Vol. 217, 2008

Polymers for Fuel Cells II
Volume Editor: Scherer, G. G.
Vol. 216, 2008

Polymers for Fuel Cells I
Volume Editor: Scherer, G. G.
Vol. 215, 2008

Photoresponsive Polymers II
Volume Editors: Lee, K. S., Marder, S.
Vol. 214, 2008

Photoresponsive Polymers I
Volume Editors: Lee, K. S., Marder, S.
Vol. 213, 2008

Polyfluorenes
Volume Editors: Scherf, U., Neher, D.
Vol. 212, 2008

Chromatography for Sustainable Polymeric Materials
Renewable, Degradable and Recyclable
Volume Editors: Albertsson, A.-C., Hakkarainen, M.
Vol. 211, 2008

Wax Crystal Control · Nanocomposites
Stimuli-Responsive Polymers
Vol. 210, 2008

Functional Materials and Biomaterials
Vol. 209, 2007

Phase-Separated Interpenetrating Polymer Networks
Authors: Lipatov, Y. S., Alekseeva, T.
Vol. 208, 2007

Hydrogen Bonded Polymers
Volume Editor: Binder, W.
Vol. 207, 2007

Oligomers · Polymer Composites
Molecular Imprinting
Vol. 206, 2007

Polysaccharides II
Volume Editor: Klemm, D.
Vol. 205, 2006

Neodymium Based Ziegler Catalysts – Fundamental Chemistry
Volume Editor: Nuyken, O.
Vol. 204, 2006

Polymers for Regenerative Medicine
Volume Editor: Werner, C.
Vol. 203, 2006

Peptide Hybrid Polymers
Volume Editors: Klok, H.-A., Schlaad, H.
Vol. 202, 2006

Supramolecular Polymers
Polymeric Betains · Oligomers
Vol. 201, 2006

Ordered Polymeric Nanostructures at Surfaces
Volume Editor: Vancso, G. J., Reiter, G.
Vol. 200, 2006

Polyfluorenes

Volume Editors: Ullrich Scherf · Dieter Neher

With contributions by

K. Becker · N. Blouin · P.-L. T. Boudreault · S.-A. Chen · S. H. Chen
E. Da Como · F. Dias · A. C. Grimsdale · A. B. Holmes · C.-W. Huang
W. Huang · S. Kappaun · S. King · M. Knaapila · M. Leclerc
E. J. W. List · S.-J. Liu · H.-H. Lu · J. M. Lupton · B.-X. Mi
A. Monkman · K. Müllen · C. Rothe · C. Slugovc · J. U. Wallace
M. J. Winokur · W. W. H. Wong · Q. Zhao

Springer

The series *Advances in Polymer Science* presents critical reviews of the present and future trends in polymer and biopolymer science including chemistry, physical chemistry, physics and material science. It is adressed to all scientists at universities and in industry who wish to keep abreast of advances in the topics covered.

As a rule, contributions are specially commissioned. The editors and publishers will, however, always be pleased to receive suggestions and supplementary information. Papers are accepted for *Advances in Polymer Science* in English.

In references *Advances in Polymer Science* is abbreviated *Adv Polym Sci* and is cited as a journal.

Springer WWW home page: springer.com
Visit the APS content at springerlink.com

ISBN 978-3-540-68733-7   e-ISBN 978-3-540-68734-4
DOI 10.1007/978-3-540-68734-4

Advances in Polymer Science ISSN 0065-3195

Library of Congress Control Number: 2008931713

© 2008 Springer-Verlag Berlin Heidelberg

This work is subject to copyright. All rights are reserved, whether the whole or part of the material is concerned, specifically the rights of translation, reprinting, reuse of illustrations, recitation, broadcasting, reproduction on microfilm or in any other way, and storage in data banks. Duplication of this publication or parts thereof is permitted only under the provisions of the German Copyright Law of September 9, 1965, in its current version, and permission for use must always be obtained from Springer. Violations are liable to prosecution under the German Copyright Law.

The use of general descriptive names, registered names, trademarks, etc. in this publication does not imply, even in the absence of a specific statement, that such names are exempt from the relevant protective laws and regulations and therefore free for general use.

Cover design: WMXDesign GmbH, Heidelberg
Typesetting and Production: le-tex publishing services oHG, Leipzig

Printed on acid-free paper

9 8 7 6 5 4 3 2 1 0

springer.com

Volume Editors

Prof. Dr. Ullrich Scherf
Institute for Polymer Technology
Macromolecular Chemistry Group
Bergische Universität Wuppertal
Gauss-Str. 20
42097 Wuppertal, Germany
scherf@uni-wuppertal.de

Prof. Dr. Dieter Neher
Institut für Physik und Astronomie
Universität Potsdam
Karl-Liebknecht.Str. 24-25
14476 Potsdam-Golm, Germany
neher@uni-potsdam.de

Editorial Board

Prof. Akihiro Abe
Department of Industrial Chemistry
Tokyo Institute of Polytechnics
1583 Iiyama, Atsugi-shi 243-02, Japan
aabe@chem.t-kougei.ac.jp

Prof. A.-C. Albertsson
Department of Polymer Technology
The Royal Institute of Technology
10044 Stockholm, Sweden
aila@polymer.kth.se

Prof. Ruth Duncan
Welsh School of Pharmacy
Cardiff University
Redwood Building
King Edward VII Avenue
Cardiff CF 10 3XF, UK
DuncanR@cf.ac.uk

Prof. Karel Dušek
Institute of Macromolecular Chemistry,
Czech
Academy of Sciences of the Czech Republic
Heyrovský Sq. 2
16206 Prague 6, Czech Republic
dusek@imc.cas.cz

Prof. Dr. Wim H. de Jeu
Polymer Science and Engineering
University of Massachusetts
120 Governors Drive
Amherst MA 01003, USA
dejeu@mail.pse.umass.edu

Prof. Hans-Henning Kausch
Ecole Polytechnique Fédérale de Lausanne
Science de Base
Station 6
1015 Lausanne, Switzerland
kausch.cully@bluewin.ch

Prof. Shiro Kobayashi
R & D Center for Bio-based Materials
Kyoto Institute of Technology
Matsugasaki, Sakyo-ku
Kyoto 606-8585, Japan
kobayash@kit.ac.jp

Prof. Kwang-Sup Lee
Department of Polymer Science &
Engineering
Hannam University
133 Ojung-Dong
Daejeon 306-791, Korea
kslee@hannam.ac.kr

Prof. L. Leibler
Matière Molle et Chimie
Ecole Supérieure de Physique
et Chimie Industrielles (ESPCI)
10 rue Vauquelin
75231 Paris Cedex 05, France
ludwik.leibler@espci.fr

Prof. Timothy E. Long
Department of Chemistry
and Research Institute
Virginia Tech
2110 Hahn Hall (0344)
Blacksburg, VA 24061, USA
telong@vt.edu

Prof. Ian Manners
School of Chemistry
University of Bristol
Cantock's Close
BS8 1TS Bristol, UK
ian.manners@bristol.ac.uk

Prof. Martin Möller
Deutsches Wollforschungsinstitut
an der RWTH Aachen e.V.
Pauwelsstraße 8
52056 Aachen, Germany
moeller@dwi.rwth-aachen.de

Prof. Oskar Nuyken
Lehrstuhl für Makromolekulare Stoffe
TU München
Lichtenbergstr. 4
85747 Garching, Germany
oskar.nuyken@ch.tum.de

Prof. E. M. Terentjev
Cavendish Laboratory
Madingley Road
Cambridge CB 3 OHE, UK
emt1000@cam.ac.uk

Prof. Brigitte Voit
Institut für Polymerforschung Dresden
Hohe Straße 6
01069 Dresden, Germany
voit@ipfdd.de

Prof. Gerhard Wegner
Max-Planck-Institut
für Polymerforschung
Ackermannweg 10
55128 Mainz, Germany
wegner@mpip-mainz.mpg.de

Prof. Ulrich Wiesner
Materials Science & Engineering
Cornell University
329 Bard Hall
Ithaca, NY 14853, USA
ubw1@cornell.edu

Advances in Polymer Science
Also Available Electronically

For all customers who have a standing order to Advances in Polymer Science, we offer the electronic version via SpringerLink free of charge. Please contact your librarian who can receive a password or free access to the full articles by registering at:

springerlink.com

If you do not have a subscription, you can still view the tables of contents of the volumes and the abstract of each article by going to the SpringerLink Homepage, clicking on "Browse by Online Libraries", then "Chemical Sciences", and finally choose Advances in Polymer Science.

You will find information about the

– Editorial Board
– Aims and Scope
– Instructions for Authors
– Sample Contribution

at springer.com using the search function.

Color figures are published in full color within the electronic version on SpringerLink.

Preface

With this collection of short review papers we would like to present a broad overview of research on polyfluorenes and related heteroanalogues over the last two decades. The collection begins with papers on the synthesis of polyfluorenes and related polyheteroarenes, then reports photophysical properties of this class of conjugated polymers both at the ensemble and the single chain level, continues with a discussion of the rich solid state structures of polyfluorenes, and finally switches to device applications (e.g. in OLEDs). In addition, two chapters are devoted to defined oligofluorenes as low molecular weight model systems for polyfluorenes and also to degradation studies.

We feel that this up-to-date collection will be very helpful to all polymer chemists and physicists, and will also aid graduate students interested in this fascinating and still growing area of research, since such a compact overview is only now available. All articles are presented by leading scientists in their fields, insuring state-of-the-art coverage of all relevant aspects. Together with the body of references this volume is meant to assist researchers in the daily lab routine. Moreover, Advances in Polymer Science, as an established series of high quality review papers, represents a very appropriate platform for our project. We hope that this short collection will be of great value both for beginners and established research scientists in the field of polyfluorene research.

Wuppertal und Potsdam, April 2008   Ullrich Scherf
                                          und Dieter Neher

Contents

Bridged Polyphenylenes – from Polyfluorenes to Ladder Polymers
A. C. Grimsdale · K. Müllen 1

Polyfluorenes for Device Applications
S.-A. Chen · H.-H. Lu · C.-W. Huang 49

Poly(dibenzosilole)s
W. W. H. Wong · A. B. Holmes 85

Poly(2,7-carbazole)s and Related Polymers
P.-L. T. Boudreault · N. Blouin · M. Leclerc 99

Polyfluorenes with On-Chain Metal Centers
S.-J. Liu · Q. Zhao · B.-X. Mi · W. Huang 125

Fluorene-Based Conjugated Oligomers
for Organic Photonics and Electronics
J. U. Wallace · S. H. Chen 145

Polyfluorene Photophysics
A. Monkman · C. Rothe · S. King · F. Dias 187

Structure and Morphology of Polyfluorenes
in Solutions and the Solid State
M. Knaapila · M. J. Winokur 227

Optically Active Chemical Defects
in Polyfluorene-Type Polymers and Devices
S. Kappaun · C. Slugovc · E. J. W. List 273

Single Molecule Spectroscopy of Polyfluorenes
E. Da Como · K. Becker · J. M. Lupton 293

Subject Index 319

Bridged Polyphenylenes – from Polyfluorenes to Ladder Polymers

Andrew C. Grimsdale[1] · Klaus Müllen[2] (✉)

[1] School of Materials Science and Engineering, Nanyang Technological University, Nanyang Avenue, 639798 Singapore, Singapore

[2] Max-Planck-Institute for Polymer Research, Ackermannweg 10, 55128 Mainz, Germany
muellen@mpip-mainz.mpg.de

1	Introduction	2
2	**Polyfluorenes**	5
2.1	Poly(dialkylfluorene)s	6
2.2	Optical Properties of PDAFs	9
2.3	Colour Control by Copolymerisation	10
2.4	Defect Emission from PDAFs	15
2.5	Poly(diarylfluorene)s	18
2.6	Polyfluorenes with Improved Charge Injection	20
2.7	Polymers Containing Spirobifluorene	25
2.8	Polymers Containing Other Bridged Biphenylene Units	26
3	**Polyindenofluorenes**	29
3.1	Poly(tetraalkylindenofluorene)s	29
3.2	Defect Emission from PIFs	30
3.3	Poly(tetraarylindenofluorene)s	31
4	**Polymers of Higher Ladder-Type Oligophenylenes**	32
4.1	Poly(ladder-Type Tetraphenylene)s	32
4.2	Poly(ladder-Type Pentaphenylene)s	33
5	**Ladder-Type Polyphenylenes**	34
5.1	Synthesis of LPPPs	35
5.2	Optical Properties of LPPPs	37
5.3	Ladder-Type Polymers with 2-Carbon Bridges	40
6	**Conclusions**	42
	References	43

Abstract This chapter reviews the synthesis of polymers based on bridged phenylenes for use in electronic applications such as blue-light-emitting diodes and polymer lasers. We show how the optoelectronic properties may be tuned by varying the amount of bridging between one bridge per two phenylenes in polyfluorenes through to completely bridged ladder-type polyphenylenes, and by changing the nature of the substituents at the bridgeheads. Of particular importance is the suppression of undesirable long wavelength emission by controlling interchain interactions and choosing synthetic methods so as to hinder or prevent the formation of emissive defects. The electroluminescence efficiency of these materials can also be enhanced by the use of charge-transporting substituents. Copolymerisation with lower band-gap units enables tuning of the emission colour across the entire visible range.

Keywords Electroluminescence · Ladder polymers · Lasing · Light-emitting diodes · Photoluminescence

Abbreviations

^{t}Bu	*tert*-butyl
ECL	Effective conjugation length
EL	Electroluminescence
eV	Electron volt
HOMO	Highest occupied molecular orbital
IR	Infrared
ITO	Indium tin oxide
LEC	Light-emitting electrochemical cell
LED	Light-emitting diode
LPPP	Ladder-type poly(*para*-phenylene)
LUMO	Lowest unoccupied molecular orbital
Me	Methyl
Me-LPPP	Methyl-substituted LPPP
M_n	Number-averaged molecular weight
M_w	Weight-averaged molecular weight
NMR	Nuclear magnetic resonance
PDAF	Poly(dialkylfluorene)
PEDOT	Poly(3,4-ethylenedioxythiophene)
PEO	Poly(ethylene oxide)
PF	Polyfluorene
Ph	Phenyl
Ph-LPPP	Phenyl-substituted LPPP
PIF	Polyindenofluorene
PSS	Poly(styrene sulfonate)
PVK	Poly(*N*-vinyl carbazole)
TGA	Thermal gravimetric analysis
UV-VIS	Ultraviolet-visible

1
Introduction

Phenylene-based polymers are one of the most important classes of conjugated polymers, and have been the subject of extensive research, in particular as the active materials in light-emitting diodes (LEDs) [1, 2] and polymer lasers [3]. These materials have been of particular interest as potential blue emitters in such devices [4]. The discovery of stable blue-light emitting materials is a major goal of research into luminescent polymers [5]. Poly(*para*-phenylene) (PPP, Scheme 1, **1**) is a blue emitter [6], but it is insoluble and so films of PPP have to be prepared via precursor routes [7]. Substitution with long alkyl

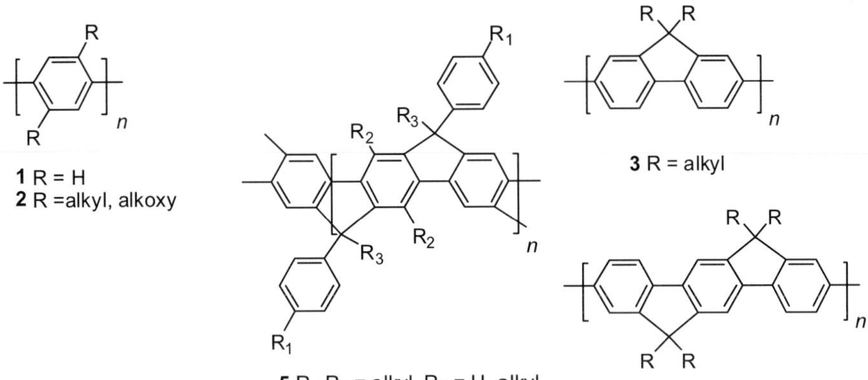

Bridged Polyphenylenes – from Polyfluorenes to Ladder Polymers

Scheme 1 Unbridged and bridged phenylene polymers

or alkoxy groups as in **2** induces solubility [8], but the steric repulsion between the sidechains causes a marked twisting of the polymer backbone [9], leading to very short conjugation lengths and a corresponding shifting of the emission into the ultraviolet. In order to obtain solubility and thus processability without undesired blue-shifting of the absorption and emission bands the substituents can be attached to short carbon bridges between some or all of the benzene rings. Some types of bridged phenylene materials that will be covered in this review are shown in Scheme 1. These range from polydialkylfluorenes (PDAFs, **3**) with a unit of two bridged benzene rings, through polymers with longer bridged units, e.g., polyindenofluorenes (PIFs, **4**) to fully bridged ladder-type polyphenylenes (LPPPs, **5**). In this introduction we present an overview of the important properties of these materials that need to be controlled in order to obtain efficient LEDs, of the general principles of how these properties may be controlled by synthetic or physical methods, and of the general synthetic methods available for the preparation of these polymers. We will then present a more detailed discussion of the synthesis and optimisation of polymer properties for each class of material.

There are several properties of luminescent materials that need to be controlled in order to make efficient LEDs and lasers. The first is the colour of the emission, which is primarily determined by the energy difference (bandgap) between the highest occupied molecular orbital (HOMO) and the lowest unoccupied molecular orbital (LUMO), but in the solid state is also affected by interactions between the molecules or polymer chains which can lead to red-shifts in the emission due to formation of aggregates. This can be controlled by manipulating both the polymer backbone and the substituents. Polyphenylenes are intrinsically blue-emitting materials with large HOMO-LUMO gaps, but as we will show, by copolymerisation with other materials it is possible to tune the emission colour across the entire visible spectrum. Even without the incorporation of comonomers it is possible to tune the emission colour over a substantial range by controlling the conjugation length through restriction of the torsion of adjacent phenylene rings. Thus, as mentioned above, substitution of PPP (1) with solubilising groups to give soluble PPPs (2) causes a blue-shift in the emission as the steric interactions between the side-chains induce increased out-of-plane twisting of the phenylene units, while the enforced coplanarity produced by the methine bridges in LPPP (5) results in a marked red-shift in the emission. The "stepladder" polymers such as PDAFs (3) and PIFs (4) produce emission spectra intermediate between those of 2 and 5. As already mentioned, solid-state interactions between polymer chains can cause red-shifts in the emission of these materials, which can be suppressed by a suitable choice of substituents. This can, however, adversely affect the charge transporting properties of the material (see below).

The second critical property to be tuned is the efficiency of the charge injection, which is determined by the energy barrier between the HOMO and the anode (for hole injection) and between the LUMO and the cathode (for electron injection), and of charge transport which is controlled by intermolecular or interchain interactions. Polyphenylenes are materials with intrinsically low-lying HOMOs (typically 5.8–6.0 eV) which create a large barrier to hole injection from the most widely used anode material, indium tin oxide (ITO), which has a work function of 4.8–5.0 eV. The LUMO values are typically around 2.2–2.5 eV, which makes electron injection from air-stable metals such as aluminium (work function 4.3 eV) difficult, thus requiring the use of more electropositive metals such as calcium (work function 2.9 eV) as cathodes. Obviously one can improve the charge injection by raising the HOMO and/or lowering the LUMO energy of the polymer, but in doing so, one reduces the size of the energy gap and so induces a red-shift in the emission colour. As a result, obtaining efficient blue emission is a particular problem.

The efficiency of devices can be increased by the incorporation of layers of charge injecting (or level-matching) materials which have energy levels intermediate between that of the emissive layer and the work function of the electrode, but the use of such layers has the disadvantages of increasing the device thickness which increases the driving voltage, and also complicates device fabrication as successive layers have to be deposited in ways such that the lower layers are not disturbed by deposition of the upper ones. Blending charge transporting materials into the emissive layer leads to problems with phase separation, giving unstable device performance. The incorporation of charge accepting units into the polymer backbone or onto the side-chains avoids both these problems. This approach has been used to successfully improve both the hole and electron accepting properties of phenylene-based polymers. Good charge transport in the solid state requires close packing of polymer chains, permitting rapid and efficient hopping of charges between chains. As mentioned above, strong inter-chain interactions can cause undesirable red-shifts in the emission bands, so that it is sometimes necessary to compromise the charge transport properties of the material to obtain the desired emission colour.

Other desirable properties are the ability to form defect-free films, preferably by solution-processing techniques, a high solid-state photoluminescence (PL) quantum efficiency, and good stability towards oxygen and light. For some applications the ability to obtain polarised light is also desirable. All of these properties are to some extent controllable by the design of the structure and the synthetic pathway. The formation of a thick, uniform, defect-free film by spin-coating or similar solution-processing methods is dependent upon many factors. The first is the molecular mass of the compound, as low molar mass materials tend to be crystalline and so do not form high-quality amorphous films, while very high molar mass polymers are difficult to dissolve. The second is the solubility. To get a good film, the material must be reasonably soluble as too dilute solutions give too thin films, and must not form aggregates in solution as these will tend to lower the film quality (uneven morphology) and may produce red-shifts in the emission. The exact PL efficiency of a given material is as yet not predictable, but the removal of fluorescence quenching defects such as halide atoms or carbonyl groups, and the suppression of inter-chain interactions leading to non-radiative decay pathways are known to assist in improving the solid state quantum efficiency of materials. The stability of a polymer towards photo-oxidation can be improved by avoiding susceptible functional groups, e.g., benzyl protons. Polarised emission is obtained by alignment of the polymer chains, which seems to be easiest for polymers which possess a liquid crystalline mesophase [10]. Circularly polarised emission has been obtained by using chiral side-chains.

There exist two general methods for the synthesis of bridged phenylene-based materials: (a) oxidative coupling of monomers, which is of strictly limited synthetic utility as only low molecular-mass materials are obtained from such methods; and (b) transition mediated polycondensations of substituted aromatic compounds and/or aryl organometallic compounds. The latter is the main method for preparing bridged phenylene-based polymers. The two main polycondensation methods used are the Suzuki polycondensation of arylhalides with arylboronic acids or esters [11], and the Yamamoto polymerisation of aryl dihalides using nickel(0) reagents [12]. Generally speaking the Suzuki method gives higher molecular masses than the Yamamoto procedure, but is synthetically more demanding. A more detailed comparison of the relative merits of the two methods will be given in the discussion of the synthesis of PDAFs (Sect. 2.1 below).

2
Polyfluorenes

Polyfluorenes (PFs), the simplest regular stepladder-type polyphenylenes, in which only every second ring is bridged, have been much studied in recent years due to their large PL quantum efficiencies and excellent chemical and thermal stability, as evidenced by the number of recent reviews [4, 13–15]. A further attractive feature of poly(9,9-dialkylfluorene)s (PDAFs) is the ready synthetic accessability of the monomers as the alkylation and halogenation of fluorene proceed smoothly and in high yields.

2.1
Poly(dialkylfluorene)s

The acidity of the bridgehead (C9) protons in fluorene makes the alkylation at this position extremely facile, with 9,9-dialkylfluorenes **6** being obtainable in high yield by treatment of fluorene **7** with an alkyl halide and base, usually under phase transfer catalysis conditions (Scheme 2). Some care is required to keep oxygen away as otherwise oxidation of the fluorenyl anion intermediate **8** leads to formation of fluorenone **9** as a byproduct. Another problem with this reaction is that the product may contain traces of the monoalkylfluorene **10**, which are difficult to detect and remove, and which as will be discussed in Sect. 2.3 can cause problems with the emission colour stability. Bromination of a dialkylfluorene **6** efficiently produces a 2,7-dibromo-9,9-dialkylfluorene **11**, which can also be made in a high yield by alklylation of 2,7-dibromofluorene. Both **6** and **11** can be used as monomers to make PDAFs.

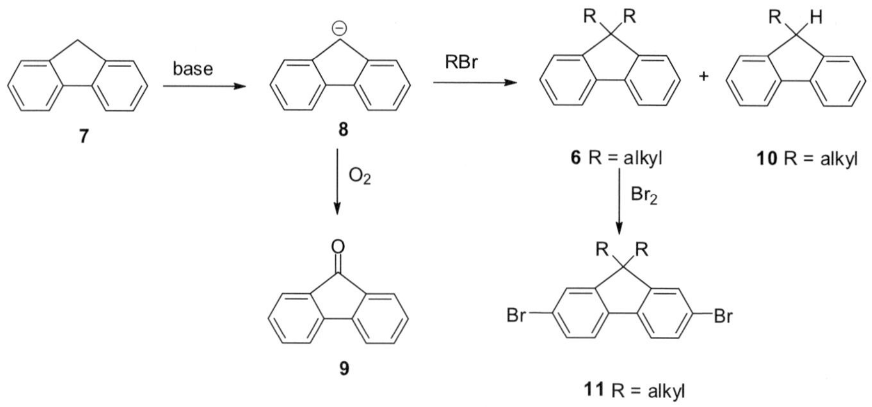

Scheme 2 Alkylation of fluorene to produce monomers for PDAFs

This ready accessability of dialkylfluorene monomers is one of the major reasons for the extensive investigation of poly(dialkylfluorene)s. Three main methods have been used for the synthesis of the polymers. PDAFs can be made by oxidative coupling of the monomer **6** with iron(III) chloride. Poly(9,9-dihexyl-2,7-fluorene) (**12**) with a molecular mass of 5000 (n = ca. 20) was made in this way by Yoshino and co-workers (Scheme 3) [16, 17], and used to make low-efficiency blue-emitting (λ_{max} = 470 nm) devices [18–20].

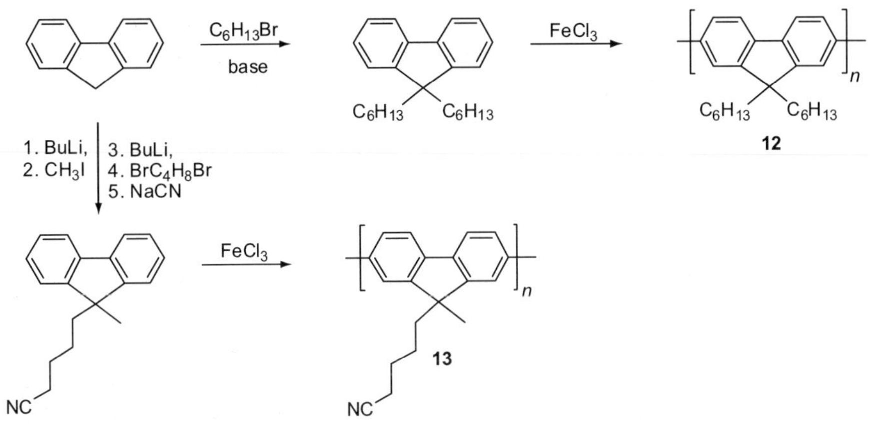

Scheme 3 Synthesis of polyfluorenes by oxidative polymerisation

These were the first blue-emitting LEDs reported using a phenylene-based polymer. The disadvantages of this method are that the degree of polymerisation is low, and the high level of defects produced due to coupling other than at the 2 and 7 positions. As a result this method is generally not used to make PDAFs. A recent exception is the synthesis of the polymer 13 with a cyanoalkyl substituent (Scheme 4) [21]. Here transition-metal mediated coupling methods such as Suzuki or Yamamoto polycondensation could not be used as the nitrile deactivates the metal catalysts by binding to the metal.

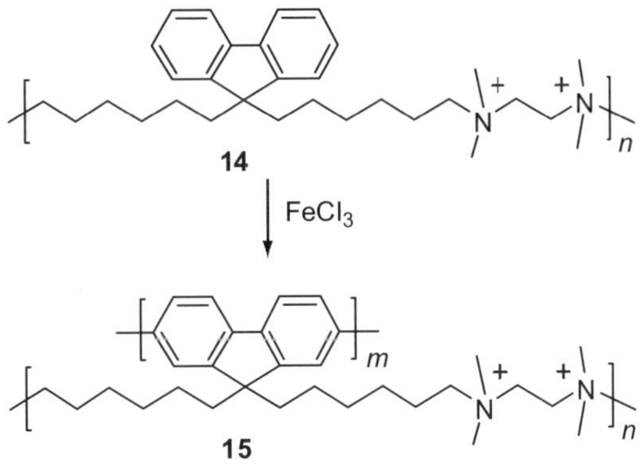

Scheme 4 Synthesis of a polyfluorene network by oxidative coupling

Oxidation of the polyionene 14 with iron(III) chloride to obtain an insoluble polyfluorene network 15 (Scheme 4) [22]. This shows violet PL (λ_{max} = ca. 410 nm), suggesting that the polyfluorene segments are rather short. Due to its insolubility this material cannot be used to make good quality films for use in LEDs, but the incorporation of a conjugated backbone within a network is one possible way to obtain isolated chains and so avoid the problems associated with interchain interactions, e.g., excimer formation.

Most syntheses of PFs use Suzuki polycondensations or Yamamoto Ni(0) couplings of dibromomonomers. A group at Dow [23–28] have developed a Suzuki cross-coupling route leading to high molecular weight PDAFs (Scheme 5), e.g., poly(9,9-dioctylfluorene) (**16**) with a $M_n > 100\,000$ g/mol was obtained after less than 24 hours reaction time. The disadvantage of this method is that exact stoichiometry of the two components is required for obtaining very high molecular weights which can be difficult to achieve experimentally owing to the well-known problems in purifying boronic acids and their derivatives [11]. Bradley and co-workers [29] used **16** prepared by this method to fabricate a blue ($\lambda_{max} = 436$ nm) LED with relatively high (0.2%) efficiency being obtained when a hole-transporting layer was used.

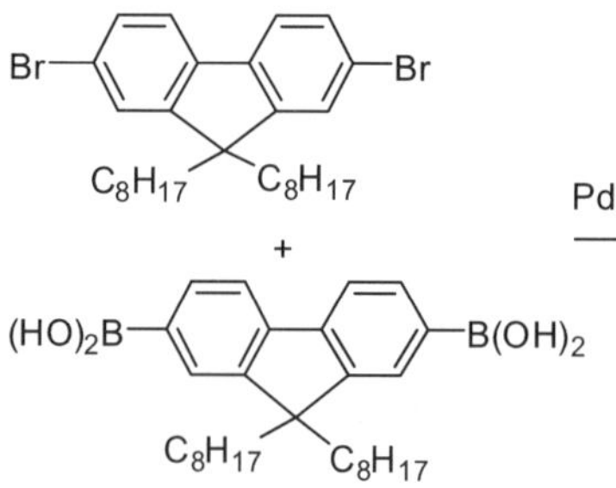

Scheme 5 Dow route to high molecular weight poly(9,9-dioctylfluorene)

In addition to the AA-BB cross-coupling method used by Dow, an AB-type Suzuki coupling of a 2-bromo-7-boronate **17** could also be used to make PDAFs. However, this method has so far been used only for preparing end-capped low molecular weight PF rods e.g., **18** as macroiniators in the synthesis of rod-coil copolymers (Scheme 6).

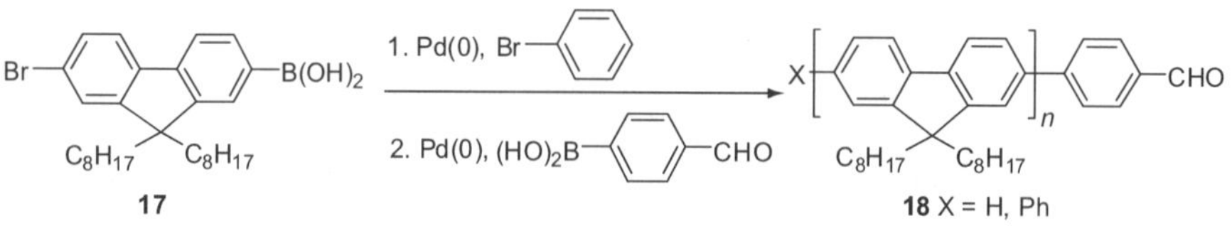

Scheme 6 Synthesis of end-capped polyfluorene by AB-type Suzuki polycoupling

Yamamoto-style polycondensations have also been used to prepare high molecular weight PFs. For example, poly[9,9-bis(2-ethylhexyl)fluorene] (**19**) with a M_n of over 100 000 g/mol has been prepared by coupling the dibromofluorene monomer **20** with bis(cycloocta-1,7-dienyl)nickel(0) and bipyridine (Scheme 7) [30]. The disadvantages of this method are that the expen-

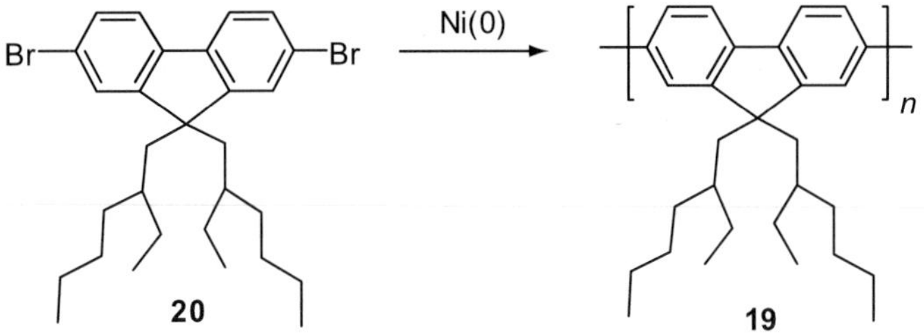

Scheme 7 PDAFS by Yamamoto polycondensation

sive nickel reagent must be used stoichiometrically, and that the molecular weights cannot match those obtained from the most efficient Suzuki couplings. The advantage is that the method is experimentally simpler than the Suzuki method.

2.2
Optical Properties of PDAFs

The emission from PDAFs is violet-blue with a primary emission maximum at about 425 nm, and a secondary peak at about 445 nm. Well-defined oligomers of **16** have been prepared, and as a result the effective conjugation length for PDAFs has been determined to be about 12 units (24 benzene rings) for absorption and 6 units for emission [31–33], indicating that the geometries of the ground and excited states are different.

PDAFs with unbranched alkyl substituents, e.g., **12** and **16**, show two thermotropic nematic liquid crystalline phases [34, 35]. Longer alkyl chains lower the transition temperatures, so that whereas for **12** the phases occur at 162–213 °C and 222–246 °C, with isotropisation occurring at 290–300 °C, for **16** the nematic phases are between 80–103 °C and 108–57 °C with isotropisation at 278–283 °C, and for the dodecyl-substituted polymer **21** (Scheme 8) they are observed at 62–77 °C and 83–116 °C with isotropisation at 116–118 °C [35]. Annealing films of the polymers at a temperature just above the second liquid crystalline transition followed by rapid cooling preserves the liquid crystalline order [34, 35]. Deposition of such a film of **16** upon a rubbing aligned polyimide alignment layer has been used to obtain polarised PL and EL [10, 36]. A rubbing-aligned layer of PPV has also been used as an alignment layer to obtain polarised EL from films of **16**, but the emission spectrum is slightly red-shifted due to some absorption by the PPV layer in the region of the main emission peak at 433 nm [37–39].

In contrast the polymer **19** with branched alkyl chains displays only one nematic phase. Thus **19** shows a phase transition at 167 °C upon heating with the reverse transition upon cooling being seen at 132 °C. Annealing of a film of **19** at 185 °C upon a rubbing-aligned polyimide layer followed by rapid

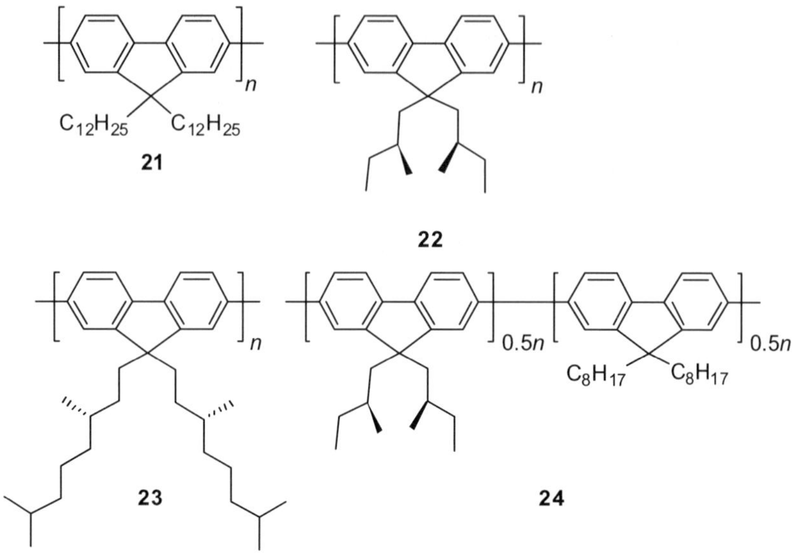

Scheme 8 Polyfluorenes with linear and branched alkyl chains

cooling produced a film which emitted polarised EL with a dichroic ratio of up to 20 [30, 36, 40, 41].

Circularly polarised PL and EL emission has been obtained from polymers, e.g., **22** and **23** (Scheme 8), bearing chiral side-chains [42–45]. The copolymer **24** also produced circularly polarised emission, but the degree of dissymmetry was much reduced [42–44]. This was attributed to the dioctylfluorene units preferring a planar conformation of the backbone while the chiral-substituted units adopt a helical structure. A combination of a helical backbone conformation and liquid crystallinity is thought essential for obtaining a high degree of circular polarisation.

2.3
Colour Control by Copolymerisation

Since PDAFs are wide bandgap materials, incorporation of other smaller bandgap chromophores as comonomers or substituents enables tuning of the emission colour due to efficient energy transfer from the fluorene to the other units. Random copolymers of fluorenes with other chromophores are readily prepared by the copolymerisation of the dihalochromophore with a dihalofluorene as exemplified in Scheme 9 for the synthesis of copolymers of dioctylfluorene with perylene dyes (1–5 mol %) [46, 47].

These materials were designed so that by efficient Förster energy transfer from the fluorene to the dye units, efficient emission across the whole visible spectrum could be obtained. Perylene dyes were chosen as the chro-

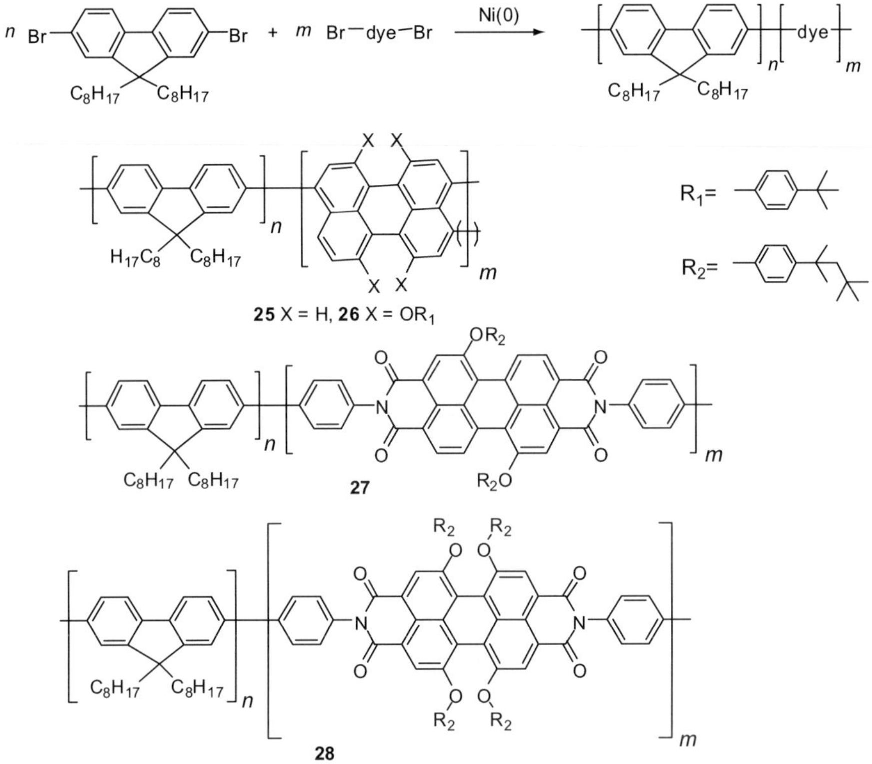

Scheme 9 Synthesis of copolymers of fluorene and perylene-based dyes

mophores due to their high solid state PL quantum yields, and their excellent thermal and photochemical stability. Four different dyes were used, with emission maxima in the green (λ_{max} = 525 nm), yellow (λ_{max} = 540 nm), orange (λ_{max} = 584 nm) and red (λ_{max} = 626 nm) respectively. The PL spectra of these materials in solution showed emission from both components roughly proportional to their relative mole ratios, but in films energy transfer occurred so that the PL spectrum was dominated by emission from the dyes. The PL quantum efficiencies of the copolymers **25–28** were only slightly lower (33–51%) than for **16** (55%). In the EL spectra only emission from the dyes was seen, which is due to a combination of energy transfer from the fluorene and the dyes acting as charge traps which leads to preferential exciton formation on them. The EL spectra closely resembled the PL emission from the dyes except for the copolymer **27** with the orange-emitting dye, where the EL spectrum was broader and red-shifted compared to the PL spectrum. External efficiencies of 0.2–0.6% were obtained from these devices, which are comparable with other polyfluorene-based devices. The emission colours are stable, unlike devices using a blend of dye and **16**, where phase separation results in a loss of half the luminescence intensity within a few minutes, accompanied by a change in emission colour as the emission from the host polymer **16** reappears as the energy transfer becomes less efficient. As all the copolymers contain at least 95 mol % of dioctylfluorene units their blends with each other and with **16** should be stable, thus enabling stable emission of practically any colour desired, including white, to be obtainable.

Blue-emitting comonomers have also been used in order to stabilise the emission from the PDAFs. Miller and co-workers have prepared copolymers **29** (Scheme 10) of dihexylfluorene with anthracene by the Yamamoto method. These show stable blue emission (λ_{max} = 455 nm) even after prolonged annealing due to trapping of the exciton at the anthracene sites and subsequent emission therefrom [33, 48, 49]. Copolymers with other acenes such as pentacene do not produce stable emission due to the ready oxidation of the acene units [50]. Other stable chromophores, e.g., cyanostilbene, have been used to tune the emission colour through exciton trapping [49, 51]. By altering the amount of stilbene the emission from the copolymer **30** could be tuned between λ_{max} = 466 nm (m = 5%) and λ_{max} = 510 nm (m = 50%).

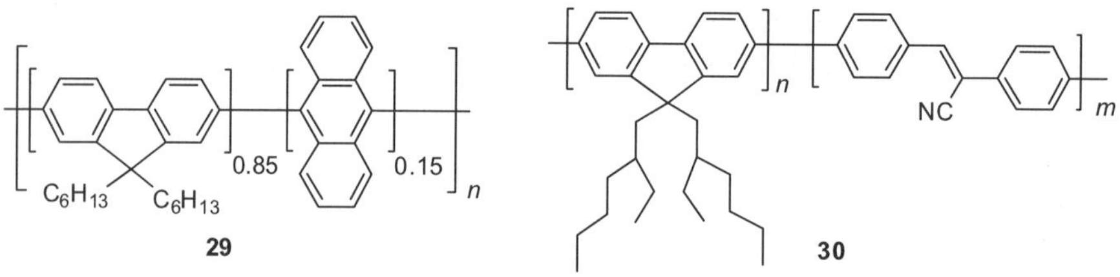

Scheme 10 Copolymers of fluorene with other stable chromophores

Efficient energy transfer was also obtained from a copolymer **31** in which the perylene was attached as an endcapping group (Scheme 11) [47]. By using a ratio of fluorene to monobromo-dye of 18 : 1 in a Yamamoto polycondensation, a polymer with M_n = 21 000 and a polydispersity of 2.1 was obtained, corresponding to about a degree of polymerisation of about 40. As with

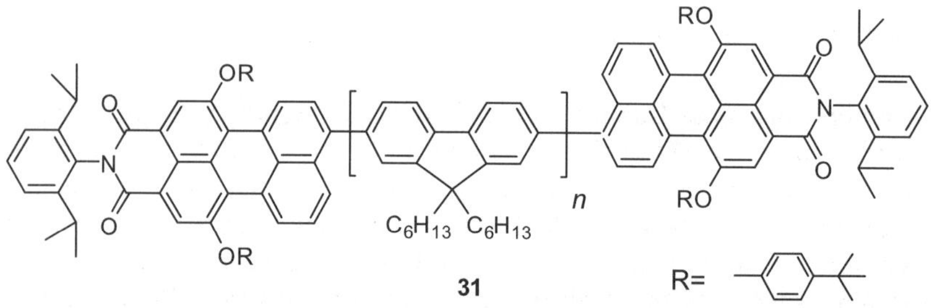

Scheme 11 Polyfluorene endcapped with perylene dyes

the random copolymers 25–28, no energy transfer was seen in solution, but in the solid state the emission was almost entirely from the dye moieties (λ_{max} = 613 nm). A problem with using the dye as an endcapping group is that due to a competing debromination reaction, complete endcapping in Yamamoto reactions is usually not obtained. In this case it was estimated that only 40 mol % of the polymer chains contained two dye units. Very efficient (up to 7.4 cd/A) green EL has been obtained by Wang and coworkers by endcapping 16 with naphthalimide dyes [52].

The copolymers described above, which have perylene dyes in the main chain or as endgroups, show energy transfer only in the solid state, due to the low mole ratio of dye units present. If the dyes are attached on the sidechain, then copolymers containing much higher mole ratios of chromophore are accessable. The copolymer 32 (Scheme 12) in which 33% of the fluorene units have dyes attached ($m:n = 2:1$) showed energy transfer in solution as well as in a thin film [47]. The emission colour differed slightly between the two states, with the emission maximum appearing at λ_{max} = 561 nm in solution with a shoulder at 599 nm, and at λ_{max} = 599 nm in the solid state. This is probably due to interaction of the chromophores in the solid phase. Stable efficient blue EL can be obtained by incorporation of 0.2 mol % of phenylene units bearing naphthalimide dyes into a PDAF [53]. At higher concentrations of the comonomer, only green emission from the dyes is seen [54].

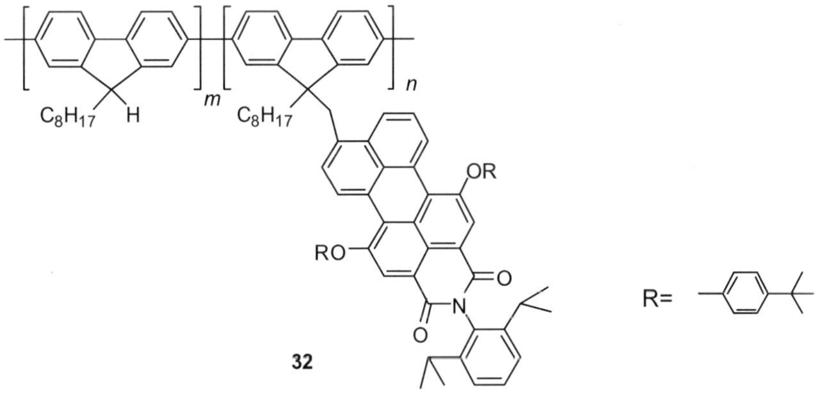

Scheme 12 Polyfluorene with dyes as side-chains

If a very low mole fraction of a red or orange-emitting chromophore is incorporated into a PDAF, then energy transfer is incomplete so that emission from both the fluorene and the comonomer is observed. By carefully balancing the amounts of emission form the fluorene and comonomer, white emission can be obtained. This approach has been used very successfully by Wang and coworkers to obtain white-emitting copolymers [55]. The highest efficiency for white emission (9 cd/A), was obtained from copolymer

33 (Scheme 13) containing 0.03 mol% of an orange emitting heterocyclic comonomer. Incorporation of low concentrations of green and red-emitting silole comonomers into a PDAF has also been used to obtain pure white emission [56].

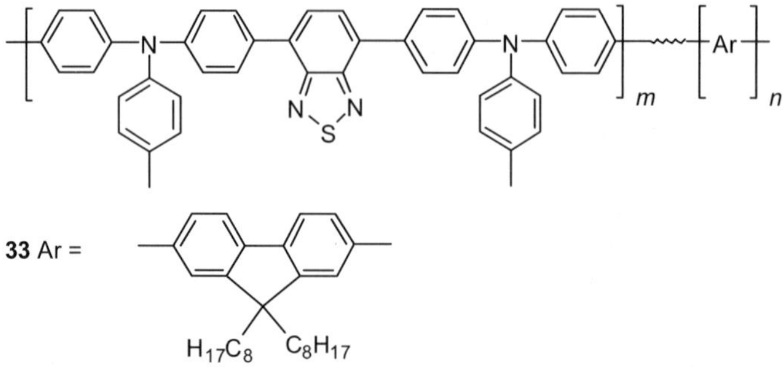

Scheme 13 High efficiency white emitting fluorene copolymer

These approaches have been adopted more recently to incorporate phosphorescent chromophores into PF in order to make use of the fact that a large proportion (up to 75%) of all excitons formed in LEDs are triplet states, whose energy can only be harvested by using phosphorescent units. The first fluorene copolymers with phosphorescent units **34–35** were made by Holmes and coworkers who added monobrominated red- or green-emitting iridium complexes to an AA-BB Suzuki polycondensation [57]. With short fluorene chains, only emission from the iridium complexes are observed, but with longer fluorene chains some blue emission is also seen. Other groups have since incorporated different phosphorescent units such as platinum [58] or zinc salen [59] units or porphyrins [60, 61].

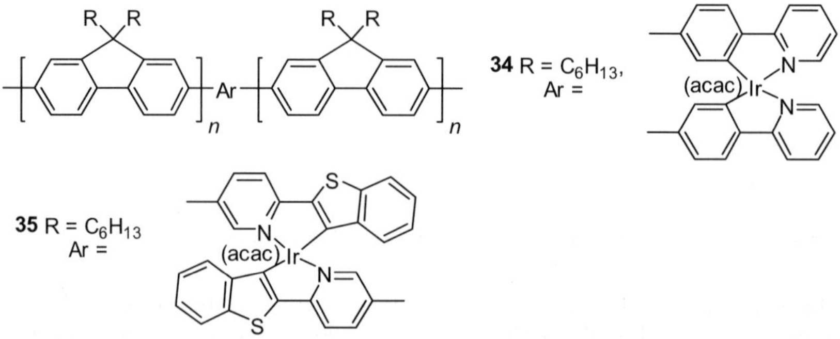

Scheme 14 PDAFs incorporating phosphorescent iridium complexes

White EL has been obtained from a copolymer containing low mole fractions of a green fluorescent (0.005–0.05 mol %) and a red phosphorescent (0.1–0.5 mol %) comonomer with efficiencies of up to 6.1 cd/A [62].

2.4
Defect Emission from PDAFs

The blue emission from PDAFs is unstable, with the appearance of a strong emission band around 530 nm after annealing or upon running an EL device [14, 15]. Initially this long wavelength emission band was believed to be due to emission from excimers [63, 64], but subsequent work has shown that it is due to emission from fluorenone defects within the PDAF chain [15].

Scherf and co-workers [65] prepared the polymer **36** (Scheme 15) with only one alkyl substituent at the 9-position by the Yamamoto method, and found that even pristine material contained fluorenone units as shown by a carbonyl stretching band at 1721 cm^{-1} in the IR absorption spectrum, and that the solid-state PL and EL spectra of **36** were dominated by a low-energy emission band centred at 533 nm due to fluorenone emission. A band at exactly the same position was seen from the corresponding dialkylfluorene polymer **37** after photooxidation in air or 30 minutes continuous operation of an LED. They accordingly suggested that the long wavelength emission band in PDAFs is due to energy transfer to ketone defect sites. The greater contribution of this band to the EL then the PL spectrum they attributed to the fluorenone units acting as electron trapping sites, thus favouring recombination at the defect sites. Further support for this view came from time-delayed PL measurements on polymer **23** in solution by Lupton et al. [66].

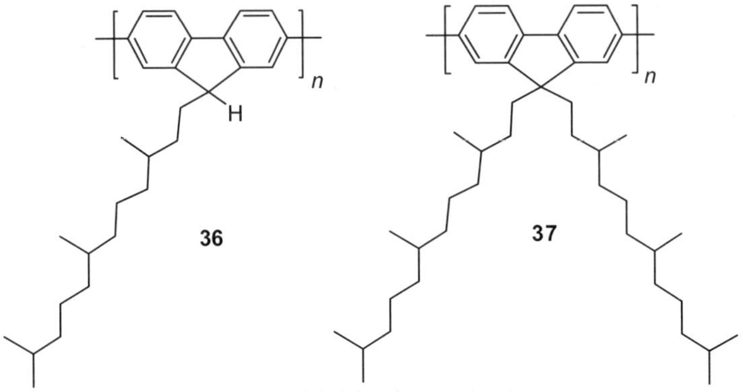

Scheme 15 Mono- and dialkylated polyfluorenes used to test the ketone defect hypothesis

To account for the formation of the defect sites in pristine **36**, they proposed that during the polymerisation, some of the alkylfluorene units were reduced by the nickel(0) to monoalkylfluorenyl anions **38**, which were then oxidised by atmospheric oxygen to the ketones **39** during the workup (Scheme 16). The oxidation of 9-fluorenyl anions by air is an established highly efficient route to fluorenones [67], making this a plausible pathway but the details of the oxidation mechanism have yet to be elucidated.

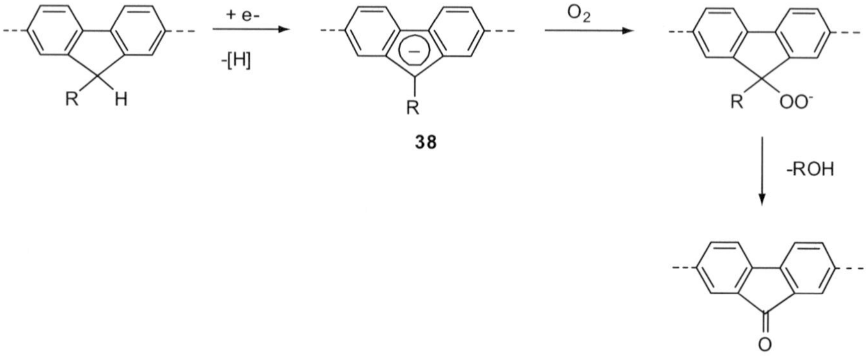

Scheme 16 Proposed mechanism for formation of ketone defects **39**

Final confirmation of this defect hypothesis was obtained by List and co-workers who found that heating a film of **37** at 200 °C in a dynamic vacuum (<10^{-4} mbar) produced no long wavelength emission band, even when the sample was simultaneously illuminated, while heating in air produced a strong band at 530 nm [68]. This demonstrated conclusively that the emission band at 530 nm from PFs is due to the formation of ketone defects by oxidation during synthesis and/or handling. This does not mean that inter-chain interactions produced during annealing of PDAFs play no part in the appearance of the long wavelength emission, as increased inter-chain interactions (aggregation) would enhance exciton migration to the defect sites.

Considerable effort has gone into developing stable blue emission from PDAFs. Fractionation of **16** to remove low molecular mass material has been reported to reduce the long wavelength emission [69]. This is at first sight a somewhat surprising result as the probability of a polymer chain containing a defect increases with increasing chain length. However, it is possible that the oligomers in the low molecular mass fraction assist in inter-chain charge and energy migration, and so removing them would reduce exciton migration to defect sites.

Stable blue emission has been obtained by Neher and co-workers from blends of the copolymer **40** with hole-transporting molecules e.g., **41** (Scheme 17) [70]. This was attributed to the dopants acting as hole traps, thus reducing the amount of charge trapping at the defect sites, and thus the emission therefrom.

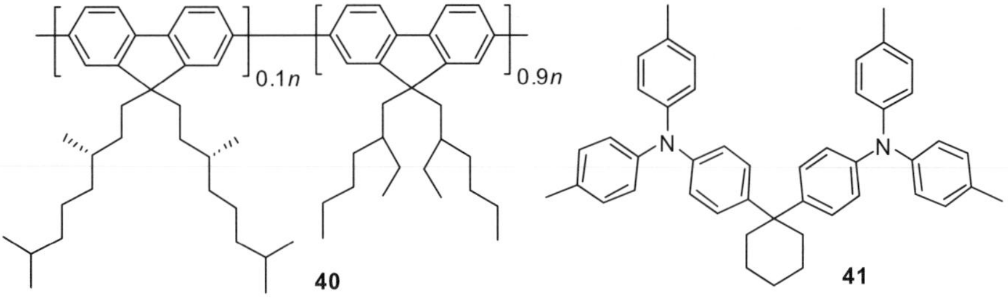

Scheme 17 Polyfluorene and typical hole-transporting material blended with it to obtain a stable blue emission

Other approaches have involved reduction of interchain interaction either by cross-linking the polymers using vinyl endgroups [71] or by attaching bulky groups, e.g., dendrimers [72, 73] or silsesquioxanes [74, 75], as end- or sidegroups. There is evidence that reduction of fluorene by calcium cathodes is implicated in the formation of ketone defects during LED operation, and this can be suppressed by introduction of a buffer layer between the polymer and the cathode [76].

The surest approach would be to ensure that the dialkylfluorene monomers contain no monoalkylfluorene impurities. Meijer and co-workers found that treating the monomers with base to remove the monoalkyl derivatives before polymerisation greatly enhanced the colour stability of the resulting polymers [77]. However, since they could not detect any impurity either before or after this treatment, indicating it was present in very low concentration, they could not tell if they had removed it completely. Recently, Grimsdale, Holmes et al., have developed a new synthesis of 9,9-dialkylfluorenes (Scheme 18) which precludes the formation of monoalkylfluorenes [78]. Polymers made from the "defect-free" monomer 42 showed stable blue emission

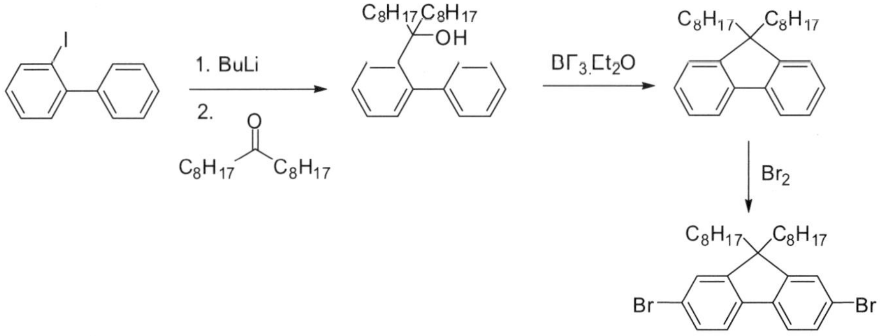

Scheme 18 Route to dialkylfluorenes without monoalkylfluorene impurities

in initial tests, though lifetime tests still need to be done to determine their absolute stability. Incorporation of even very low (0.06 mol %) amounts of a monoalkylfluorene comomoner was found to lead to formation of a green emission band from the polymers in an LED.

Another approach to avoiding defect formation is to have substitution at C9 which cannot be oxidised to form a ketone. Replacement of the carbon with a nitrogen bridge (carbazole) or a silicon (silafluorene) is covered in other chapters to which the reader is referred. Replacement of the alkyl groups at C9 with aryl groups has proved to be a successful way to obtain stable blue emission from PFs and is the subject of the next section.

2.5
Poly(diarylfluorene)s

A general synthesis of polyfluorenes with aryl substituents at the 9-position has been developed (Scheme 19) [73]. This synthetic route precludes formation of the bridgehead hydrogens which are oxidised to ketones and the bulk reduces inter-chain interactions, and hence exciton diffusion to any defect sites that may be formed. As aryl groups, unlike alkyl groups, generally cannot be directly substituted onto the 9-position of fluorenes (for an exception see Scheme 29 below), the monomers were prepared by the addition of aryllithium reagents to the biphenyl-2-carboxylic acid methyl ester **43**, followed by ring closure. The ring closures were performed by dissolving the carbinol **44** in hot acetic acid, and then adding a small amount of concentrated sulfuric acid. This produced a dark blue or lilac colour due to the formation of the cation, followed by the appearance of a white precipitate of the desired 9,9-diarylfluorenes **45**. These were then polymerised using nickel(0). The ring closure can also be performed with Lewis acids such as BF_3 [79].

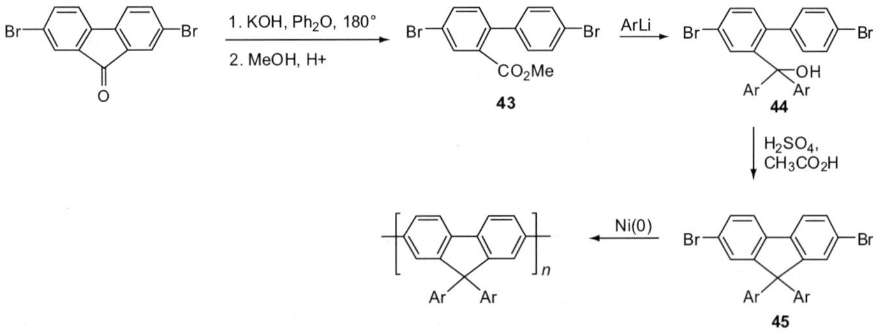

Scheme 19 Synthetic route to poly(9,9-diarylfluorene)s

The diphenylfluorene polymer **46** (Scheme 20) was virtually totally insoluble. Characterisation by MALDI-TOF of a soluble fraction, obtained by

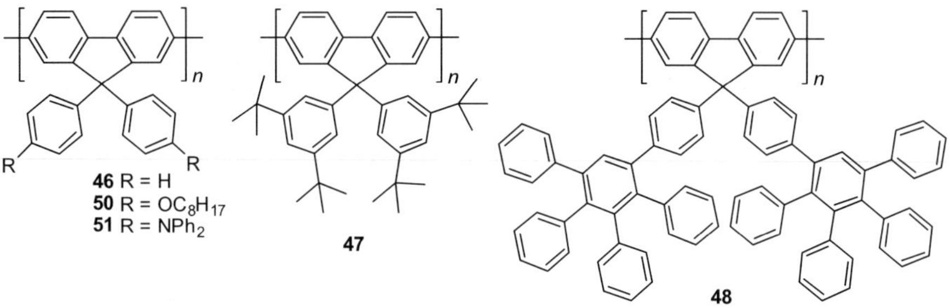

Scheme 20 Poly(9,9-diarylfluorene)s

prolonged Soxhlet extraction with toluene, suggested that it consisted mainly of oligomers up to the hexamer ($n = 6$), though chains of up to 15 fluorene units were detected [164]. The insoluble fraction contained chains with degrees of polymerisation of up to 25, which suggests that some chain addition occurred even after the material was no longer soluble in the reaction medium.

In contrast the materials **47** with di-*tert*-butylphenyl and **48** with first generation dendron side-chains were processable from toluene [73]. The degrees of polymerisation for these polymers were rather low, with the dendronized polymer **48** having a M_n of only 10 000 g/mol ($n = 10$) probably due to the bulkiness of the substituents. Polymer **48** was also found to be exceptionally thermally stable, with no significant mass loss being observed in thermal gravimetric analysis (TGA) until 570 °C (cf. 463 °C for **89** [73]).

Stable blue PL emission has been obtained from films of both **47** and **48**, with no sign of an emission band at 530 nm even after annealing for 24 h in air at 100 °C. This shows that aryl substituents are much stabler towards oxidation than alkyl groups as annealing of **16** under these conditions produces a strong defect emission band. The dendronised polymer **48** shows blue EL with no significant defect emission even after several minutes operation [46, 80]. Studies have shown that the bulky dendrimers in **48** affect the photophysics of the polymer [81–84]. One effect is to increase the lifetime of triplet excited states which may reduce the efficiency of LEDs due to increasing singlet-triplet quenching [84], so that the use of phenyl substituents rather than dendrimers to suppress the long wavelength emission would appear to be advisable.

A second way to attach aryl groups at C9 is by acid-catalysed condensation of dibromofluorenone **49** with activated aromatics such as phenol or triphenylamine (Scheme 21). This method has been used to make polymers **50** [85] and **51** [86], which both show stable blue EL.

One effect of aryl substituents on the physical properties of polyfluorenes is to inhibit chain packing as demonstrated by the lack of liquid crystalline

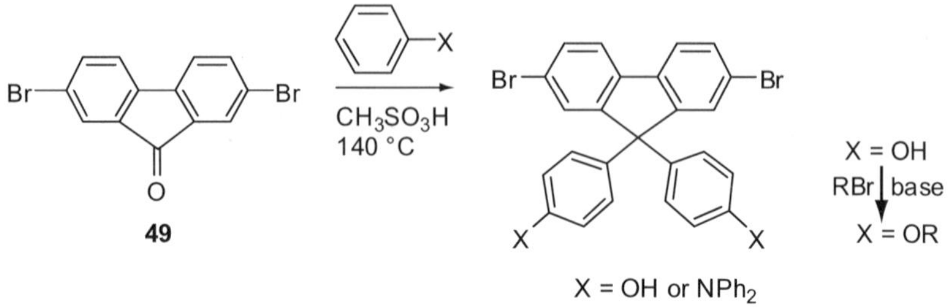

Scheme 21 Addition of highly activated arenes to fluorenone

phases, and the absence of any signs of organization in the polymer films when studied by atomic force microscopy (AFM) [87].

2.6
Polyfluorenes with Improved Charge Injection

A second problem with PDAFs is that of charge injection. The energy levels of the HOMO and LUMO orbitals of **16** have been determined to be 5.8 eV and 2.12 eV by cyclic voltammetry [88]. This means there are large barriers to charge injection from electrodes such as calcium (work function 2.9 eV) and ITO (4.7–5.0 eV). The value for the HOMO-LUMO gap of 3.7 eV determined by electrochemistry is much larger than the optical bandgap of 2.95 eV calculated from the onset of absorption, which suggests that one or both of the electrochemically derived orbital energy levels are inaccurate. As the efficiency of devices using calcium electrodes seem to be limited by hole injection [89], most probably it is the LUMO value which is at fault (too high).

A number of approaches have been adopted to overcome the problem of obtaining efficient charge injection. Satisfactory electron injection seems to be obtained by use of calcium as a cathode, and a composite LiF/Ca/Al cathode is reported to give better electron injection than a simple calcium cathode [90], but there usually remains a large barrier to hole injection due to the difference between the HOMO and ITO energy levels. As a result hole-injecting layer are used in PDAF devices in order to obtain good efficiency. A second method is to blend charge-transporting materials into the emissive polymer, though this raises the problems of phase separation. Neher and co-workers have examined blends of the copolymer **41** (Scheme 17) with hole-transporting small molecules [70]. Not only did this increase the efficiency and luminance of the devices, but as mentioned above it also suppressed the green emission from the ketone defects. No phase separation leading to device degradation was observed by them during the characterisation of their devices, but the long-term stability of such blends is uncertain.

Another approach is to incorporate charge-transporting groups into the polymer, either as endgroups, as units in the polymer chain or as sidechains. The IBM group [91] have compared the efficiency, using both calcium and aluminium cathodes, of triblock polymers (Scheme 22) containing an anthracene-dialkylfluorene copolymer as the emissive block with hole-transporting poly(vinyl triphenylamine) (**52**) or electron transporting poly(vinyl oxadiazole) (**53**) blocks at each end with the simple copolymer **29** (Scheme 10).

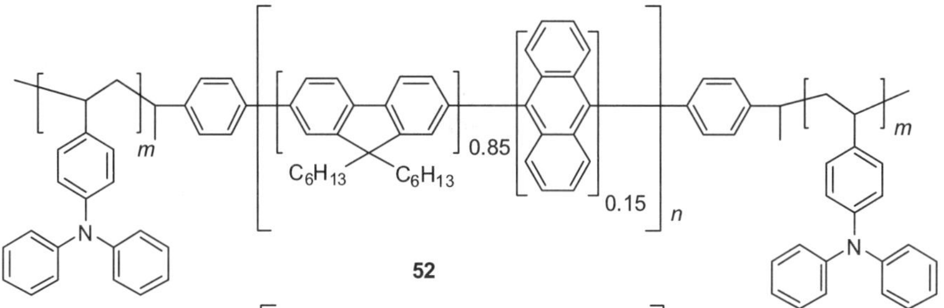

Scheme 22 Triblock copolymers containing both emissive and charge-transporting blocks

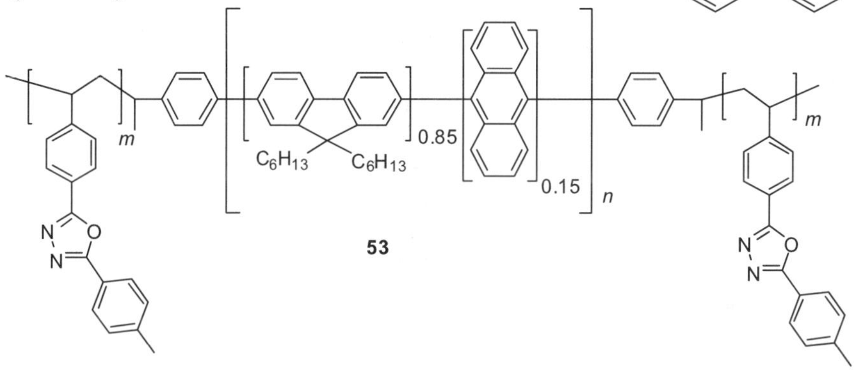

In single-layer devices with calcium cathodes the EL efficiency of the hole-transporting triblock **52** was nearly double that of **29** (0.35% versus 0.02%), but that of the electron transporting triblock **53** was slightly lower (0.014%). This shows that with calcium electrodes, hole injection is the limiting factor for efficiency. With aluminium electrodes, electron injection by contrast became the limiting factor so that the highest efficiency (0.01%) was obtained with **53**, although the efficiency of **52** (0.008%) was still much higher than for **29** (0.001%). The triblock **52** showed a strong orange-red emission band due to emission from the triphenylamine units in the EL, but not in the PL spectrum. The highest efficiency (0.54%) was obtained from a double-layer device of **52** with a hole-transporting layer and a calcium cathode. Again, some emission from the triphenylamine units was seen, though it was much weaker than in the single-layer device.

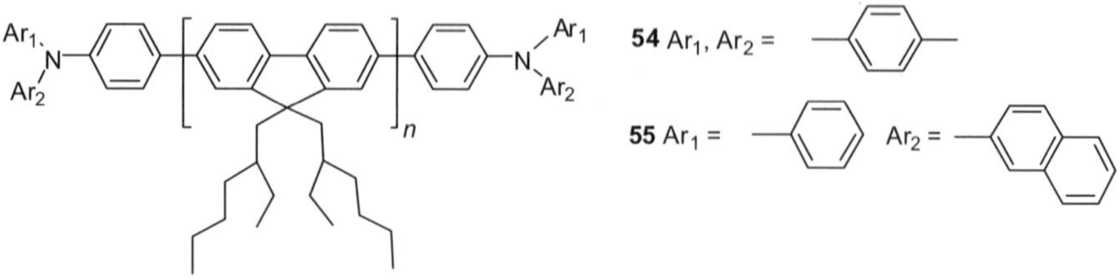

Scheme 23 Polyfluorenes endcapped with hole-transporting groups

Scherf, Neher and co-workers prepared liquid crystalline polyfluorenes, **54** and **55** (Scheme 23), by adding bromo-substituted triarylamines as endcapping reagents to a Yamamoto polymerisation [92]. By varying the amount of endcapper they were able to control the molecular weight and polydispersity of the polymer, so that for **54** 2 mol % of endcapper gave a polymer with M_n = 102 000 and M_w/M_n = 1.4, while 9 mol % of endcapper produced a polymer with M_n = 12 000 and M_w/M_n = 2.6. NMR analysis showed that the endcapping was incomplete so that 2 mol % of endcapper produced 1.8% of triarylamine endgroups and 9 mol % resulted in 8.3% incorporation of the endgroups. Films of these polymers deposited on an aligned polyimide layer produced polarised emission with better efficiencies (up to 0.75 cd/A) than for the homopolymer **19**. Even better efficiencies (up to 2.7 cd/A) have been obtained from devices using an emissive layer of **54** (2 mol % endcapper) and a series of three cross-linked hole-transporting layers of varying work function [93].

The Dow group have prepared altenating copolymers of dialkylfluorene with hole-transporting triphenylamines (Scheme 24), e.g., **56**, by Suzuki coupling of fluorene bisboronic acids with dibromotriarylamines [25–28, 94]. Blue emission has been reported from some of these polymers, e.g., **57** [94] (λ_{max} = 481 nm). It would appear, however, that these materials are primarily intended to be used as hole-transporting rather than emissive materials.

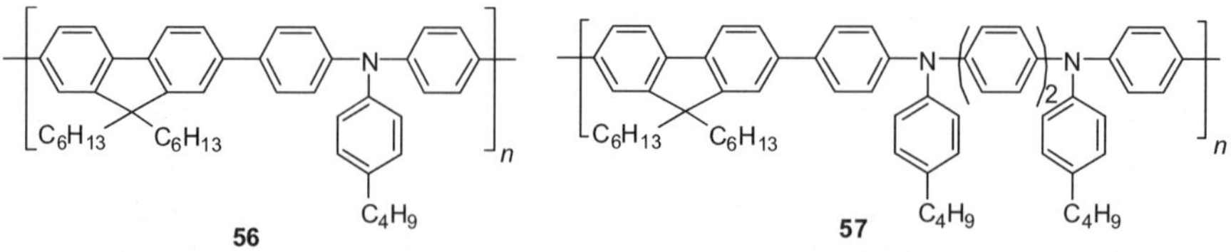

Scheme 24 Copolymers of fluorene with hole-transporting triarylamines

Xia and Advincula have prepared copolymers **58** containing hole-transporting carbazole units by Yamamoto copolymerisation (Scheme 25) [165]. Cyclic voltammetry showed that the HOMO energy level increased from 5.8 eV to 5.6 eV with 10 mol % carbazole and to 5.5 eV with 30 mol % car-

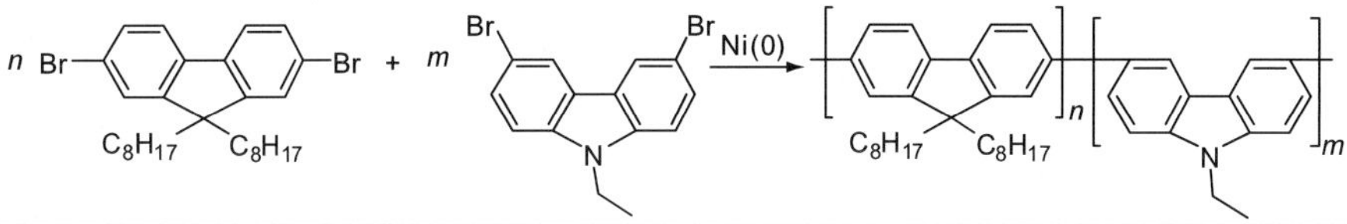

Scheme 25 Synthesis of fluorene-carbazole copolymers

bazole. They also found that films of these copolymers showed stabler blue PL than the homopolymer **16** with the green ketone emission band appearing only slowly upon annealing at 200 °C.

The IBM group have investigated the incorporation of charge-transporting groups into the copolymer **29** [33]. Introduction of hole-transporting triphenylamine units produced a slight red-shift (λ_{max} = 462 nm) in the EL spectrum, while electron-accepting diphenylsulfone units caused a small blue-shift (λ_{max} = 445–449 nm) in the emission, in both cases with a decrease in the EL efficiency. By contrast, the copolymer **59** (Scheme 26) containing 10 mol % of both groups showed stable blue EL (λ_{max} = 460 nm) with nearly twice the efficiency of **29**.

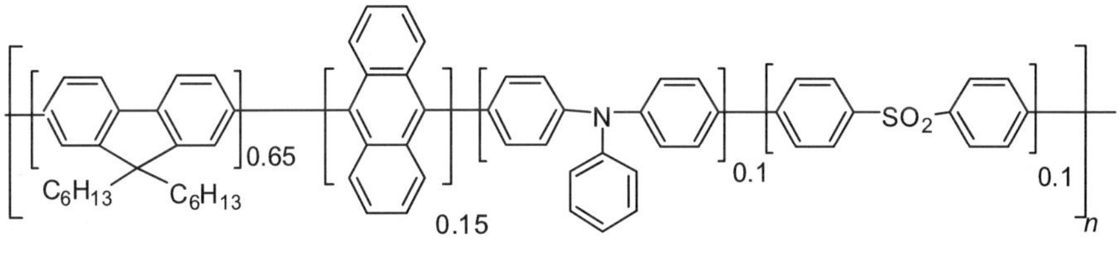

Scheme 26 Copolymer of fluorene with both hole and electron-transporting units

Müllen and co-workers prepared the polymer **51** in which hole-transporting triphenylamine units are introduced at the 9-position by Friedel-Crafts alkylation of triphenylamine with 9-fluorenyl cations produced from dibromofluorenone under the reaction conditions (Scheme 21) [86]. The degree of polymerisation is not very high (n = 14) due to the limited solubilising power of the substituents. The HOMO of **51** has been determined by cyclic voltammetry to be at 5.5 eV, so the barrier to hole injection from ITO is much smaller than for PDAFs. Polymer **51** shows stable blue EL (λ_{max} = 428 nm), but the overall EL efficiency is lower than for PDAFs despite the better hole-accepting properties, probably because of the lower PL efficiency of **51** (22%, cf. 55% for **16** [29]). The use of a PVK hole-injecting layer does not improve the efficiency as it does with PDAFs, indicating hole injection into **51** is superior.

Müllen and co-workers have also prepared the blue-emitting (λ_{max} = 450 nm) polyketal **60**, films of which can be converted by exposure to

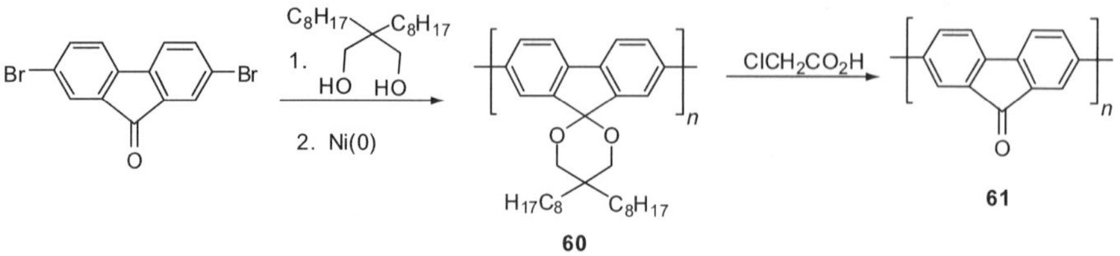

Scheme 27 Precursor route to polyfluorenone

dichloroacetic acid vapour to the orange-emitting (λ_{max} = 580 nm) polyfluorenone **61** (Scheme 27) [95]. The carbonyl groups enhance the electron accepting properties of **61**, and this polymer shows useful electron-transporting properties, though the acidic residues from the conversion are a potential source of problems for electronic applications [96].

Copolymers **62** (Scheme 28) of fluorene and fluorenone display green EL (λ_{max} = 535 nm) with the highest EL efficiency of 0.1% being observed for the copolymer with 3 mol % of fluorenone [97].

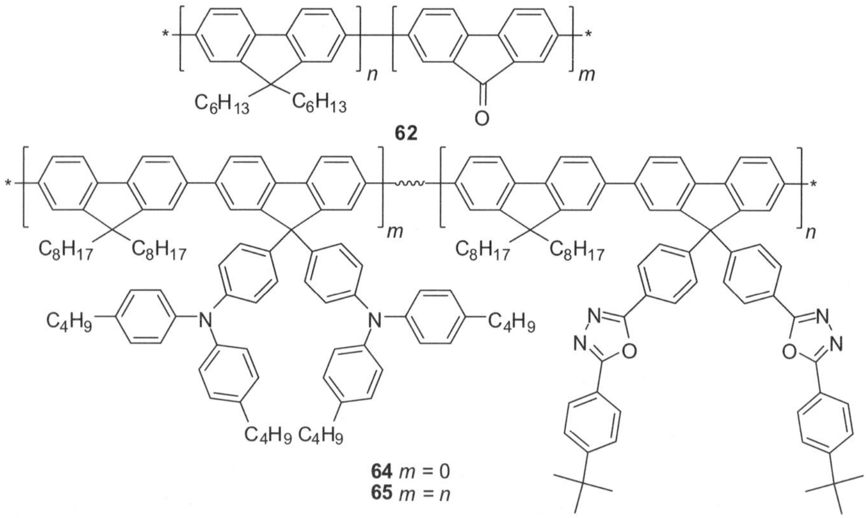

Scheme 28 Fluorene copolymers with electron-accepting units at C9

A route has been developed by Shu, Jen, and co-workers for the synthesis of a fluorene monomer **63** bearing two aryloxadiazole groups at the bridgehead by nucleophilic substitution of 4-fluorobenzonitrile with the fluorenyl anion, followed by conversion of the nitriles to oxadiazoles via tetrazole intermediates (Scheme 29) [98]. The EL efficiency of the alternating copolymer **64** in a single-layer device is considerably higher than for the PDAF homopolymer

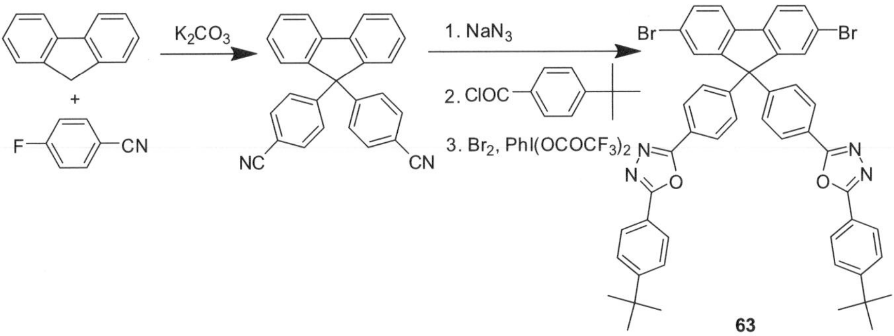

Scheme 29 Synthesis of a fluorene with oxadiazole substituents

16 (0.52% vs 0.2%). The copolymer **65** (Scheme 28) bearing both hole- and electron-accepting groups has an even higher EL efficiency (1.21%) [99].

2.7
Polymers Containing Spirobifluorene

Spirobifluorenes have been investigated by Salbeck [100] and found to be promising materials for use in blue LEDs, and so some effort has been made to incorporate these units into polymers (Scheme 30). As poly(diarylfluorene)s have proven to be stable blue emitters it comes as no surprise to find that polymers containing spirobifluorenes such as **66** [101] and **67** [102] also produce stable blue emission. A fluorene-spirobifluorene alternating copolymer **68** has been made and was found to give stabler emission than the fluorene homopolymer **16**, but green emission was still observed upon heating in air at 150 °C [103].

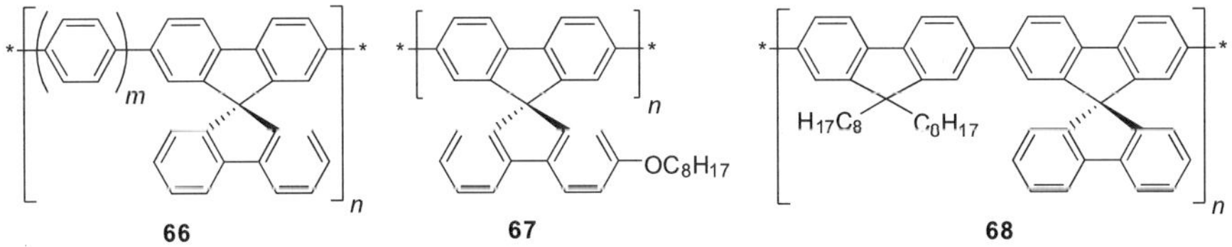

Scheme 30 Polymers containing spirobifluorene units

Other spiro-linkages have also been introduced at the C9-position of fluorenes in attempts to stabilize the emission (Scheme 31). Polymers, e.g., **69**, with spirocycloalkyl groups still produce green emission upon annealing [104], whereas the polymer **70** with aryl groups at C9 displays stable blue PL and EL [105].

Meerholz and co-workers [106] have used Suzuki cross-coupling of a spirobifluorene bisboronate with various dibromo-comonomers to prepare

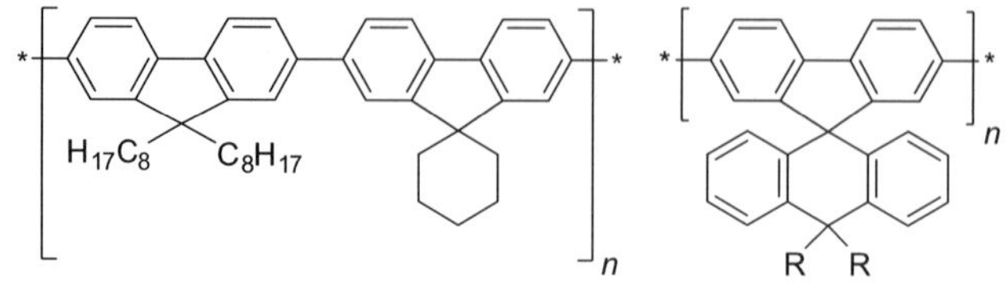

Scheme 31 Polymers containing other spiro-linked fluorenes

crosslinkable copolymers 71–73 (Scheme 32) which respectively produced blue (λ_{max} = 457 nm), green (λ_{max} = 507 nm), and red (λ_{max} = 650 nm) EL. Illumination of films of these materials with ultraviolet light in the presence of a photoacid induced cationic polymerisation of the oxetane rings to crosslink the materials. Comparison of the EL properties of crosslinked 71 with a similar non-crosslinkable polymer showed that the crosslinking had no significant effect upon the luminescence of the materials. They used these materials to make a pixelated red, green, and blue display in which each of the materials was patterned by irradiation through a mask.

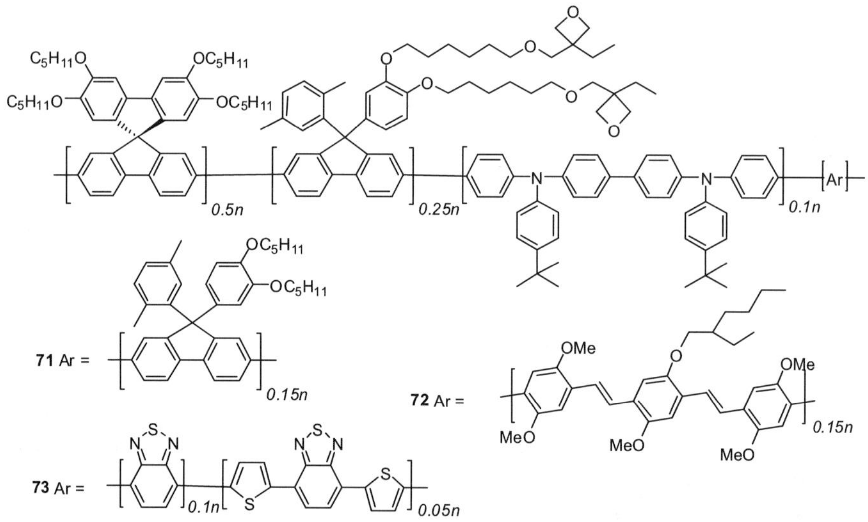

Scheme 32 Cross-linkable copolymers containing spirobifluorenes

2.8
Polymers Containing Other Bridged Biphenylene Units

Poly(9,10-dihydrophenanthren-2,7-diyl)s 74 (Scheme 33) which are analogous to polyfluorenes, but with ethane instead of methine bridges have been prepared by Yamamoto and co-workers [107]. The PL emission from 74

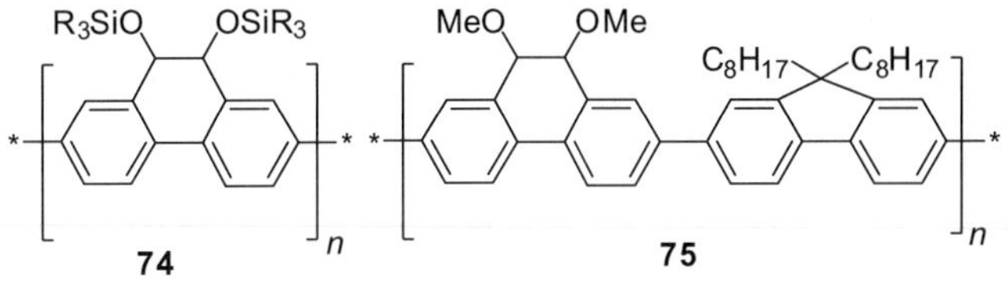

Scheme 33 Polymers based on 9,10-dihydrophenathrene

(λ_{max} = 431 nm) and the alternating copolymer **75** with dioctylfluorene is similar to that from polyfluorenes (λ_{max} = 428 nm).

Poly(tetrahydropyrene)s have two ethane bridges across each biphenyl unit, so that the torsion angle between the benzene rings is 20°. Poly(2,7-dioctyl-4,5,9,10-tetrahydropyrene) **76** has been made from a 2,2'-bis(1''-alkenyl)biphenyl as shown in Scheme 34. The molecular mass of the polymer (M_n = 20 000 g/mol) corresponds to about 46 monomer units (92 phenylene rings). The emission from **76** is blue in solution (λ_{max} = 425 nm) but red-shifts to blue-green (λ_{max} = 457 nm) in the solid-state which is attributed to aggregation. Blue-green EL has been obtained from **76**. Interestingly the effective conjugation length for absorption in **76** is about 20 benzene rings [120], which is longer than for LPPP (**5**) (11 rings), but comparable with the values for PDAFs (24 rings) and PIFs (18–21 rings).

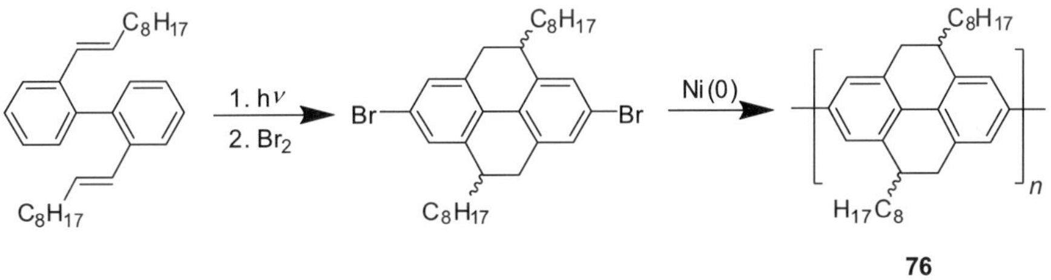

Scheme 34 Synthesis of poly(2,7-dioctyl-4,5-9,10-tetrahydropyrene)

When longer carbon chains are used to bridge biphenyls the biphenyl units become twisted reducing the conjugation [108, 109]. The introduction of 30–40 mol % of such units into fluorene copolymers **77–80** (Scheme 35) has been shown to suppress the long wavelength emission seen for the fluorene homopolymers [110]. The best EL efficiency (0.24 cd/A) was seen for **77** which

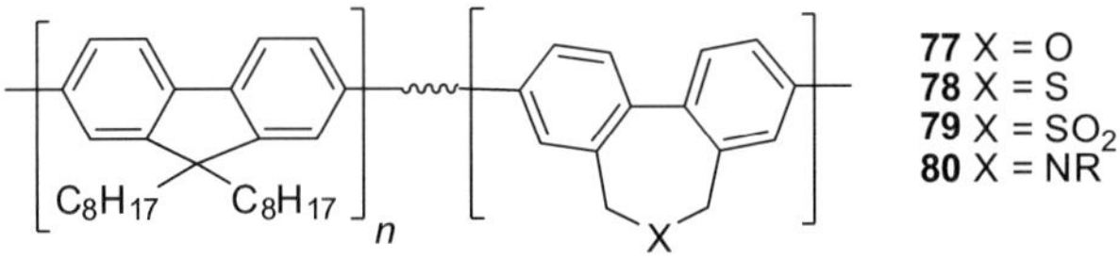

77 X = O
78 X = S
79 X = SO_2
80 X = NR

Scheme 35 Polymers containing biphenyls with larger bridges

is however, significantly below that of **16** (0.51 cd/A). These effects are attributed to disruption of the chain packing which not only hinders exciton migration to emissive defects, but also charge transport.

Phenanthrene is a biphenyl with an ethene bridge. Poly(2,7-phenanthrylene)s **81** have been prepared by polymer-analogous McMurry coupling of precursor poly(2-acyl-*p*-phenylene)s **82** (Scheme 36) [111].

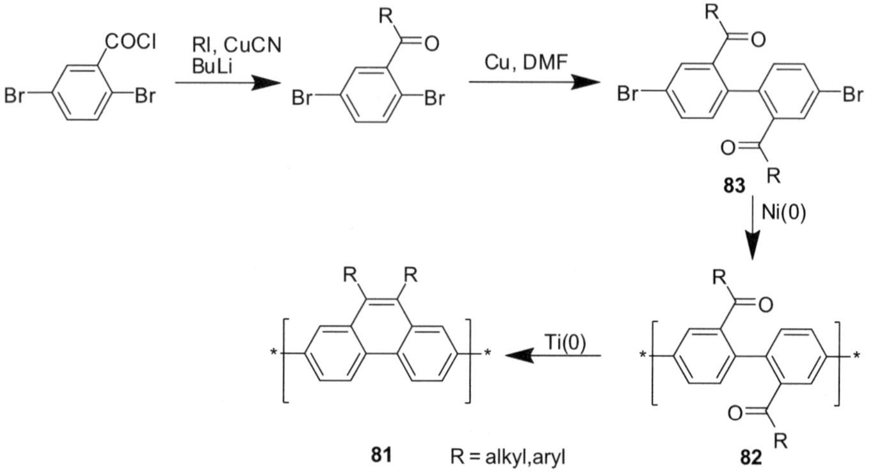

Scheme 36 Synthesis of poly(2,7-phenanthrylene)s

The coupling has to be performed in this way because McMurry coupling of dibromodiacylbiphenyl **83** proceeds with dehalogenation. Halogenation of the resulting 9,10-dialkyl(diaryl)phenanthrenes **84** occurs at the 3 and 6 positions so providing a route to the poly(3,6-phenanthrylene)s **85** (Scheme 37).

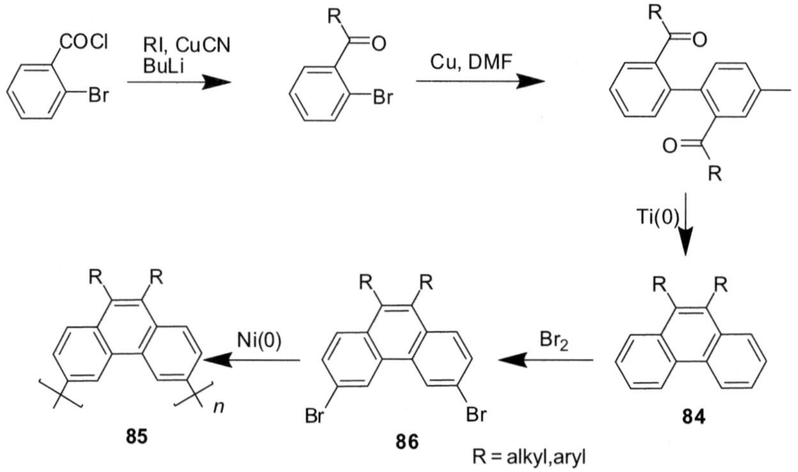

Scheme 37 Synthesis of poly(3,6-phenanthrylene)s

It was found that the Yamamoto polycoupling of **86** produced a large amount of a cyclic trimer, but coupling **86** with the corresponding diboronate gave high molecular weight polymers. While the 2,7-polymers **81** are analogues of PPP, the 3,6-polymers **85** are PPPV-analogues. These materials show similar optical properties however, with blue PL and EL (λ_{max} = 400–425 nm) being obtained from both of them.

3
Polyindenofluorenes

3.1
Poly(tetraalkylindenofluorene)s

Poly(tetraalkylindenofluorene)s (PIFs, Scheme 38) which are intermediate in structure between PDAFs and LPPP have been prepared by Müllen and co-workers by Yamamoto coupling [112]. In solution these polymers showed strong blue PL with maxima around 430 nm, which made them attractive candidates for use in blue LEDs. By extrapolating the absorption and emission maxima of oligomers the effective conjugation lengths were determined to be n = 5–6 (15–18 benzene rings) for emission and n = 6–7 for absorption. The phase behaviour of PIFs **87** and **88** resembles that of the corresponding PDAFs **16** and **19**, but with higher phase transition temperatures. Thus the octyl polymer **87** shows two nematic phases with transition temperatures at 250 and 290 °C (reverse transitions at 270 and 140 °C), while the ethylhexyl polymer **88** exhibits only a single nematic phase with a transition at 290 °C (220 °C for the reverse transition).

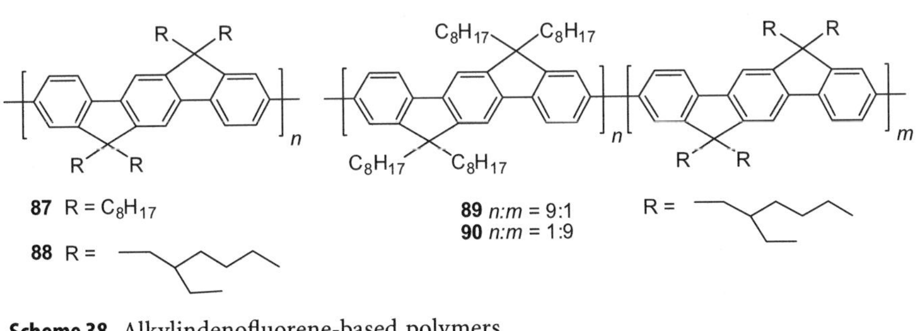

87 R = C_8H_{17}

88 R =

89 $n{:}m$ = 9:1
90 $n{:}m$ = 1:9

R =

Scheme 38 Alkylindenofluorene-based polymers

Unfortunately, as with PDAFs, obtaining stable blue emission in the solid state has presented difficulties. The tetraoctylpolymer **87** was a green emitter in the solid state, with the PL and EL spectrum being dominated by a broad emission band at 560 nm [113]. Films of the polymer **88** with ethylhexyl substituents show blue PL (λ_{max} = 429 450 nm). The EL from **88** was initially blue, but rapidly red-shifted, with a simultaneous loss of emission intensity, so that the devices had a half-life (time for intensity to drop to half its intial value) of less than 1 hour. The green emission from **87** was stable, with an estimated half-life of 5000 h. Copolymers **89** and **90** showed intermediate behaviour.

3.2
Defect Emission from PIFs

The source of this green emission might have been either excimers or defects. The amount of green emission in the solid state spectra was shown to correlate well with the presence of long ordered structures due to π-stacking in the film morphology revealed by atomic force microscopy (AFM) studies [113]. These results were consistent with the green emission arising from aggregates. Certainly the greater solid state PL efficiency for **88** (36%) and the copolymers **89** (40%) and **90** (50%) compared with **87** (24%) was consistent with the bulkier branched side-chains reducing interchain interactions and so reducing the possibility of non-radiative decay. An alternative explanation was that the emission arrives from ketone defects **91** (Scheme 39) which would be expected to show green emission slightly reshifted compared to that from the fluorenone defects in PDAFs at 530 nm. The above EL results would then be explicable in terms of the relative ability of excitons to diffuse to defect sites in the polymer films. Accordingly a model compound **92** was prepared, whose emission spectrum was found to closely match the green emission band in the polymers [114]. Further when the monomer for **87** was prepared with rigorous efforts to obtain complete alkylation, it was found that the polymer was a blue emitter in the solid state, though the emission was slightly red-shifted compared with **88**. Accordingly the green emission band can now be confidently assigned to a defect.

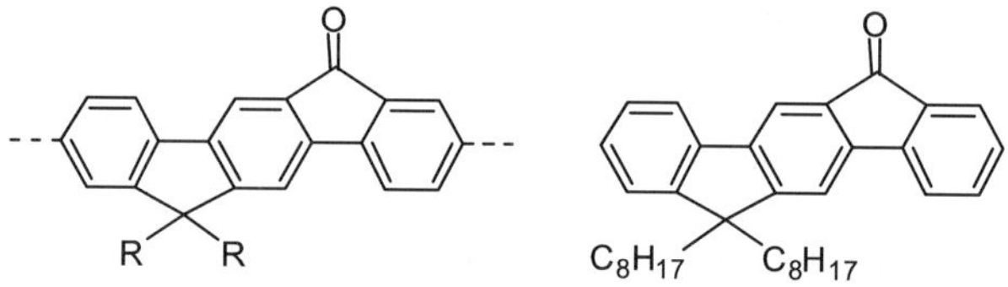

Scheme 39 Proposed emissive defect in polyindenofluorenes and a model compound

As with PDAFs, a copolymer of PIF with anthracene **93** (Scheme 40, $n:m = 85:15$) shows stable blue EL ($\lambda_{max} = 445$ nm), probably due to exciton confinement [115].

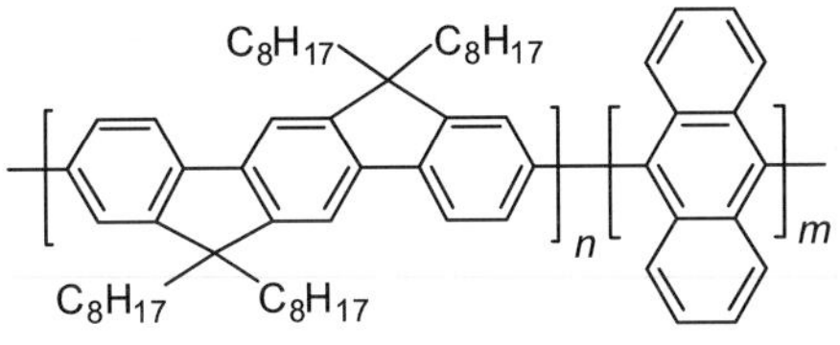

Scheme 40 Indenofluorene-anthracene copolymer which produces a stable blue emission

3.3
Poly(tetraarylindenofluorene)s

Since aryl substituents have been shown to greatly enhance the stability of blue emission from PDAFs, the effect of such substituents in PIFs has also been investigated. The synthesis of the tetraarylmonomers **94** (Scheme 41) follows a route similar to that used for the diarylfluorenes **45** (see Sect. 2.5 above). Suzuki coupling of a dibromoterephthalate [116] with commercially available 4-trimethylsilylbenzeneboronic acid gave the terphenyl **95** in high (92%) yield. Treatment of this with 4 equivalents of an aryllithium followed by electrophilic displacement of the silyl group with bromine and then ring-closure produced tetraarylindenofluorene monomers **94** in good (70%) yield. These were then polymerised with nickel(0) [117].

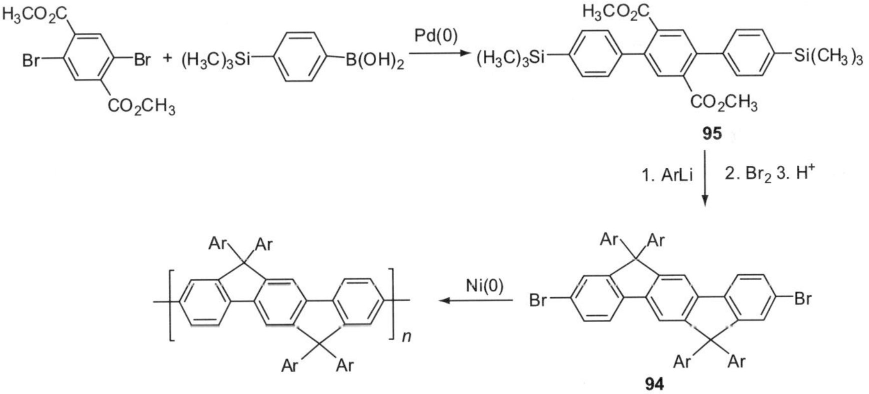

Scheme 41 Route to poly(tetraarylindenofluorene)s

Due to the limited solubilising power of the substituents, only oligomers ($n = 2$–6) of the *tert*-butylphenyl polymer **96** (Scheme 42) were obtained. The octylphenyl groups in **97** provided much better solubility, so that this polymer was obtained with $M_n = 66\,400$ g/mol, $M_w/M_n = 3.86$ (measured against a polyphenylene standard), corresponding to a degree of polymerisation of about $n = 66$ (ca. 200 phenylene rings).

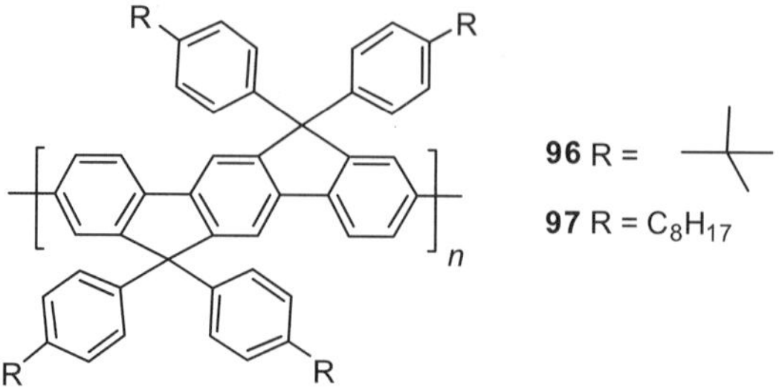

Scheme 42 Poly[tetra(4-alkylphenyl)indenofluorene]s

The PL emission from **97** is blue both in solution (λ_{max} = 428 nm) and in thin films (λ_{max} = 434 nm), with no sign of the green emission band seen for **87**. Some unoptimised LEDs have been constructed using **97**, which produce stable blue emission, though with only moderate device efficiency [117]. Some further device optimisation, e.g., by using charge transporting layers, is still required to get high efficiency. Blending 0.3 wt % of the ketone **92** into **97** has been found to totally suppress the blue emission, demonstrating how even very low levels of defects can dominate the emission from conjugated polymers [114]. A polymer **98** (Scheme 43) with spirolinkages at both bridgeheads has been made and reported to show stable blue emission (λ_{max} = 445 nm) [118].

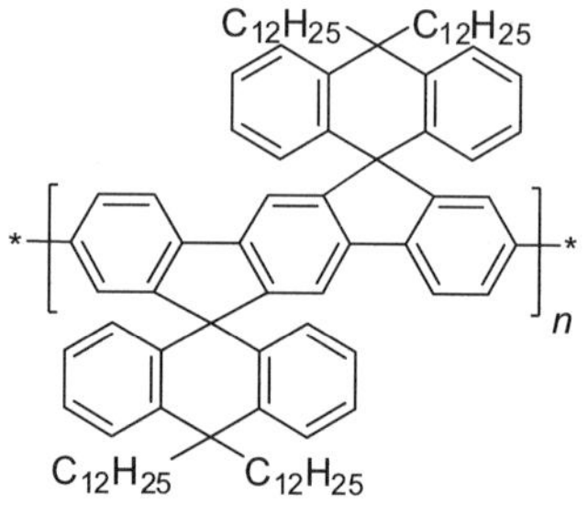

Scheme 43 Polyindenofluorene with double spirolinkages

4
Polymers of Higher Ladder-Type Oligophenylenes

4.1
Poly(ladder-Type Tetraphenylene)s

The emission from PIFs is still violet-blue, rather than a pure blue, and so to obtain a purer blue emission colour, a poly(ladder-type tetraphenylene) **99** has been made (Scheme 44) [119]. This polymer emits blue PL

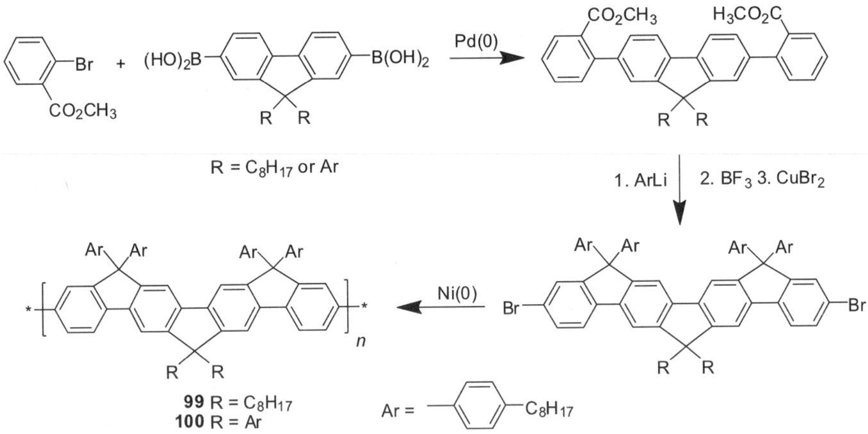

Scheme 44 Synthesis of poly(ladder-type tetraphenylene)s

(λ_{max} = 442 nm) in both the solution and thin film. The EL spectrum exhibits an extra band in the green at 510 nm, which does not appear to be due to defects from oxidation, as heating **99** in air produces a band at 575 nm, but is thought to arise from an interaction between the polymer and the metal cathode. A fully arylated polymer **100** has been prepared by the same route [120], whose stability towards oxidation is expected to be higher, but EL measurements have yet to be completed.

4.2
Poly(ladder-Type Pentaphenylene)s

The gap between PF and LPPP has finally been bridged by the preparation of ladder-type pentaphenylenes **101–102** by a route analogous to that for arylin-

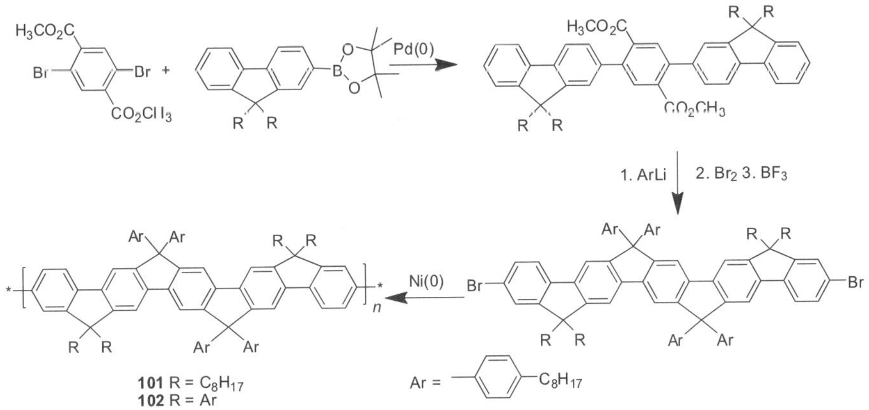

Scheme 45 Synthesis of poly(ladder-type pentaphenylene)s

denofluorenes but replacing a fluorene boronate for the phenylboronate in the Suzuki coupling (Scheme 45) [121].

The emission from **101** is blue with an emission maximum at 445 nm and a very small Stokes shift and well-resolved spectra very similar to those for LPPP **3a**. The rigidity of polymer **101** is further exemplified by its unusually high persistence length of 25 nm [122]. The blue emission from **101** is much stabler than from **16**, but experiments showed that a long wavelength emission band appeared rapidly upon heating in air. By contrast the all-aryl polymer **102** was remarkably stable, with no sign of degradation even after prolonged heating at 200 °C in air [123]. This indicates the green emission band from **101** arises from oxidation of the alkylated bridgeheads, possibly indicating the presence of some bridgehead hydrogens, due to incomplete alkylation.

5
Ladder-Type Polyphenylenes

An obvious approach to overcoming the problem of phenylene-phenylene torsion in substituted PPPs is to tether adjacent rings together with short alkyl bridges to make a ladder-type polymer. Ladder-type polymers are of considerable scientific interest as they are intermediate between linear and three-dimensional materials [124]. They can be prepared in two ways: (a) by iterative multi-centre condensation or addition (e.g., Diels–Alder cycloaddition) reactions; (b) by polymer-analogous conversion of suitably functionalised single-stranded precursors. A major feature of the second method is that the polymer analogous reactions must proceed quantitatively to avoid formation of defects in the final polymer. Though ribbon-like polyacenes can be prepared by polycycloaddition methods [124], linear ladder-type PPPs are only accessible through the conversion of single-stranded PPPs. If methine or ethene bridges are used the phenylene backbone is forced to be coplanar, but use of ethane or longer alkane bridges allows some torsion between adjacent phenylene rings (Scheme 46). For an ethane bridge the torsion angle is predicted to be about 20° [125]. Thus some degree of control over the optical properties can be achieved by varying the type of bridge(s) used, as the more coplanar the polymer the greater the expected degree of conjugation, and thus the longer the wavelengths of the absorption and emission maxima. The

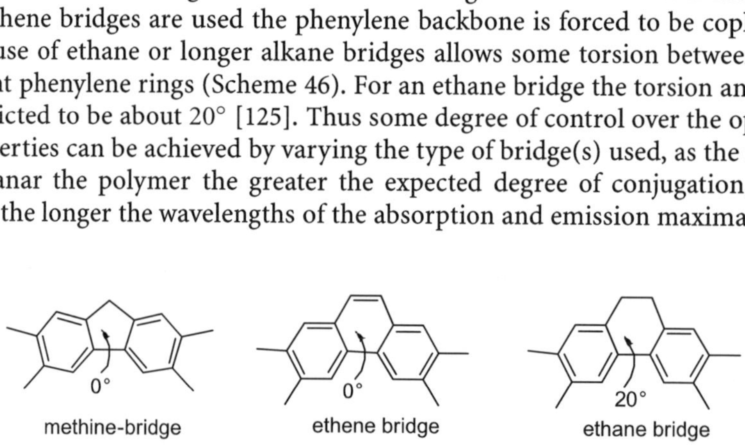

methine-bridge 0°   ethene bridge 0°   ethane bridge 20°

Scheme 46 Bridges and associated torsion angles in ladder polymers methine and ethene bridges also impart greater rigidity to the structure and thus reduce the Stokes shift between the absorption and emission maxima.

5.1 Synthesis of LPPPs

Scherf and Müllen prepared (Scheme 47) the ladder-type polyphenylene (LPPP, **5**) with methine bridges [126–129], via a poly(diacylphenylene-*co*-phenylene) precursor copolymer **103** obtained by an AA-BB type Suzuki polycondensation. The key step is the polymer analogous Friedel-Crafts ring-closing reaction on the polyalcohol **104**, obtained by the reduction of **103**. This was found to proceed quickly and smoothly upon addition of borontrifluoride to a solution of **104** in dichloromethane. The reaction appeared to be complete by both NMR and MALDI-TOF analysis, indicating the presence of less than 1% of defects due to incomplete ring closure. LPPPs with num-

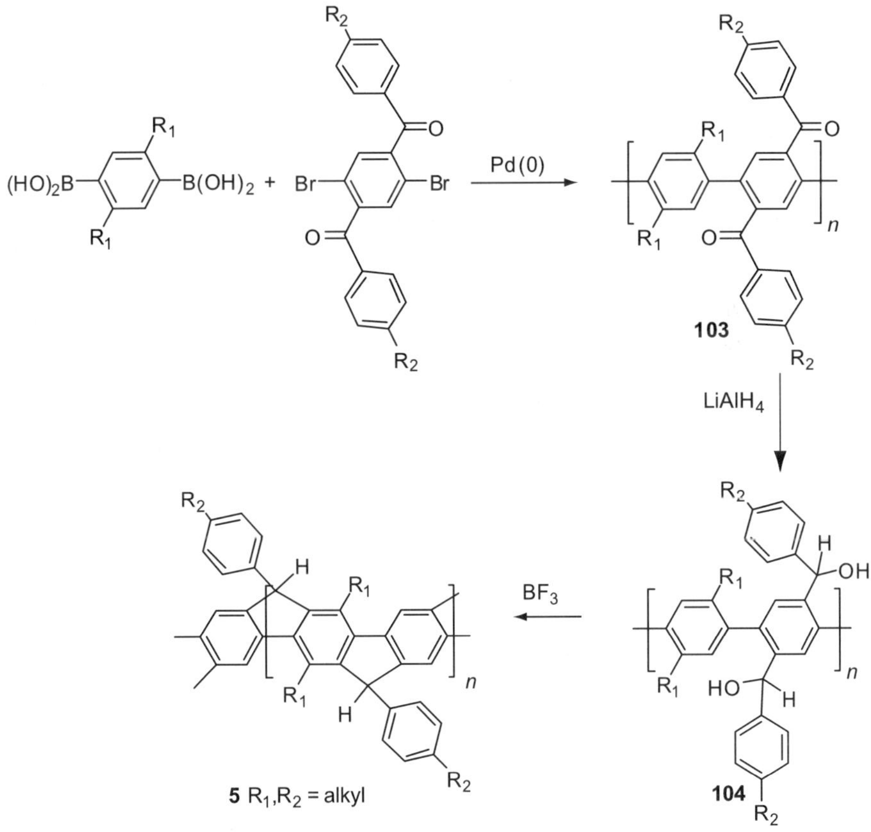

Scheme 47 Synthesis of ladder-type PPP

ber average molecular weights (M_n) of up to 50 000 g/mol have been obtained corresponding to about 150 phenylene rings.

A chiral LPPP **105** (Scheme 48) containing cyclophane units has been prepared by using a resolved cyclophane bisboronic acid [130, 131]. This is a candidate for obtaining circularly polarised EL.

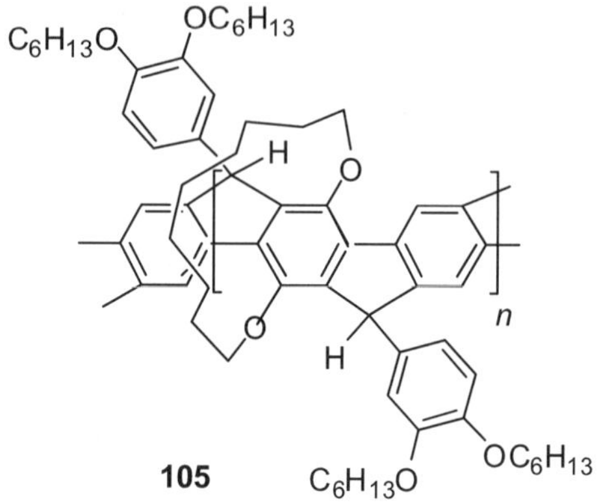

Scheme 48 Chiral ladder-type PPP

The hydrogen at the methine bridges can be replaced with a methyl group to give Me-LPPP (**106**) (Scheme 49) [132]. This is achieved by treating the precursor polymer **103** with methyl lithium, followed by ring-closure of the resulting polyalcohol **107** with boron-trifluoride as in the preparation of **5**.

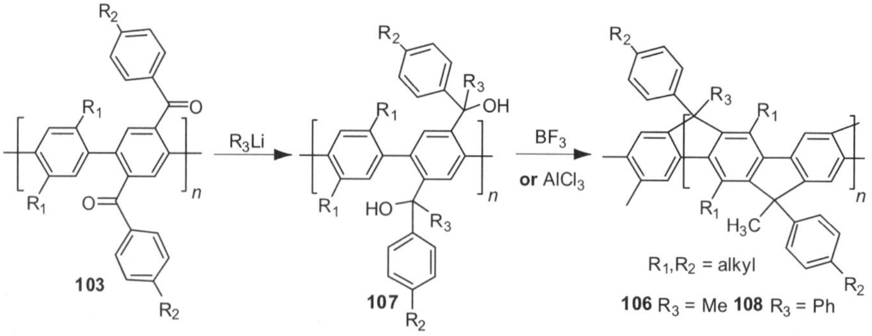

Scheme 49 Synthesis of Me-LPPP and Ph-LPPP

By using phenyllithium instead of methyl lithium Ph-LPPP (**108**) can be made [124]. In this case it has been found that $AlCl_3$ needs to be used instead of BF_3 as the Lewis acid to achieve complete ring closure.

While complete ring closure is indicated by mass spectral and NMR analysis, characterization of **5** by X-ray and neutron scattering [133], and by dynamic light scattering experiments [134] suggested this polymer has a worm-like structure with a persistence length of 6.5 nm, rather than the rigid ribbon-like structure expected for a fully ladder-type material. This suggests there are considerable numbers of defects in the structure, so that polymer 5 is best represented as a series of rigid ribbon-like segments linked by single bonds. The persistence length of 6.5 nm for 5 is actually shorter than the values obtained for step-ladder polymers such as polyfluorenes (vide supra) [122].

5.2
Optical Properties of LPPPs

The planarisation of the PPP backbone in LPPP (5) has been found to lead to better vibrational resolution in both absorption and emission spectra and to a much smaller Stokes shift [135]. The absorption maximum is at 440–450 nm, which is considerably bathochromically shifted with regard to single-stranded PPP. The absorption band also shows an unusually sharp absorption edge. The PL of 5 is an intense blue colour in solution with a maximum at 450–460 nm. The Stokes shift is thus only about 150 cm^{-1}. Such a small value is a clear indication of how the rigidity of the polymer hinders deformation in going from the ground to the excited state. A further result of this rigidity is that the PL quantum efficiencies in solution of LPPPs are very high (up to 90%) as non-radiative decay pathways are seriously reduced [136].

The emission from the substituted LPPPs 106 and 108 is slightly red-shifted compared to 5 with emission maxima at 461 nm. Single molecule spectroscopy studies on 106 have provided a new understanding of the nature of this and other conjugated polymers [137]. Whereas a molecule of low molar mass 106 (M_w = 25 kDa, corresponding to ca. 62 phenylene units) produced only a single emission peak at 459 nm which matched the emission maximum observed in ensemble measurements, a larger molecule (M_w = 67 kDa, ca. 165 phenylene units) produced up to 5 emission peaks at wavelengths between 450 and 461 nm. By performing a large number of single molecule experiments it was established that there was a linear correlation between the average number of chromophores and the chain length. This indicates that the polymer 106 consists of a series of linked chromophores, each of which can emit separately. This is consistent with the model described above, derived from light-scattering and other studies, of 5 as a series of ladder-type segments linked together by single bonds. The size of these chromophores has been determined by comparing their emission with that of well-defined oligomers. Extrapolation from the absorption maxima of bridged oligomers (Scheme 50) ranging from 3 rings (109) to 7 rings (110) has previously been used to estimate the effective conjugation length (ECL) for absorption in LPPPs as being about 11–12 benzene rings [125]. When a ladder-type undecamer 111 was made its absorption and emission were found to be slightly blue-shifted compared with Me-LPPP (106) suggesting

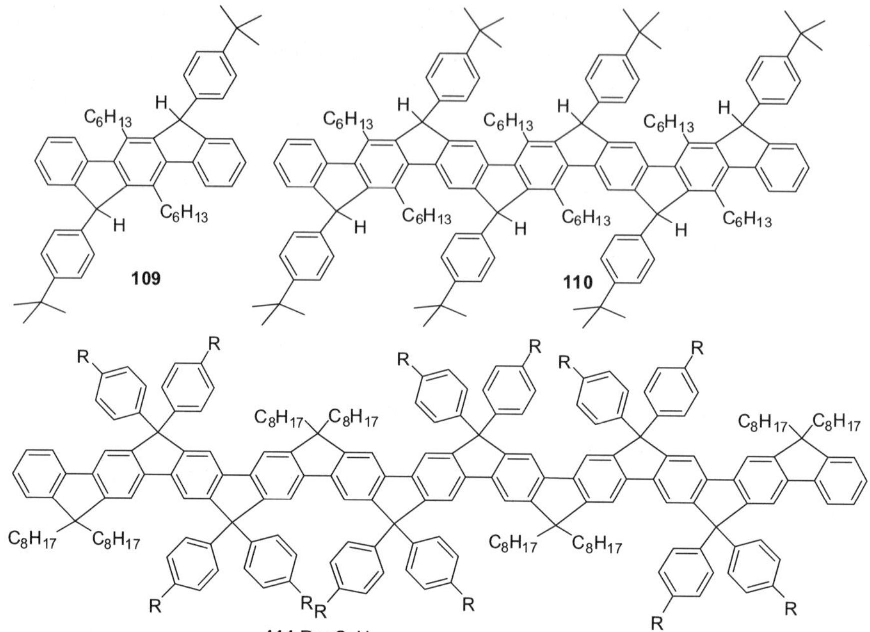

Scheme 50 Ladder-type oligophenylenes

a larger value for the ECL [137]. Single molecule spectroscopy studies on **111** demonstrated that the PL emission from the oligomer at 451 nm matches that from the smallest chromophores on the polymer, so that the emission maxima for the larger chromophores must correspond to longer segments of up to ca. 16 benzene rings, with the emission maximum at 459 nm seen in the bulk sample corresponding to a chromophore of about 15 phenylenes. The effective conjugation length for emission in LPPPs is thus around 15 benzene rings, which is similar to that for polyfluorenes and polyindenofluorenes (see above). The ECL for absorption and emission of **5** are also clearly very similar as would be expected for a rigid ladder-type structure in which the ground state and excited state geometries are constrained to be very similar.

By comparison, the effective conjugation length for poly(tetrahydropyrene)s **76** (see Sect. 2.8) which have an estimated 20° torsion angle between adjacent phenylene rings, was found to be about 19 phenyl rings. Thus, contrary to expectation it has been found that increased planarisation of the aromatic π-system leads to a decrease and not an increase in the effective conjugation length in PPP derivatives [125].

While the PL from solutions of LPPP **5** is blue, in thin films the emission is dominated by a broad, featureless band in the yellow (λ_{max} = 600 nm) [138–141]. The relative intensity ratio of the blue and yellow bands is strongly dependent upon the method used to prepare the films and varies with solvent and film thickness. The blue band disappears completely upon annealing a film of **3** at 150 °C. As a result, LEDs using **5** show yellow EL [142]. The efficiencies from single-layer devices are 0.4% with calcium cathodes and 0.02% with aluminium cathodes. Double layer devices using PPV as a hole-transporting layer show 0.6% and 0.04% efficiency with calcium and aluminium cathodes respectively. Blue emission (λ_{max} = 450–460 nm) has been observed from LEDs using **5** but has been found to be unstable, with the yellow band rapidly appearing [140].

Originally this yellow emission band was attributed to excimers from aggregates formed by π-stacking of the polymer chains. Evidence supporting this came from photophysical experiments [143, 144]. Also consistent with this hypothesis is the obtaining of pure blue EL from blends of **5** (1 wt %) in PVK [142]. The efficiency was 0.15% with a calcium cathode. The emission was found to turn white after only a few tens of minutes of device operation, which was attributed to formation of excimers due to the Joule heat produced by the passing of electricity through the device.

As mentioned above, the emission from **106** is slightly red-shifted compared with **5**, with an emission maximum at λ_{max} = 461 nm and a secondary peak at λ_{max} = 491 nm, so that the emission colour is blue-green. Unlike **5**, Me-LPPP shows almost identical emission from films and solutions, with PL quantum efficiencies of over 90% in solution and up to 60% in the solid state. There is a broad emission band centred at 560 nm, which has been attributed to emission from aggregates [145]. This band is much weaker than the yellow band from LPPP (**5**), and the emission does not change upon annealing. LEDs using **106** produce blue-green emission with EL efficiencies of up to 4% [146–150]. These high emission efficiencies make Me-LPPP a particularly promising material for use in organic solid-state lasers [3]. Optically pumped lasing has been observed from films of **106** in both waveguide and "distributed feedback" configurations by the groups of Leising and Lemmer [151–155].

Subsequently, Lupton [156] reported that the broad emission feature centred at 560 nm could be detected in the delayed fluorescence from both films and dilute solutions of Me-LPPP. This is not consistent with the suggestion that the long wavelength emission from ladder-type PPPs comes from aggregates, as dilute solutions should not contain any aggregates. He therefore proposed that the long wavelength emission band originates from defects on the polymer chains.

Convincing evidence has been produced that long wavelength emission from polyfluorenes is due to fluorenone (**39**) units (see Sect. 4.1 below), so a probable structure for the defect in LPPPs is a ketone as in **112** (Scheme 51). As the ketone in **112** has more extended conjugation than in fluorenone (**39**) one would expect the emission from it to be bathochromically shifted, which is consistent with the long-wavelength emission from LPPPs occurring at 560–600 nm, and that from polyfluorenes at about 530 nm.

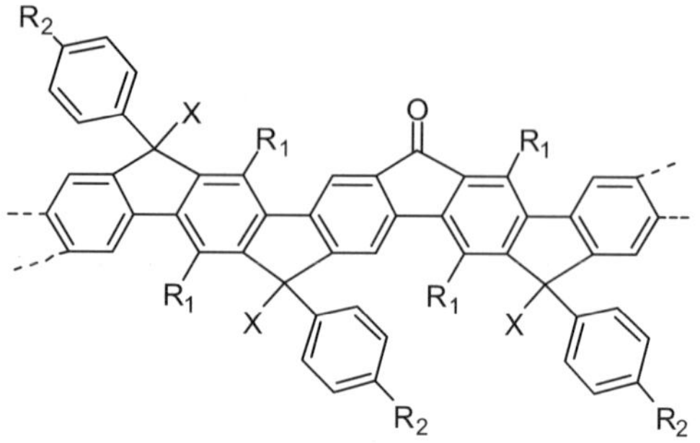

112 R$_1$,R$_2$ = alkyl, X = H, CH$_3$

Scheme 51 Proposed emissive defect in ladder-type PPPs

The ketone **112** presumably arises from oxidation of the methine bridge by oxygen from the air. The difference between the emission spectra in solution and the solid state would then reflect the more efficient energy transfer to the defect sites in the latter due to the increased intermolecular interactions. The much lower intensity of the defect band and the greater emission stability of Me-LPPP over LPPP can be explained as being due to the greater difficulty in oxidising the methyl-substituted methine bridge producing a much lower level of defect sites. If a fluorene bisboronate is used in the synthesis of LPPPs instead of benzene bisboronate the blue-green EL from the resulting polymers is reportedly stabler than the emission from **5**, presumably because fewer bridges are being made during the polymer analogous ring closure reaction and so fewer defects are produced [157].

That even low levels of defects can produce strong emission is exemplified by the case of Ph-LPPP (**108**). The PL emission from **108** is very similar to that from **106** with maxima at 460 and 490 nm. However, the EL spectrum shows an additional long wavelength band. This is not a broad featureless band as seen for the defect emission from **5** or **106**, but one with well-resolved maxima at 600 and 650 nm. Photophysical investigation of this emission showed the feature at 600 nm to be emission from a triplet exciton (phosphorescence) with a vibronic shoulder at 650 nm [158]. Elemental analysis of the polymer showed it contained 80 ppm of palladium (cf. <2 ppm in **106**). It was therefore proposed that residues of the palladium catalyst used to make the precursor polymer **103** reacted with the phenyllithium and the polymer to introduce covalently bound palladium centres onto the polymer chain. These then act as sites for phosphorescent emission.

5.3
Ladder-Type Polymers with 2-Carbon Bridges

The polymers discussed above have methine bridges. A ladder polymer **113** with dihydroxyethane bridges has been made by Forster and Scherf

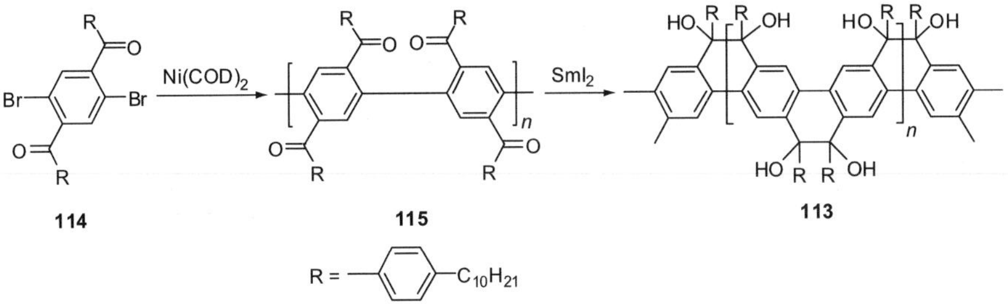

Scheme 52 Synthesis of a ladder-type polyphenylene with dihydroxyethane bridges

(Scheme 52) [159]. Yamamoto polycondensation of a dibromodibenzoylbenzene **114** gave the poly(diacylphenylene) **115**, which was coupled with samarium(II) iodide coupling to give **113**. This polymer shows strong blue-green fluorescence in solution (λ_{max} = 459 nm) and the solid-state (λ_{max} = 482 nm). The red-shift in the solid state PL indicates that **113** is less rigid than the LPPPs with methine bridges, but still shows only weak geometrical changes in going from the ground to the excited state. No long wavelength emission band is seen in the PL spectrum of **113**.

Treatment of the poly(dibenzoylphenylene)s **115** and **116** with boron sulfide gives polymers **117** and **118** with ethene bridges (Scheme 53) [160, 161]. These also show blue-green emission (λ_{max} = 478 nm and 484 nm respectively) with some long wavelength emission in the solid state which has been attributed to aggregates. Their EL efficiency is reported to be very low (<0.1%) [162].

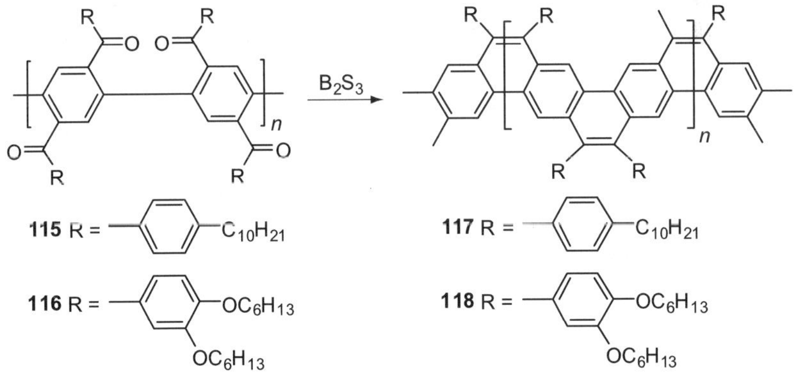

Scheme 53 Synthesis of ladder-type polyphenylene with ethene bridges

A similar polymer **119** was prepared by Goldfinger and Swager by an acid-catalysed cyclisation of a PPP precursor **120** with alkyne sidechains (Scheme 54) [163]. There is no report on the emission from this material, but

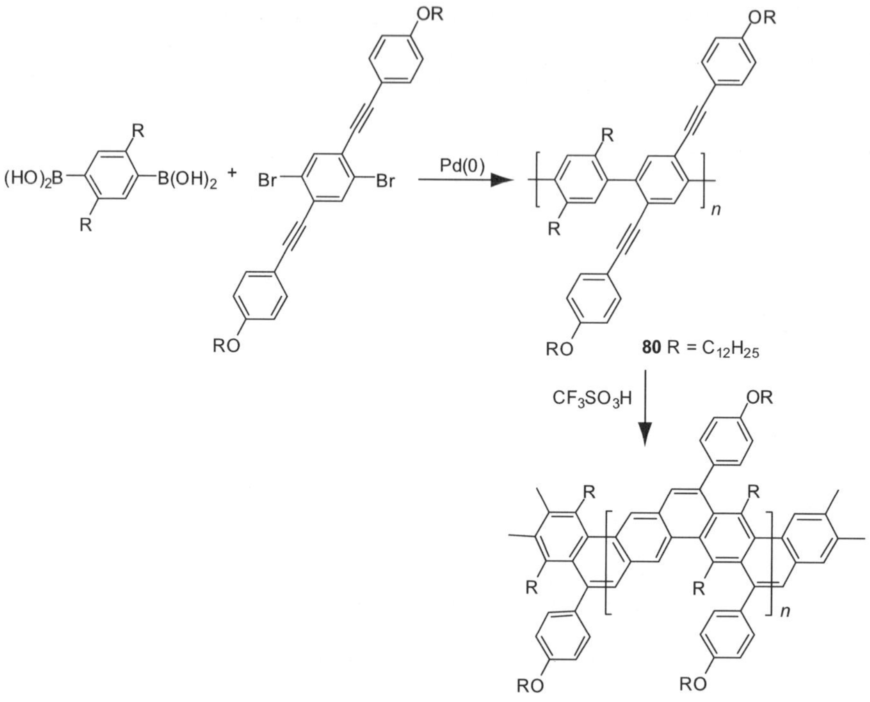

Scheme 54 Ladder-type polyphenylene by alkyne cyclisation

the absorption edge was reported to be at 478 nm, suggesting it should be a blue-green or green-emitter.

6
Conclusions

As shown in this review, the properties of carbon-bridged phenylene-based materials so as to maximise their potential as active materials in LEDs or polymer lasers can be controlled by deliberate synthetic design. By changing the number and type of bridges between some or all of the phenylene units the effective conjugation length of the polymer may be controlled, while the interactions between the chains and the injection of charges may be regulated by careful selection of substituents. (The emission and electrochemical properties of phenylenes can also be tuned by using heteroatoms in the bridges, as is described in other chapters.) By these means it is possible to minimise interchain interactions which lead to loss of luminescence efficiency and/or undesirable red-shifts in the emission spectrum. The goal of obtaining stable blue emission now appears to be attainable if steps are taken to minimise the formation of emissive defects in the polymers and the diffusion of excitons to them. The emission colour can also be tuned efficiently over the whole visible spectrum by incorporation of suitable chromophores. With ongoing interdisciplinary research efforts, the fabrication of polyphenylene-based high efficiency full-colour LED-based displays with long lifetimes may soon be possible. The prospects for polyphenylene-based lasers also appear good, though much work remains to be done in this field.

Acknowledgements We gratefully acknowledge the funding for research into emissive polyphenylenes at Mainz provided by the European Commission through Project NAIMO (Integrated Project Number NMP4-CT-2004-500355).

References

1. Kraft A, Grimsdale AC, Holmes AB (1998) Angew Chem Intl Ed 37:403
2. Mitschke U, Bäuerle P (2000) J Mater Chem 10:1471
3. McGehee MD, Heeger AJ (2000) Adv Mater 12:1655
4. Grimsdale AC, Müllen K (2006) Adv Polym Sci 199:1
5. Kim DY, Cho HN, Kim CY (2000) Prog Polym Sci 25:1089
6. Grem G, Leditzky G, Ullrich B, Leising G (1992) Adv Mater 4:36
7. Gin DL, Conticello VP (1996) Trends Polym Sci 4:217
8. Schlüter AD, Wegner G (1993) Acta Polymer 44:59
9. Park KC, Dodd LR, Levon K, Kwei TK (1996) Macromolecules 29:7149
10. Grell M, Bradley DDC (1999) Adv Mater 11:895
11. Schlüter AD (2001) J Polym Sci A: Polym Chem 39:1533
12. Yamamoto T (1992) Prog Polym Sci 17:1153
13. Leclerc M (2001) J Polym Sci, Part A: Polym Chem 39:2867
14. Neher D (2001) Macromol Rapid Commun 22:1365
15. Scherf U, List EWJ (2002) Adv Mater 14:477
16. Fukuda M, Sawada K, Yoshino K (1989) Jpn J Appl Phys 28:L1433
17. Fukuda M, Sawada K, Yoshino K (1993) J Polym Sci, Polym Chem 31:2465
18. Ohmori Y, Uchida K, Muro K, Yoshino K (1991) Jpn J Appl Phys 30: L1941
19. Ohmori Y, Uchida M, Morishima C, Fujii A, Yoshino K (1993) Jpn J Appl Phys 32:L1663
20. Uchida M, Ohmori Y, Morishima C, Yoshino K (1993) Synth Met 57:4168
21. Liu B, Chen Z-K, Yu W-L, Lai Y-H, Huang W (2000) Thin Solid Films 363:332
22. Advincula R, Yia C, Inaoka S (2000) Polym Prepr 41(1):846
23. Woo EP, Inbasekaran M, Shiang W, Roof GR (1997) PCT Intl Pat Appl: WO 97/05184
24. Inbasekaran M, Wu W, Woo EP (1998) US Patent 5777070
25. Bernius M, Inbasekaran M, O'Brien J, Wu W (2000) Adv Mater 12:1737
26. Bernius M, Inbasekaran M, Woo E, Wu W, Wujkowski L (2000) J Mater Sci; Mater Electron 11:111
27. Bernius M, Inbasekaran M, Woo E, Wu W, Wujkowski L (2000) Thin Solid Films 363:55
28. Inbasekaran M, Woo E, Bernius M, Wujkowski L (2000) Synth Met 111–112:397
29. Grice AW, Bradley DDC, Bernius MT, Inbasekaran M, Wu WW, Woo EP (1998) Appl Phys Lett 73:629

30. Grell M, Knoll W, Lupo D, Meisel A, Miteva T, Neher D, Nothofer H-G, Scherf U, Yasuda A (1999) Adv Mater 11:671
31. Klaerner G, Miller RD (1998) Macromolecules 31:2007
32. Lee SH, Tsutsui T (2000) Thin Solid Films 363:76
33. Miller RD, Klaerner G, Fuhrer T, Kreyenschmidt M, Kwak J, Lee V, Chen W-D, Scott JC (1999) Nonlinear Optics 20:269
34. Grell M, Bradley DDC, Inbasekaran M, Woo EP (1997) Adv Mater 9:798
35. Teetsov J, Fox MA (1999) J Mater Chem 9:2117
36. Grell M, Bradley DDC, Whitehead KS (2000) J Korean Phys Soc 36:331
37. Whitehead KS, Grell M, Bradley DDC, Inbasekaran M, Woo EP (2000) Synth Met 111–112:181
38. Whitehead KS, Grell M, Bradley DDC, Jandke M, Strohriegl P (2000) Appl Phys Lett 76:2946
39. Whitehead KS, Grell M, Bradley DDC, Jandke M, Strohriegl P (2000) Proc SPIE 3939:172
40. Miteva T, Meisel A, Grell M, Nothofer H-G, Lupo D, Yasuda A, Knoll W, Kloppenburg L, Bunz UHF, Scherf U, Neher D (2000) Synth Met 111–112:173
41. Nothofer H-G, Meisel A, Miteva T, Neher D, Forster M, Oda M, Lieser G, Sainova D, Yasuda A, Lupo D, Knoll W, Scherf U (2000) Macromol Symp 154:139
42. Oda M, Meskers SCJ, Nothofer HG, Scherf U, Neher D (2000) Synth Met 111–112:575
43. Oda M, Nothofer H-G, Lieser G, Scherf U, Meskers SCJ, Neher D (2000) Adv Mater 12:362
44. Nothofer H-G, Oda M, Neher D, Scherf U (2000) Proc SPIE 4107:19
45. Tang H, Fujiki M, Motonaga M, Torimitsu K (2001) Polym Prepr 42(1):440
46. Becker S, Ego C, Grimsdale AC, List EWJ, Marsitzky D, Pogantsch A, Setayesh S, Leising G, Müllen K (2002) Synth Met 125:73
47. Ego C, Marsitzky D, Becker S, Zhang J, Grimsdale AC, Müllen K, MacKenzie JD, Silva C, Friend RH (2003) J Am Chem Soc 125:437
48. Klärner G, Davey MH, Chen W-D, Scott JC, Miller RD (1998) Adv Mater 10:993
49. Klärner G, Lee J-I, Davey MH, Miller RD (1999) Adv Mater 11:115
50. Tokito S, Weinfurtner K-H, Fujikawa H, Tsutsui T, Taga Y (2001) Proc SPIE 4105:69
51. Lee J-I, Klaerner G, Davey MH, Miller RD (1999) Synth Met 102:1087
52. Cao J, Zhou Q, Cheng Y, Geng Y, Wang L, Ma D, Jing X, Wang F (2005) Synth Met 152:237
53. Liu J, Min C, Zhou Q, Cheng Y, Wang L, Ma D, Jing X, Wang F (2006) Appl Phys Lett 88:083505
54. Liu J, Tu G, Zhou Q, Cheng Y, Geng Y, Wang L, Ma D, Jing X, Wang F (2006) J Mater Chem 16:1431
55. Liu J, Zhou Q, Cheng Y, Geng Y, Wang L, Ma D, Jing X, Wang F (2006) Adv Funct Mater 16:957
56. Wang F, Wang L, Chen J, Cao Y (2007) Macromol Rapid Commun 28:2012
57. Sandee AJ, Williams CK, Evans NR, Davies JE, Boothby CE, Köhler A, Friend RH, Holmes AB (2004) J Am Chem Soc 126:7041
58. Galbrecht F, Yang XH, Nehls BS, Neher D, Farrell T, Scherf U (2005) Chem Commun, p 2378
59. Peng Q, Xie M, Huang Y, Lu Z, Cao Y (2005) Macromol Chem Phys 206:2373
60. Hou Q, Zhang Y, Yang RQ, Yang W, Cao Y (2005) Synth Met 153:193
61. Zhuang W, Zhang Y, Hou Q, Wang L, Cao Y (2006) J Polym Sci (A), Polym Chem 44:4174
62. Jiang J, Xu Y, Yang W, Guan R, Liu Z, Zhen H, Cao Y (2006) Adv Mater 18:1769

63. Bliznyuk VN, Carter SA, Scott JC, Klärner G, Miller RD, Miller DC (1999) Macromolecules 32:361
64. Lee J-I, Klärner G, Miller RD (1999) Synth Met 101:126
65. List EWJ, Guentner R, Scanducci de Freitas P, Scherf U (2002) Adv Mater 14:374
66. Lupton JM, Craig MR, Meijer EW (2002) Appl Phys Lett 80:4489
67. Pei J, Ni J, Zhou X-H, Cao X-Y, Lai Y-H (2002) J Org Chem 67:4924
68. Gaal M, List EJW, Scherf U (2003) Macromolecules, p 4236
69. Weinfurtner K-H, Fujikawa H, Tokito S, Taga Y (2000) Appl Phys Lett 76:2502
70. Sainova D, Miteva T, Nothofer H-G, Scherf U, Glowacki I, Ulanski J, Fujikawa H, Neher D (2000) Appl Phys Lett 76:1810
71. Klärner G, Lee J-I, Lee VY, Chan E, Chen J-P, Nelson A, Markiewicz D, Siemens R, Scott JC, Miller RD (1999) Chem Mater 11:1800
72. Marsitzky D, Vestberg R, Blainey P, Tang BT, Hawker CJ, Carter KR (2001) J Am Chem Soc 123:6965
73. Setayesh S, Grimsdale AC, Weil T, Enkelmann V, Müllen K, Meghdadi F, List EJW, Leising G (2001) J Am Chem Soc 123:946
74. Lee J, Cho H-J, Jung B-J, Cho NS, Shim H-K (2004) Macromolecules 37:8523
75. Takagi K, Kunii S, Yuki Y (2005) J Polym Sci (A), Polym Chem 43:2119
76. Gong X, Iyer PK, Moses D, Bazan GC, Heeger AJ, Xiao SS (2003) Adv Funct Mater 13:325
77. Craig MR, de Kok MM, Hofstraat JW, Schenning APHJ, Meijer EW (2003) J Mater Chem 13:2861
78. Cho SY, Grimsdale AC, Jones DJ, Watkins SE, Holmes AB (2007) J Am Chem Soc 129:11910
79. Jacob J, Oldridge L, Zhang J, Gaal M, List EJW, Grimsdale AC, Müllen K (2004) Curr Appl Phys 3:339
80. Pogantsch AF, Wenzl FP, List EWJ, Leising G, Grimsdale AC, Müllen K (2002) Adv Mater 14:1061
81. List EWJ, Pogantsch A, Wenzl FP, Kim C-H, Shinar J, Loi MA, Bongiovanni G, Mura A, Setayesh S, Grimsdale AC, Nothofer HG, Müllen K, Scherf U, Leising G (2001) Mat Res Soc Symp Proc 665:C5.47.1
82. Lupton JM, Schouwink P, Keivanidis PE, Grimsdale AC, Müllen K (2003) Adv Funct Mater 13:154
83. Pogantsch A, Gadermaier C, Cerullo G, Lanzani G, Scherf U, Grimsdale AC, Müllen K, List EJW (2003) Synth Met 139:847
84. Pogantsch A, Wenzl FP, Scherf U, Grimsdale AC, Müllen K, List EJW (2003) J Chem Phys 119:6904
85. Lee J-H, Hwang D-H (2003) Chem Commun, p 2836
86. Ego C, Grimsdale AC, Uckert F, Yu G, Srdanov G, Müllen K (2002) Adv Mater 14:809
87. Surin M, Hennebicq E, Ego C, Marsitzky D, Grimsdale AC, Müllen K, Brédas JL, Lazzaroni R, Leclère P (2004) Chem Mater 16:994
88. Janietz S, Bradley DDC, Grell M, Giebeler C, Inbasekaran M, Woo EP (1998) Appl Phys Lett 73:2453
89. Gross M, Muller DC, Nothofer H-G, Scherf U, Neher D, Meerholz K (2000) Nature 405:861
90. Brown TM, Friend RH, Millard IS, Lacey DJ, Burroughes JH, Cacialli F (2001) Appl Phys Lett 79:174
91. Chen JP, Markiewicz D, Lee VY, Klaerner G, Miller RD, Scott JC (1999) Synth Met 107:203

92. Miteva T, Meisel A, Knoll W, Nothofer HG, Scherf U, Müller DC, Meerholz K, Yasuda A, Neher D (2001) Adv Mater 13:565
93. Müller DC, Braig T, Nothofer H-G, Arnoldi M, Gross M, Scherf U, Nuyken O, Meerholz K (2000) Chem Phys Chem 1:207
94. Bernius MT, Inbasekaran M, Woo EP, Wu W, Wujkowski L (1999) Proc SPIE 3797:129
95. Uckert F, Setayesh S, Müllen K (1999) Macromolecules 32:4519
96. Uckert F, Tak Y-H, Müllen K, Bässler H (2000) Adv Mater 12:905
97. Kulkarni AP, Kong X, Jenekhe SA (2004) J Phys Chem B 108:8689
98. Wu F-I, Reddy S, Shu C-F, Liu MS, Jen AK-Y (2003) Chem Mater 15:269
99. Shu C-F, Dodda R, Wu F-I, Liu MS, Jen AK-Y (2003) Macromolecules 36:6698
100. Salbeck J (1996) Ber Bunsenges Phys Chem 100:1666
101. Kreuder W, Lupo D, Salbeck J, Schenk H, Stehlin T (1996) EU Patent Appl EP 707 020
102. Wu Y, Li J, Fu Y, Bo Z (2004) Org Lett 6:3485
103. Yu W-L, Pei J, Huang W, Heeger AJ (2000) Adv Mater 12:828
104. Grisorio R, Mastronilli P, Nobile CF, Romanazzi G, Surana EP, Acierno D, Amendola E (2005) Macromol Chem Phys 206:448
105. Vak D, Chun C, Lee CL, Kim J-J, Kim D-Y (2004) J Mater Chem 14:1342
106. Müller CD, Falcou A, Reckefuss N, Rojahn M, Wiederhirn V, Rudati P, Frohne H, Nuyken O, Becker H, Meerholz K (2003) Nature 421:829
107. Yamamoto T, Asao T, Fukumoto H (2004) Polymer 45:8085
108. Lu P, Zhang H, Zheng Y, Ma Y, Zhang G, Chen X, Shen J (2003) Synth Met 135–136:205
109. Lu P, Zhang H, Shen F, Yang B, Li D, Ma Y, Chen X, Li J, Tama N (2003) Macromol Chem Phys 204:2274
110. Lim S-F, Friend RH, Rees ID, Li J, Ma Y, Robinson K, Holmes AB, Hennebicq E, Beljonne D, Cacialli F (2005) Adv Funct Mater 15:981
111. Yang C, Scheiber H, List EJW, Jacob J, Müllen K (2006) Macromolecules 39:5213
112. Setayesh S, Marsitzky D, Müllen K (2000) Macromoledules 33:2016
113. Grimsdale AC, Leclère P, Lazzaroni R, MacKenzie JD, Murphy C, Setayesh S, Silva C, Friend RH, Müllen K (2002) Adv Funct Mater 12:729
114. Keivanidis PE, Jacob J, Oldridge L, Sonar P, Carbonnier B, Baluschev S, Grimsdale AC, Müllen K, Wegner G (2005) ChemPhysChem 6:1650
115. Marsitzky D, Scott JC, Chen J-P, Lee VY, Miller RD, Setayesh S, Müllen K (2001) Adv Mater 13:1096
116. Lambda JJS, Tour JM (1994) J Am Chem Soc 116:11723
117. Jacob J, Zhang J, Grimsdale AC, Müllen K, Gaal M, List EJW (2003) Macromolecules 36:8240
118. Vak D, Lim B, Lee S-H, Kim D-Y (2005) Org Lett 7:4229
119. Mishra AK, Graf M, Grasse F, Jacob J, List EJW, Müllen K (2006) Chem Mater 18:2879
120. Laquai F, Mishra AK, Ribas MR, Petrozza A, Jacob J, Akcelrud L, Müllen K, Friend RH, Wegner G (2007) Adv Funct Mater 17:3231
121. Jacob J, Sax S, Piok T, List EJW, Grimsdale AC, Müllen K (2004) J Am Chem Soc 126:6987
122. Somma E, Loppinet B, Fytas G, Setayesh S, Jacob J, Grimsdale AC, Müllen K (2004) Colloid Polym Sci 282:867
123. Jacob J, Grimsdale AC, Müllen K, Sax S, Gaal M, List EJW (2005) Macromolecules, p 9933
124. Scherf U (1999) J Mater Chem 9:1853
125. Grimme J, Kreyenschmidt M, Uckert F, Müllen K, Scherf U (1995) Adv Mater 7:292
126. Scherf U, Müllen K (1991) Makromol Chem Rapid Commun 12:489

127. Scherf U, Müllen K (1992) Macromolecules 25:3546
128. Scherf U, Müllen K (1995) Adv Polym Sci 123:1
129. Scherf U, Müllen K (1997) ACS Symp Ser 672:358
130. Fiesel R, Huber J, Scherf U (1996) Angew Chem, Intl Ed Engl 35:2111
131. Fiesel R, Huber J, Scherf U (1998) Enantiomer 3:383
132. Scherf U, Bohnen A, Müllen K (1992) Makromol Chem 193:1127
133. Hickl P, Ballauff M, Scherf U, Müllen K, Lindner P (1997) Macromolecules 30:273
134. Petekidis G, Fytas G, Scherf U, Müllen K, Fleischer G (1999) J Polym Sci (A), Polym Chem 37:2211
135. Graupner W, Grem G, Meghdadi F, Paar C, Leising G, Scherf U, Müllen K, Fischer W, Stelzer F (1994) Mol Cryst Liq Cryst 256:549
136. Stampfl J, Graupner W, Leising G, Scherf U (1995) J Lumin 63:117
137. Schindler F, Jacob J, Grimsdale A, Scherf U, Müllen K, Lupton JM, Feldmann J (2005) Angew Chem Intl Ed. 44:1520
138. Grem G, Martin V, Meghdadi F, Paar C, Stampfl J, Sturm J, Tasch S, Leising G (1995) Synth Met 71:2193
139. Grem G, Leising G (1993) Synth Met 55–57:4105
140. Hüber J, Müllen K, Saalbeck J, Schenk H, Scherf U, Stehlin T, Stern R (1994) Acta Polymer 45:244
141. Leising G, Grem G, Leditzky G, Scherf U (1993) Proc SPIE 1910:70
142. Grüner J, Wittmann HF, Hamer PJ, Friend RH, Huber J, Scherf U, Müllen K, Moratti SC, Holmes AB (1994) Synth Met 67:181
143. Mahrt RF, Siegner U, Lemmer U, Hopmeier M, Scherf U, Heun S, Göbel EO, Müllen K, Bässler H (1995) Chem Phys Lett 240:373
144. Köhler A, Grüner J, Friend RH, Müllen K, Scherf U (1995) Chem Phys Lett 243:456
145. Haugeneder A, Lemmer U, Scherf U (2002) Chem Phys Lett 351:354
146. Leising G, Köpping-Grem G, Meghdadi F, Niko A, Tasch S, Fischer W, Pu L, Wagaman MW, Grubbs RH, Althouel L, Froyer G, Scherf U, Huber J (1995) Proc SPIE 2528:307
147. Leising G, Ekström O, Graupner W, Meghdadi F, Moser M, Kranzelbinder G, Jost T, Tasch S, Winkler B, Athouel L, Froyer G, Scherf U, Müllen K, Lanzani G, Nisoli M, DeSilvestri S (1996) Proc SPIE 2852:189
148. Leising G, Tasch S, Meghdadi F, Athouel L, Froyer G, Scherf U (1996) Synth Met 81:185
149. Tasch S, Niko A, Leising G, Scherf U (1996) Appl Phys Lett 68:1090
150. Tasch S, Niko A, Leising G, Scherf U (1996) Mat Res Soc Symp Proc 413:71
151. Leising G, Tasch S, Brandstätter C, Graupner W, Hampel S, List EWJ, Meghdadi F, Zenz C, Schlichting P, Rohr U, Geerts Y, Scherf U, Müllen K (1997) Synth Met 91:41
152. Leising G, List EWJ, Zenz C, Tasch S, Brandstätter C, Graupner W, Markart P, Meghdadi F, Kranzelbinder G, Niko A, Resel R, Zojer E, Schlichting P, Rohr U, Geerts Y, Scherf U, Müllen K, Smith R, Gin D (1998) Proc SPIE 3476:76
153. Stagira S, Zavelani-Rossi M, Nisoli M, DeSilvestri S, Lanzani G, Zenz C, Mataloni P, Leising G (1998) Appl Phys Lett 73:2860
154. Kallinger C, Hilmer C, Haugeneder A, Perner M, Spirkl W, Lemmer U, Feldmann J, Scherf U, Müllen K, Gombert A, Wittwer V (1998) Adv Mater 10:920
155. Riechel S, Kallinger C, Lemmer U, Feldmann J, Gombert A, Wittwer V, Scherf U (2000) Appl Phys Lett 77:2310
156. Lupton JM (2002) Chem Phys Lett 365:366
157. Qiu S, Lu P, Liu X, Lu FS, Liu L, Ma Y, Shen J (2003) Macromolecules 36:9823

158. Lupton JM, Pogantsch A, Piok T, List EWJ, Patil S, Scherf U (2002) Phys Rev Lett 89:7401
159. Forster M, Scherf U (2000) Macromol Rapid Commun 21:810
160. Chmil K, Scherf U (1993) Makromol Chem, Rapid Commun 14:217
161. Chmil K, Scherf U (1997) Acta Polym 48:208
162. Kirstein S, Cohen G, Davidov D, Scherf U, Klapper M, Chmil K, Müllen K (1995) Synth Met 69:415
163. Goldfinger MB, Swager TM (1994) J Am Chem Soc 116:7895
164. Trimpin S, Grimsdale AC, Räder HJ, Müllen K (2002) Anal Chem 74:3777
165. Xia C, Advincula RC (2001) Macromolecules 34:5854

Polyfluorenes for Device Applications

Show-An Chen (✉) · Hsin-Hung Lu · Chih-Wei Huang

Chemical Engineering Department, National Tsing-Hua University,
30013 Hsinchu, Taiwan, Republic of China
sachen@che.nthu.edu.tw

1	Introduction	51
2	Concepts for Performance Improvement of Polyfluorene Devices	51
2.1	Background of Electronic States of PFs	51
2.2	Adjustment of Chain Structures in PFs	54
2.2.1	Chemical Modifications	54
2.2.2	Physical Manipulation	63
2.3	Sources of Undesirable Green Emission in PF-based PLEDs	64
3	PFs with Various Emission Colors	65
3.1	Green and Red Emitters	65
3.1.1	Chemical Strategy via Fluorescent Moiety Incorporated in Polymer Backbone, Side Chain, or Chain End	65
3.1.2	Chemical Strategy via Phosphorescent Moiety Incorporated in Polymer Side Chain and Backbone	70
3.2	White Emitters	72
3.2.1	Physical Blending System	72
3.2.2	Chemical Strategy: Single-Polymer Approach	73
4	Alteration of Interfaces Between PFs and Electrodes	75
4.1	Hole-Transporting Layer	75
4.2	Cathode Materials and Electron-Injection/Hole-Blocking Layer	78
5	Summary	81
	References	82

Abstract This article mainly reviews the approaches that have been proposed to improve the device performance of polyfluorene (PF)-based polymer light-emitting diodes (PLEDs). Chemical modifications on main chains, side chains, and chain ends of PFs via the incorporation of charge-transport moieties can reduce the hole- and electron-injection barriers; while physical manipulation on main-chain structures of PFs, poly(9,9-di-n-octylfluorene) (PFO), after dipping in mixed solvent/non-solvent (tetrahydrofuran/methanol) can generate β-phase with extended conjugation length, leading to a promoted charge balance and stable pure blue emission. Hole- and electron-injection barriers can be effectively lowered by the insertion of hole-transport and electron-injection layers, respectively. The recent development of PFs with various emission colors, via physical blending or chemical modification, are presented for a comprehensive understanding of PFs for device applications. The deliberate choice of cathode material with work function matching the lowest unoccupied molecular orbital (LUMO) levels of PFs is another efficient method for increasing electron flux, and is also discussed in this review.

Keywords Device performance · Electron-transport moiety · Hole-transport moiety · Polyfluorenes · Polymer light-emitting diode

Abbreviations

Alq$_3$	Tris-(8-hydroxyquinoline) aluminum
Ca(acac)$_2$	Calcium acetylacetonate
CIE	Commission Internationale de l'Eclairage
CN	Cyano
CV	Cyclic voltammetry
Cz	Carbazole
DBT	4,7-Di-2-thienyl-2,1,3-benzothiadiazole
DOF	9,9-Dioctylfluorene
E	Electric-field
Ea	Electron affinity
Eg	Optical gap energy
EL	Electroluminescence
ETM	Electron-transport moiety
F-TBB	1,3,5-Tris-(4′-fluorobiphenyl-4-yl)benzene
FETs	Field-effect transistors
FI	Electric field induction
HOMO	Highest occupied molecular orbital
HTL	Hole-transport layer
HTM	Hole-transport moiety
Ip	Ionization potential
ITO	Indium tin oxide
LEDs	Light-emitting diodes
LUMO	Lowest unoccupied molecular orbital
NTSC	National Television System Committee
OCz	9-Octylcarbazole
OXD	Oxidazole
PEDOT:PSS	Poly(styrene sulfonic acid)-doped poly(3,4-ethylenedioxy-thiophene)
PEO20	PEO doped with 20 wt. % Cs$_2$CO$_3$
PFO	Poly(9,9-di-n-octylfluorene)
PFs	Polyfluorenes
PLED	Polymer light-emitting diode
PLQEs	Photoluminescence quantum efficiencies
poly-TPD	Polymeric triphenyldiamine derivative
PVK	Poly(N-vinylcarbazole)
SAM	Self-assembly monolayer
SPF	Spiropolyfluorene
TAA	Triarylamine
TAZ	Triazole
TOF	Time-of-flight
TPA	Triphenylamine
TPBI	1,3,5-Tris-(N-phenylbenzimidazol-2-yl)benzene
TPD-Si$_2$	4,4′-Bis[(p-trichlorosilylpropylphenyl)phenylamino]biphenyl
TSC	Thermally stimulated current
UV	Ultraviolet-visible
V_{bi}	Built-in voltage

η_e	Fraction of emitted photons that pass out of the device
η_r	Fraction of electron/hole pairs that recombine to form excitons
μ_p	Hole mobility
Φ_{EL}	External EL quantum efficiency
Φ_{PL}	Photoluminescent efficiency
χ	Fraction of hole-electron recombinations resulting in singlet excitons
$(FO)_n$	Oligo(9,9-dioctylfluorenyl-2,7-diyl)

1
Introduction

The applications of conjugated polymers in light-emitting diodes (LEDs) [1, 2], field-effect transistors (FETs) [3, 4], and plastic solar cells [5, 6] have attracted great attention recently. In polymer light-emitting diodes (PLEDs), polyfluorenes (PFs) are promising candidates as blue emitters due to their high photoluminescence quantum efficiencies (PLQEs) as solid films [7], their excellent solubility and film-forming ability, and the ease of controlling their properties via facile substitution in the 9,9-position of the fluorene unit [8–10]. In addition, chemical modifications on main chain, side chain, and chain end of PFs allow elaborate tuning of emission color covering the whole visible range (blue, green, yellow, red, and white), enhancement of device performance, and improvement of long-term operational stability. On the other hand, various types of chain conformation and chain stacking of PFs permit relatively easy physical manipulation for identical purposes as the chemical methods. In this review, we focus on the application of PFs in electroluminescent devices and reveal approaches for an enhancement of device performance. The basic electronic properties of PFs are first introduced to give a guideline on the effects of chemical modification and physical manipulation of PF structure on device performance. After that, we review the recent developments of PFs with various emission colors for a comprehensive understanding of PFs in device applications. Finally, the topic of alteration of the interface between PF and electrode is discussed.

2
Concepts for Performance Improvement of Polyfluorene Devices

2.1
Background of Electronic States of PFs

The fundamental properties of electronic states include energy levels of the highest occupied molecular orbital (HOMO) and lowest unoccupied molecular orbital (LUMO) as well as mobilities of hole and electron carriers. These allow one to identify the minor charge carrier in PFs so that a strategy can be planned for promoting device performance. The factors governing the fluorescent external electroluminescence (EL) quantum efficiency Φ_{EL} can be depicted by the equation, $\Phi_{EL} = \eta_r \chi \Phi_{PL} \eta_e$ [11], where η_r is the fraction of electron/hole pairs that recombine to form excitons, χ the fraction of hole–electron recombinations resulting in singlet excitons, Φ_{PL} the photoluminescent efficiency, and η_e the fraction of emitted photons that passes out of the device. However, χ can be promoted to a higher value in cases where: (1) triplet excitons so generated can be recovered to become emissive in the visible light region, and (2) charge trapping and/or energy transfer occur in multiple emissive species system.

For poly(9,9-di-n-octylfluorene) (PFO), which is deemed a model polymer in PFs, Janietz and coworkers reported its HOMO (equivalent to ionization potential, Ip) and LUMO (equivalent to electron affinity, Ea) levels as 5.8 and 2.12 eV below the vacuum level, respectively, as determined from cyclic voltammetry (CV) measurements of PFO thin solid film [11] as shown in Fig. 1a. However, the HOMO and LUMO levels provide a larger band gap of 3.68 eV than the value of 2.95 eV determined from the onset position in the ultraviolet-visible (UV) absorption spectrum of PFO film. If the band gap is taken as 2.95 eV and the HOMO level as 5.8 eV, the LUMO level of PFO

can be determined as 2.85 eV, quite close to the work function of a calcium electrode (2.9 eV). They recommended the use of the LUMO level determined by CV measurement (2.12 eV) because current density–electric field data exhibit an injection-limited nature for the device with a configuration of indium tin oxide (ITO)/PFO/Ca. But, this is still a controversial issue because, as shown in Fig. 1b, the hole-only current [ITO/poly(styrene sulfonic acid)-doped poly(3,4-ethylenedioxy-thiophene) (PEDOT:PSS)/PFO/Au] is two to three orders of magnitude smaller than the electron-only current (ITO/Ag/PFO/Ca) reported by Boudenbergh et al. [13], implying that a LUMO value of 2.85 eV seems more accurate because this value (very close to the

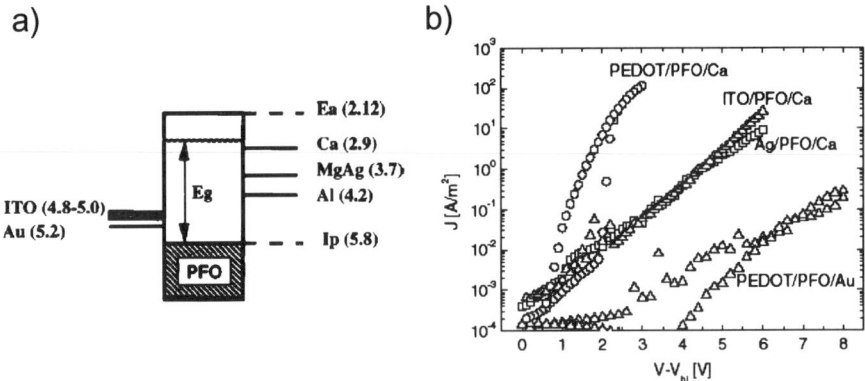

Fig. 1 a Schematic diagram showing the Ip and Ea values of PFO relative to the work functions of common electrode materials used in PLEDs. The figures in *brackets* are the respective energies in eV. The optical gap energy E_g (2.95 eV) is also shown (taken from [12]). **b** Current density–voltage characteristics of a PEDOT/PFO/Au (hole-only device) and Ag/PFO/Ca (electron-only device). Bipolar device of ITO/PFO/Ca and PEDOT/PFO/Ca are also shown (taken from [13]). Note that *PEDOT* is the abbreviation of PEDOT:PSS here

work function of the Ca cathode) can generate a negligible electron-injection barrier to cause the observed larger electron current.

Bradley and coworkers reported the hole mobility of PFO to range from 3×10^{-4} to 4.2×10^{-4} cm^2 V^{-1} s^{-1} for an electric field ranging from 4×10^4 to 5×10^5 V cm^{-1} (see Fig. 2a) via the thick-film ($\sim$3 µm) time-of-flight (TOF) technique [14]. Meanwhile, PFO film after alignment on polyimide film (pre-

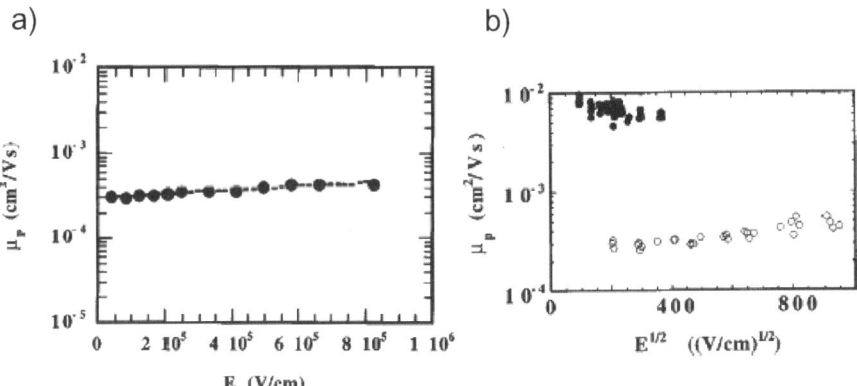

Fig. 2 a Field dependence of the hole mobility (μ_p) in spin-coated PFO film at 300 K (taken from [14]). **b** Field dependence of the room-temperature hole mobilities for an aligned, quenched PFO film (*filled circles*) and for an isotropic spin-coated PFO film (*empty circles*) (taken from [15])

viously rubbed by nylon cloth) under thermal annealing at 200 °C gives a higher hole mobility of 5×10^{-3} to 9×10^{-3} cm^2 V^{-1} s^{-1} for electric fields of 1.6×10^5 to 1×10^4 V cm^{-1} than isotropic PFO film, as shown in Fig. 2b [15]. On the other hand, electron mobility of PFs was scarcely reported due to the dispersive nature of electron current versus time in TOF measurements [14]. However, based on the configuration of FET, Friend and coworkers reported the electron mobility of PFO as 5×10^{-3} cm^2 V^{-1} s^{-1}, which is almost one order higher than its hole mobility of 3×10^{-4} cm^2 V^{-1} s^{-1}, also measured by the configuration of FET [16].

2.2
Adjustment of Chain Structures in PFs

2.2.1
Chemical Modifications

In this section, we review the effects of structure modifications for PFs with alkyl and aryl side chains or with charge-transport moieties on device performance. Spiropolyfluorenes (SPFs) having the structure of two phenylene units perpendicularly connected by a tetrahedrally bonded carbon atom are also discussed.

2.2.1.1
PFs with Alkyl and Aryl Side Chains

The first single-layer PF-based PLED with polymer **1** as emitting layer was reported by Yoshino and coworkers [17]. For this device with ITO and magnesium–indium alloy (Mg:In) as anode and cathode, respectively, its main EL emission peak locates at 470 nm, but no data on efficiency was given. However, the authors pointed out that electron injection is difficult because of the large energy barrier (1 eV) between **1** (LUMO 2.9 eV) and the Mg:In cathode (work function of Mg 3.9 eV), while the hole-injection barrier [0.2 eV, as estimated from the work function of ITO (5.5 eV) and the HOMO level of **1** (5.7 eV)] is relatively low as compared to the electron-injection barrier (1 eV). Bradley and coworkers reported that the device with PFO as emit-

1

ting layer (200 nm) and polymeric triphenyldiamine derivative (poly-TPD, 60 nm) as hole-transport layer (HTL) exhibits a blue emission with the main peak at 436 nm and reaches a luminance of 600 cd m^{-2} and current efficiency of 0.25 cd A^{-1} at the drive voltage 20 V [18]. The incorporation of poly-TPD (HOMO 5.28 eV) in the device introduces a new step in between ITO (work function 5 eV) and PFO (HOMO 5.8 eV) for better hole injection.

Carter and coworkers studied how side-chain branching in PFs affects device performance with and without an additional HTL of cross-linkable polymer 2 [19]. They found that the device efficiency is affected more by the position of the exciton recombination zone than by variations of polymer morphology induced by side-chain branching, which mainly controls the relative emission between vibrational energy levels and has a minimal effect on polymer charge transport properties. For double-layer devices (ITO/PEDOT:PSS/2/3, 4, or 5/Ca), a typical brightness of 100 cd m^{-2} at 0.8 MV cm^{-1}, maximum luminance of 10 000 cd m^{-2} at 1.5 MV cm^{-1}, and device efficiencies between 1.3 and 1.8 cd A^{-1} for 3 and 5 branching can be achieved.

Lee and Hwang synthesized a PF with aryl side chains 6 via the Yamamoto coupling reaction [20]. The device fabricated therewith (ITO/PEDOT:PSS/6/Ca/Al) emits blue light with suppressed long tail emission but gives low performance with maximum efficiency of 0.03 cd A^{-1} and maximum luminance of 820 cd m^{-2}. Note that the device could bear considerably high current density (>1.5 A cm^{-2}).

Yu et al. first reported the introduction of spiro links onto part of the PF backbone [21]. The resulting spiro-funcationalized PF copolymer, **7**, exhibits narrower EL emission spectra than polyalkylfluorene and the device [ITO/PEDOT:PSS/poly(N-vinylcarbazole) (PVK)/7/Ca/Al] exhibits low efficiency (0.54%), low luminance (15 cd m^{-2}), and high turn-on voltage (7 V). However, the full SPF, **8**, synthesized using an AB-type monomer via Suzuki coupling gives no apparent green emission, even after heat treatment, though no device performance data were provided [22]. Another SPF with branched dialkyl side chains, **9**, prepared by Vak et al. [23] was found to give pure blue emission with Commission Internationale de l'Eclairage (CIE) coordinates (0.17, 0.12) from the device ITO/PEDOT:PSS/9/ LiF/Ca/Ag and exhibited a turn-on voltage of 6 V, maximum luminance of 1600 cd m^{-2} and external quantum efficiency of 0.20% (0.19 cd A^{-1}).

2.2.1.2
Incorporation of Hole-Transport Moiety (HTM)

To obtain better device performance, achieving charge balance in PLED is crucial. The imbalance in charge carriers is due to the high barrier for hole injection and the discrepancy in charge carrier mobilities. Therefore, facilitating hole injection via introducing HTMs onto the PFs would be the key factor to improve the device performance.

Meerholz and coworkers incorporated the HTM (triphenyldiamine derivative) into the backbone of SPF to elevate the device efficiency through promoting hole injection/transport properties [24]. The cross-linkable oxetane-functionalized SPF derivatives **10–12** were also synthesized to realize full color display via spin-coating processes. The resulting EL devices (ITO/PEDOT:PSS/10, 11, or 12/Ca/Ag) showed maximum efficiencies of 2.9, 7.0, and 1.0 cd A^{-1} for blue, green, and red emissions, respectively.

Li et al. synthesized an alternating 9,9-dioctylfluorene/9-octylcarbazole (DOF/OCz) copolymer **13** with a (DOF:OCz) molar ratio of 3:1 [25]. The device [ITO/13/1,3,5-tris-(4'-fluorobiphenyl-4-yl)benzene (F-TBB)/tris-(8-hydroxyquinoline) aluminum (Alq$_3$)/LiF/Al, where F-TBB and Alq$_3$ are hole-blocking and electron-transport layers, respectively] exhibits maximum device luminance of 350 cd m^{-2} (27 V) and luminance efficiency of 0.72 cd A^{-1} at a practical brightness of 100 cd m^{-2}. The results are better than those

10 (blue), Ar: 15% (B1)/ 25% (B2)/ 10% (HTM)
11 (green), Ar: 25% (B2)/ 15% (G1)/ 10% (HTM)
12 (red), Ar: 25% (B2)/ 10% (G2)/ 5% (R1)/ 10% (HTM)

of the PFO-based device (maximum device luminance and luminance efficiency were 160 cd m^{-2} at 27 V and 0.30 cd A^{-1} at 100 cd m^{-2}, respectively). The higher device performance of **13** is attributed to a more efficient hole injection from the ITO anode into the emitting layer because the HOMO level of **13** is reduced to 5.39 eV and is closer to the ITO work function (4.8 eV) than PFO (HOMO level 5.63 eV). On the other hand, the **13**-based device emits blue light with two vibronic peaks at 423 and 443 nm and a low intensity in the region of 500–600 nm.

For the case of PFs with HTM on the side chain, the incorporation of triphenylamine (TPA) into PF side chains without the alkyl spacer **14** was carried out by Müllen and coworkers [26] to enhance blue PLED performance.

Here, TPA acts as soluble group and prevents the formation of aggregates and, in addition, can lower the hole-injection barrier between PF and ITO anode due to the closer Ip of TPA (5.3 eV) to the work function of the ITO anode (5 eV). However, for the device configuration (ITO/PEDOT:PSS/14/Ba), this TPA-containing PF, **14**, did not improve device performance as compared to PFO or **3**. This was attributed to the lower PLQE of **14** film (22%) than that of PFO film (50%) and the hole-trap property of TPA. In addition, its EL spectrum exhibited a pale blue emission (main peak at 428 nm) with CIE coordinates of (0.184, 0.159).

Our group has incorporated the HTM carbazole (Cz) into PF side chains by connecting this group with flexible decyl spacers. The device based on the polymer **15** with the configuration ITO/PEDOT:PSS/15/Ca/Al exhibits a turn-on voltage of 3.3 V, maximum efficiency of 1.28 cd A^{-1}, and maximum brightness of 5079 cd m^{-2}. The device emits sky blue light with two characteristic peaks at 420 and 460 nm, as for PFO, and an additional strong broad peak at 525 nm [27]. After further investigation [28], we found that the Cz groups in **15** play multiple roles as:

1. Hole transporters at the polymer–PEDOT:PSS interface to provide easier hole injection
2. Hole trapping sites (in the form of Cz/Cz dimers) able to catch electrons to form blue-emitting excitons
3. A source of green emission from an electroplex formed via electric field-mediated interaction of a Cz/Cz radical cation with an electron in the nearby PF backbone

On the other hand, Scherf and coworkers used N,N-bis(4-methylphenyl)-N-phenylamine as the end-capper to yield the polymer **16** (end-capper concentration 3 mol %); its single-layer device (ITO/PEDOT:PSS/16/Ca/Al) exhibited maximum luminance of 1600 cd m^{-2} and efficiency of 1.1 cd A^{-1}, with CIE coordinates (0.15, 0.08) at 8.5 V [29]. This efficiency is higher than that of non-end-capped polymer **3** (that is, bromine exists at each chain end) by one order of magnitude and was attributed to an efficient hole trapping at the

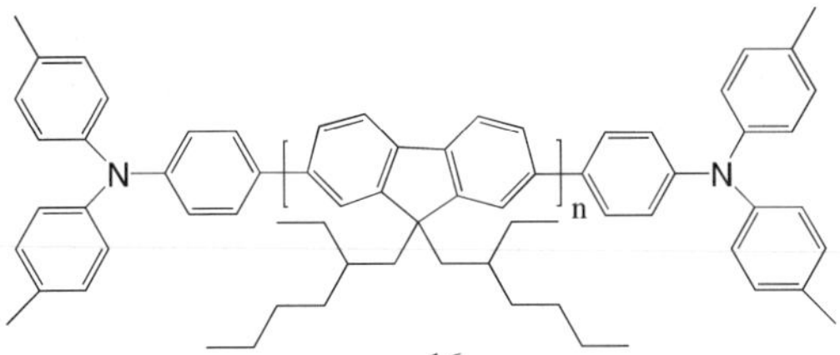

16

end-capper groups [due to the lower HOMO level of triarylamine (TAA) end-capper (5.48 eV) than 3 (5.88 eV)], resulting in more chances for holes and electrons to recombine on the main chain than at sites with less efficient emissions, such as aggregates and excimer-forming sites.

2.2.1.3
Incorporation of Electron-Transport Moiety (ETM)

For ETM incorporated in PF main chain, Jen and coworkers [30] presented copolymers with units of 9,9-dihexylfluorene and 2,5-dicyanobenzene and reported that the device based on the cyano (CN)-containing copolymer **17** as the emitting layer shows a low turn-on voltage (3 V) and better EL efficiency (0.5%) and high brightness (5430 cd m^{-2}) than those (0.044% and 717 cd m^{-2}) of the homopolymer **1** device with the same configuration (ITO/PEDOT:PSS/emitting polymer/Ca). The better device performance was attributed to the improved electron injection from the calcium cathode to the polymer because the CN electron-withdrawing functionality can lead to a significant increase of Ea from 2.12 eV of **1** to 2.98 eV of **17**, that is, close to the work function of the calcium cathode (2.9 eV). However, the EL spectrum of **17** is red-shifted (main emission peak at 477 nm) with respect to that of **1** (428 nm).

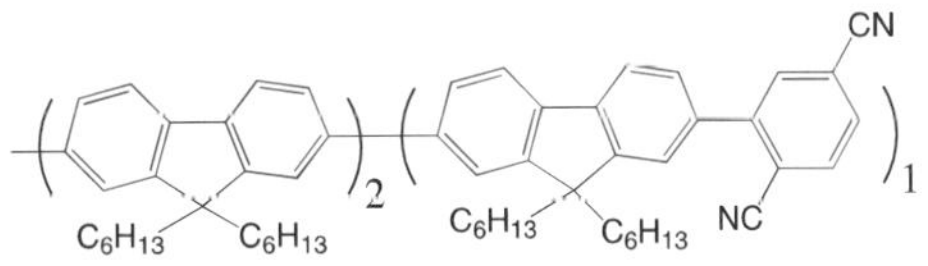

17

For PFs with HTM grafting as side chain, the alternating copolymer **18** with electron-deficient moiety (4-*tert*-butylphenyl-1,3,4-oxadiazole) functionalized fluorene and monomer of PFO was synthesized by Shu and coworkers [31]. The device with the configuration ITO/PEDOT:PSS/**18**/Ca/Ag showed improved performance: turn-on voltage of 5.3 V (defined as voltage needed for brightness of 1 cd m^{-2}), maximum brightness 2770 cd m^{-2} at 10.8 V, and maximum external quantum efficiency 0.52% at 537 cd m^{-2} relative to that of PFO (corresponding performance: 8 V, 600 cd m^{-2} at 20 V, and 0.2% at 17 V). This resulted from a better electron injection (LUMO 2.47 eV for **18** and 2.12 eV for PFO [12]; work function of Ca 2.9 eV) and transport in **18**. However, only little improvement was achieved, which could be due to the similar Ip between PFO (5.8 eV) and 4-*tert*-butylphenyl-1,3,4-oxadiazole (5.76 eV), leading to a significantly large hole-injection barrier from the anode (5.2 eV) to the emitting polymer. In addition, the EL spectrum of the device peaks at 428 nm and has no undesirable excimer/aggregate emission.

Another electron-deficient moiety, quinoline, was introduced into PF **19** by Su et al. [32]. However, the EL device performance (with the structure ITO/PEDOT:PSS/**19**/1,3,5-tris-(*N*-phenylbenzimidazol-2-yl)benzene (TPBI)/ Mg:Ag) remains low, with the maximum external quantum efficiency of 0.8%, maximum luminance of 1121 cd m^{-2} and high turn-on voltage of 7.2 V. Therefore, the introduction of electron-deficient moieties (either 4-*tert*-butylphenyl-1,3,4-oxadiazole or quinoline) cannot provide efficient EL performance, probably due to the large hole-injection barrier.

Our group reported that capping both ends of the PFO chain with electron-deficient moieties (oxidazole, OXD, and triazole, TAZ), which can induce a minor amount of long conjugating length species (regarded as the β-phase),

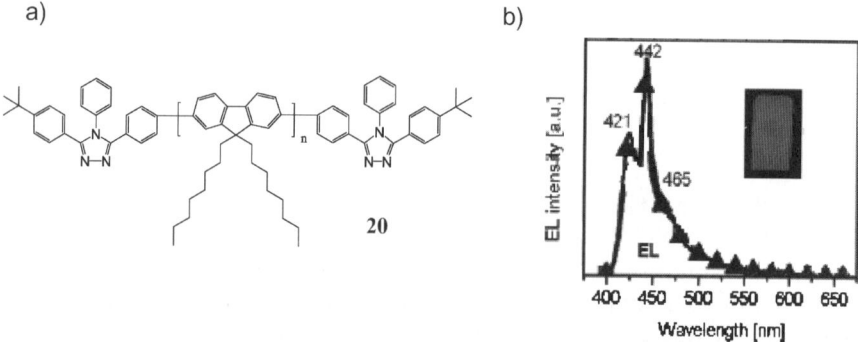

Fig. 3 a Chemical structure of polymer **20**. **b** EL spectra of spin-cast films of **20**. The *inset* shows picture of emission from **20**-based device operated at 4 V (taken from [33])

allowed an occurrence of incomplete energy transfer from the amorphous matrix to the β-phase and consequently resulted in a stable pure blue emission upon cyclic operations and promoted device efficiency [33]. For the case with TAZ (ITO/PEDOT:PSS/20/CsF/Al), the EL spectrum (in Fig. 3b) has a peak at 421 nm from the amorphous matrix and one at 442 nm from the β-phase, and gives the CIE coordinates (0.165, 0.076). Its maximum luminance efficiency (1.67 cd A^{-1}) and brightness (4000 cd m^{-2}) are better than those with *tert*-butylphenyl as end-capper, (0.74 cd A^{-1} and 2500 cd m^{-2}, respectively).

2.2.1.4
PFs with both HTM and ETM

In addition to the incorporation of a single charge-transport moiety on PFs, as mentioned above, simultaneously grafting electron-rich TAA and electron-deficient OXD groups on PF side chains was reported by Shu and coworkers [34]. This side-chain modification can lead to variations in both HOMO and LUMO levels. This polymer, **21**, showed maximum external quantum efficiency of 1.21% and luminance of 4080 cd m^{-2} at the device structure ITO/PEDOT:PSS/21/Ca/Ag. The enhanced efficiency was about twice that of the device with **18** as the emitting layer. This resulted from the better charge injection for **21** since its HOMO and LUMO levels (5.3 and 2.54 eV, respectively) are closer to the work functions of the ITO/PEDOT:PSS anode ($\sim$5.2 eV) and Ca cathode (2.9 eV) than **18** (HOMO 5.76 eV, LUMO 2.47 eV) and more efficient charge recombination (however, no evidence was given to support this claim).

On the other hand, Shu and coworkers reported that the incorporation of TAA and OXD in PF main chain and side chains, respectively, provided an improvement in the efficiency and purity of blue emission (based on the device ITO/PEDOT/22/TPBI/Mg:Ag) to 2.07 cd A^{-1} and a CIE value of

$x + y = 0.29$, respectively, relative to those of **21** (corresponding performance: 1.66 cd A^{-1} and CIE $x + y = 0.3$) [35]. The HOMO and LUMO energy levels of **22** are 5.20 and 2.38 eV, respectively. No doubt the hole-injection ability of **22** can be improved as compared to that of **21** (HOMO 5.3 eV) because the *lower* HOMO energy level of **22** is better matched to the work function of the ITO/PEDOT:PSS electrode (5.2 eV) than that of **21**. But, the lower LUMO level of **22** (2.38 eV) than that of **21** (2.54 eV) cannot improve the electron-injection ability because the LUMO level of TPBI is 2.7 eV [36]. Therefore, we think that the reasons for the enhanced performance of the **22**-based device are still not clear and are in need of further investigation.

Another example of bipolar PF, **23**, was also reported by Shu and coworkers [37]. They claimed that the bulky hole-transport Cz-derivative provides more effective hole injection than **18**. The blue EL device, ITO/PEDOT:PSS/**23**/TPBI/Mg:Ag, exhibits a maximum external quantum efficiency of 1.0%, maximum luminance of 1070 cd m^{-2}, and turn-on voltage of 4.5 V.

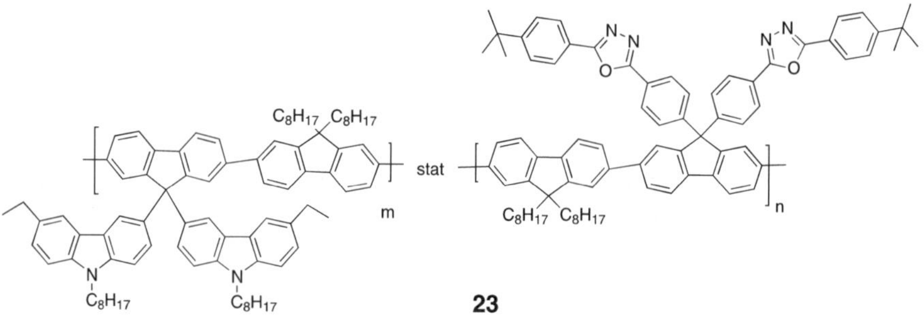

23

Shu and coworkers synthesized another kind of PF, **24**, with hindered dipolar pendent groups, TAA linked with OXD via a conjugated bridge, which was claimed to provide efficient charge transport/injection properties [38]. The

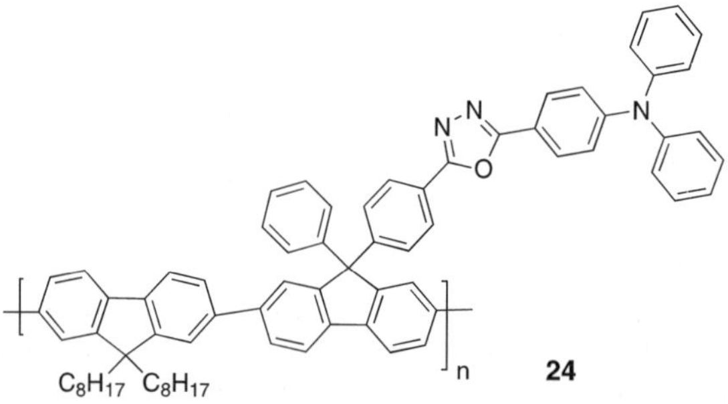

24

resulting EL device with the structure ITO/PEDOT:PSS/24/TPBI/Mg:Ag provides a maximum external quantum efficiency of 1.4%, maximum luminance of 2080 cd m^{-2}, and turn-on voltage of 5 V.

2.2.2
Physical Manipulation

Physical methods include tuning of chain conformation [39–42] and manipulation of supermolecular structure [43]. Because of its highly coplanar backbone, PFO can be physically transformed into a variety of supermolecular structures [39–42], such as crystalline phases (i.e., α and α' phases) and non-crystalline phases (i.e., amorphous, nematic, and β-phase; for β-phases it has an extended conjugation length of about 30 repeat units as calculated from wide-angle X-ray diffraction measurements) [44]. However, studies on the effects of tuning chain conformation on EL are scarce, but reports on effects of manipulating supermolecular structures on PL behaviors of PFO are extensive [40–42, 44].

As mentioned in Sect. 2.2.1.3 [33], we proposed that a trace amount of β-phase, induced by the use of an electron-deficient moiety (TAZ) as an end-capper for PFO, can improve device performance to give a better blue purity. Following the idea of β-phase formation, we further proposed a novel simple physical method to generate β-phase at a content of up to 1.32% in a PFO film spin-coated on a substrate (the remaining part is amorphous phase) by immersing it in a mixed solvent/non-solvent (tetrahydrofuran/methanol) for a few seconds [45]. The device based on PFO with 1.32% β-phase (ITO/PEDOT:PSS/emitting polymer/CsF/Al) has a dramatically enhanced device efficiency and an improved blue-color purity of 3.85 cd A^{-1} (external quantum efficiency, 3.33%) and CIE of $x + y = 0.283$ (less than the limit of

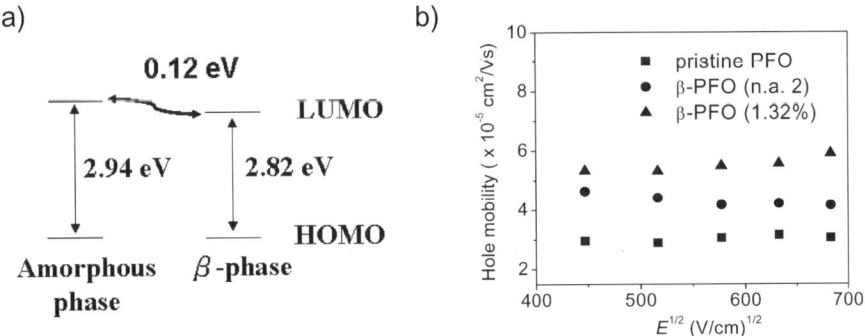

Fig. 4 a Energy-level diagram for the amorphous and β-phases. **b** Electric-field (E)-dependent hole mobilities of pristine PFO (as spin-coated without dipping), β-PFO (n.a. 2) (*n.a.* 2 denotes trace β-phase content), and β-PFO (1.32%) (taken from [45])

0.3 for pure blue), relative to that without such treatment, 1.26 cd A^{-1} (1.08%) and $x + y = 0.323$. Such a high efficiency (the highest among the reported pure-blue-emitting devices) results from the special functionalities of the β-phase: electron-trapping (see Fig. 4a) and promoted hole mobility (see Fig. 4b as revealed from thin-film TOF measurements). The energy levels in Fig. 4a were determined from UV absorption spectra and CV and are supported by thermally stimulated current (TSC) data. Because holes are minor charge carriers in pristine PFO, these two characteristics will lead to more balanced hole and electron fluxes and thus higher device performance. In addition to the improved purity of blue emission, the β-phase can also enhance the stability of blue emission during cyclic operation because: (1) linear alkyl side chains of the β-phase chains located beside fluorene units [46] can hinder neighboring PFO main chains from the intimate contact needed to form green-emitting field-induced excimers [47], and (2) efficient Förster energy transfer from the amorphous to the β-phase can also prevent formation of excimers.

2.3
Sources of Undesirable Green Emission in PF-based PLEDs

An important issue concerning whether PFs can act as pure-blue emitters is undesirable green emission (480–650 nm), and investigations on such additional lower energy emission have been intensive recently. Early reports attribute it to the aggregate/excimer emissions due to the stiff and planar chain geometry of the PF backbone [48]. As reported by Weinfurtner et al. [49], oligomers with molecular weight of less than 10 000 Da can give this green emission (main peak at 510 nm) due to the ease of aggregate formation. In addition, polar end group-enhanced aggregation emission at 507 nm for PFO without end-capping (the end groups are boron ester on one end and bromine groups on the other end) was demonstrated by us to be another reason for green emission [50]. However, several research groups directly attributed this phenomenon to the chemical defects on the PF main chain (i.e., keto defect or fluorenone) that emit green light at 535 nm [51–55]. Recently, we found that electric field induction (FI) accompanied by side-chain motion in PFs can lead to formation of excimers, which contribute to a growth of the green component peaking at 485 and 520 nm in the EL spectrum [47]. We also revealed that higher polarity of the side chain in PFs can cause a more pronounced FI effect. For polymer 25 having cross-linkable highly polar oxtane groups at the ends of the side chains, the green component can even dominate the entire emission after a few cycles of device operation. Lowering the content of the cross-linkable commoner in the copolymers with DOF from 50 to 25 mol %, or even cross-linking 25, only moderately suppresses the formation of FI excimers. This phenomenon also occurs for PFO with non-polar octyl side chains, especially after cyclic operations with higher electric fields

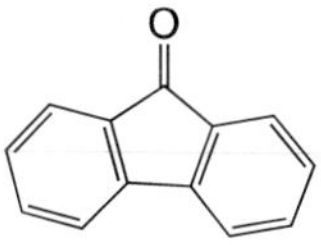

Fluorenone

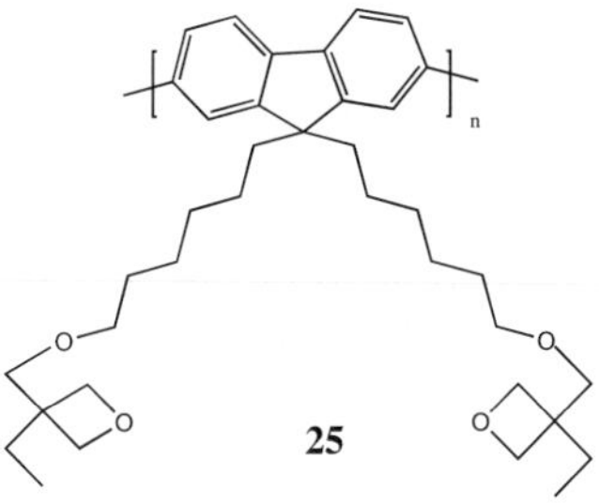

25

(1×10^6 V cm^{-1}). Thus, we must emphasize that there are several sources of the undesired green emission.

3
PFs with Various Emission Colors

3.1
Green and Red Emitters

Because the red-, green-, and blue-emitting polymers are essential in the fabrication of full-color PLED displays, fluorene copolymers that emit colors covering the entire visible range have been extensively studied and their syntheses are mainly via the Suzuki or Yamamoto routes [56]. However, more efforts have been made on the development of green- and red-emitting PF copolymers because PF homopolymer itself emits blue light. In this section, we review typical cases based on chemical incorporation of a fluorescent or phosphorescent moiety on PF for color tuning.

3.1.1
Chemical Strategy via Fluorescent Moiety Incorporated in Polymer Backbone, Side Chain, or Chain End

Heeger and coworkers adopted bithiophene as comonomer to copolymerize with fluorene and found that the EL emission can be tuned to emit green light with a peak at 493 nm and a shoulder at 515 nm [57]. The HOMO and LUMO energy levels of copolymer **26** are 5.30 and 2.85 eV, respectively, as estimated

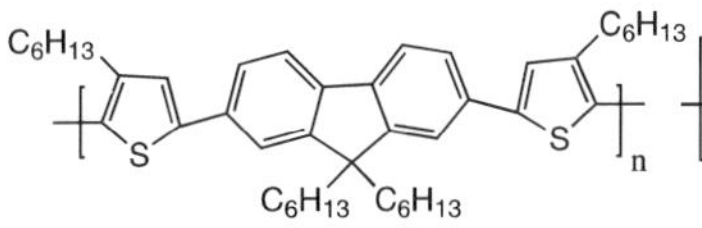

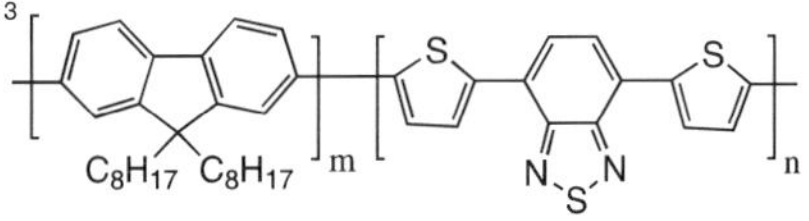

26

27: m = 85, n = 15

from the oxidation and reduction onset potentials of CV measurements. The single layer device (ITO/**26**/Ca) emitted green light at about 21 V. By employing a PVK layer (90 nm) between the ITO anode and the emitting polymer film as hole-injection layer, the turn-on voltage for light output of the double-layer device was reduced to about 8 V, and the maximum external quantum efficiency increased from 0.05 to 0.6%.

After that, Cao and coworkers synthesized a series of copolymers derived from DOF and narrow-band-gap 4,7-di-2-thienyl-2,1,3-benzothiadiazole (DBT) with various feed ratios of DOF/DBT [58]. The highest quantum efficiency and brightness were only 1.4% and 259 cd m^{-2}, respectively, from the device based on the copolymer with 15% DBT content (ITO/PEDOT:PSS/**27**/Ba/Al). As DBT content increases from 1 to 35%, the EL emission peak shifts from 628 nm to 674 nm, indicating that the copolymers are promising candidates as pure-red emitters though the performance needs to be further improved.

Another PF copolymer **28**, composed of alternating DOF and thieno-[3,2-*b*]thiophene, was reported by Lim et al. [59]. The incorporation of electron-rich thieno-[3,2-*b*]thiophene moiety in the polymer backbone led to lower HOMO (5.38 eV) and higher LUMO (2.4 eV) levels than those of PFO (HOMO 5.8 eV, LUMO 2.12 eV [12]). The device ITO/PEDOT:PSS/**28**/LiF/Al exhibited a pure green emission with a peak at 515 nm and CIE coordinates of (0.29, 0.63), which is very close to the standard green required by the National Television System Committee (NTSC) (0.26, 0.65). In addition, the maximum brightness and efficiency of this device were 970 cd m^{-2} and 0.32 cd A^{-1}, respectively.

28

Jen and coworkers reported random copolymers with 5 and 10 mol % of electron-deficient dithienosilole chromophore (**29** and **30**, respectively) [60]. The device with **30** as the emitting layer and an in-situ polymerized bis-tetraphenylenebiphenyldiamine-perfluorocyclobutane, **31**, as the HTL (ITO/**31**/**30**/Ca) can emit green light (λ_{max} = 500 nm) with the low turn-on voltage of 4 V, maximum brightness of 25 900 cd m^{-2}, and maximum external quantum efficiency of 1.64%, which are much better than the device with **29** as the emitter. This improvement was attributed to both an improved charge injection and to charge recombination in this higher dithienosilole-containing copolymer **30** because its HOMO and LUMO levels (5.5 and

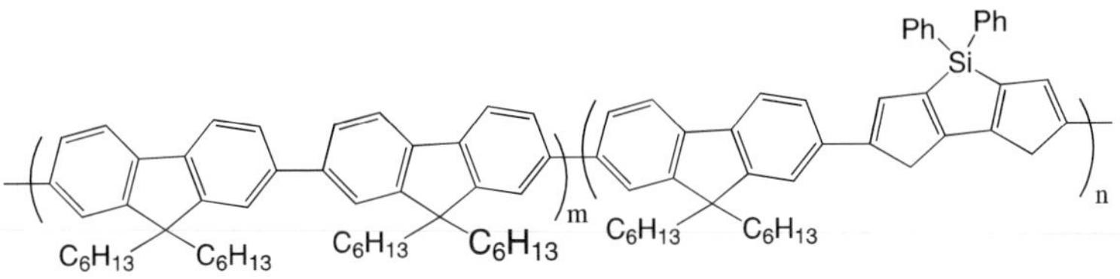

29: m = 19, n = 1
30: m = 9, n = 1

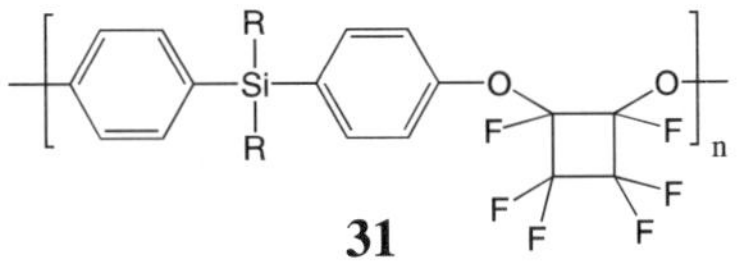

31

R =

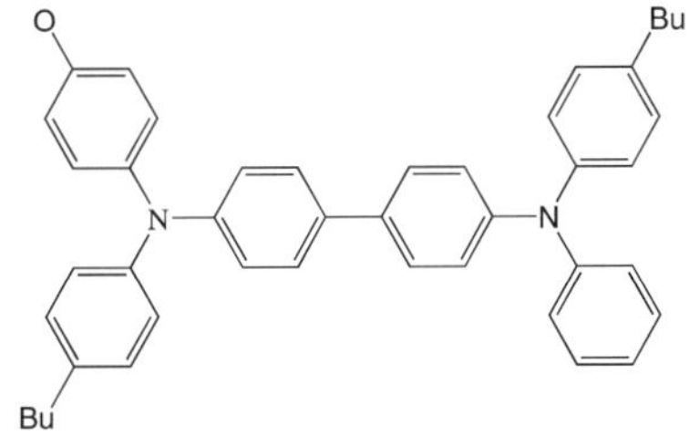

2.32 eV, respectively) are closer to the HOMO level of **31** (5.32 eV) and work function of Ca (2.9 eV) than those of **29** (5.68, and 2.18 eV, respectively).

Müllen and coworkers [61] incorporated perylene dyes to PF chains as:

1. Comonomers in the main chain
2. End-capper at the chain ends
3. Pendant side groups to tune emission color covering the whole visible range

For case (1), copolymers with various chemical structures of perylene can exhibit green and red light. For example, based on ITO/PEDOT:PSS/emitting polymer/Ca/Al, EL emission from **32** showed green light with a peak at 520 nm [CIE coordinates (0.362, 0.555)], and the emission from **33** was deep red [CIE coordinates (0.636, 0.338)] with a maximum at 675 nm. The maximum EL efficiencies for the PLEDs using **32** and **33** were 0.9 and 1.6 cd A^{-1}, respectively. For case (2), polymer **34** exhibited a narrow red photoluminescence (PL) emission with a maximum at 613 nm in the film state via an almost complete energy-transfer mechanism from fluorene units to perylene end-cappers, but the energy transfer was not seen in solution. For case (3), copolymer **35** containing 33 mol % of fluorene units with pendant dyes as a film exhibited a PL emission maximum at 599 nm. Even in solution, efficient energy transfer to the pendant groups was clearly apparent, as evidenced by the PL emission of a peak at 561 nm and a shoulder at 599 nm, due to the higher concentration of perylene units in the polymer than case (2).

Cho et al. prepared **36** by incorporating low-band-gap thiophene together with CN-substituted vinylene, which gives pure red EL emission [CIE coordinates (0.66, 0.33)] that is almost identical to the standard red (0.66, 0.34) demanded by the NTSC [62]. Another similar copolymer **37** also exhibited a pure red emission [CIE coordinates (0.63, 0.38)], and its maximum luminance and maximum external quantum efficiency were approximately 3100 cd m^{-2} at 4.6 V, and 0.46% at 4 V, respectively, which are better than those of **36**. The difference between **36** and **37** is the position of CN groups on the vinylene unit. The CN group in the latter was at β-position, and this configuration caused a strong steric interaction of CN groups with the alkoxy side chains of the inner phenylene ring, which shortens the conjugation length of the molecules and therefore results in a blue-shift in the emission (630 nm) relative to the former (675 nm).

Dow Chemical Company presented a series of PF copolymers, consisting of various charge-transport moieties and low band-gap chromophores on the backbone, for tuning the emission color of PLED to cover the whole visible range [63]. They claimed that their blue-, green-, and red-emitting copolymers (without disclosing chemical structures and ratios of comonomers) can exhibit acceptable device performance, as shown in Table 1.

Table 1 Characteristics of device performance based on ITO/PEDOT:PSS/copolymer/Ca/Al configuration (taken from [63])

	Green	Red	Blue
EL λ_{max} [nm]	534	648	476
CIE (x, y) [a]	(0.40, 0.58)	(0.68, 0.32)	(0.16, 0.16)
Efficiency [lm W^{-1}, cd A^{-1}]			
200 cd m^{-2}	7.64, 6.92	1.43, 1.50	1.31, 2.82
1000 cd m^{-2}	8.31, 8.84	0.68, 1.2	0.90, 2.60
4000 cd m^{-2}	7.36, 10.01	–, –	0.48, 2.00
10 000 cd m^{-2}	5.62, 10.25	–, –	–, –

[a] Measured at 200 cd m^{-2}

3.1.2
Chemical Strategy via Phosphorescent Moiety Incorporated in Polymer Side Chain and Backbone

In addition to the incorporation of fluorescent chromophores in PF, phosphorescent moieties have been widely used to copolymerize with fluorene or as pendent group in PFs for color tuning (via the selection of ligands chelated to heavy metals) and performance improvement. The ability to enhance device performance via phosphors is because strong spin–orbit coupling (resulting from the inclusion of heavy metal atoms in the phosphor structure) can efficiently utilize triplet excitons for electroluminescence and theoretically there are three times as many as triplet as singlet excitons.

Our group reported a new route for the design of electroluminescent polymers by grafting high-efficiency phosphorescent iridium complexes as dopants and charge-transport moieties onto alky side chains of fully conjugated polymers for PLED with single layer/single polymers [27]. As shown in Fig. 5a, the polymer system studied involves PF as the base-conjugated polymer, Cz as the charge-transport moiety and a source for green emission by forming an electroplex with the PF main chain (as revealed above), and cyclometalated Ir complexes as phosphorescent dopants. The devices prepared therewith (ITO/PEDOT:PSS/**38** or **39**/Ca/Al) can emit red light with the high efficiency of 2.8 cd A^{-1} at 7 V and 65 cd m^{-2} from **38** (with high Ir complex content of 1.3 mol % in the feed), and can emit the light with a broad band containing blue, green, and red peaks (2.16 cd A^{-1} at 9 V) from **39** (with low Ir complex content of 0.8 mol % in the feed) as shown in Fig. 5b. The inset of Fig. 5b shows that the HOMO and LUMO (the triplet state) energy levels of the red Ir complex lay between those of the main chain, which permits both hole and electron trapping during the electric field excitation.

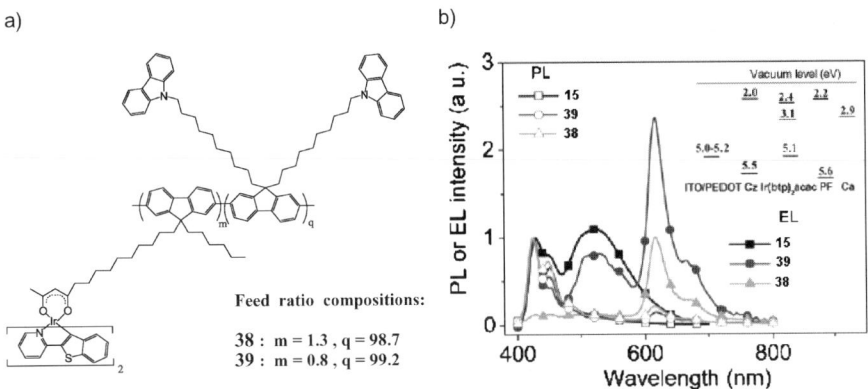

Fig. 5 a Chemical structures of Ir-complex and Cz-moiety grafted PFs. **b** PL and EL spectra of the polymers. Also shown is the band diagram, in which the LUMO for Ir(btp)$_2$acac has the energy level for singlet state at 2.4 and triplet state at 3.1 eV (taken from [27])

Holmes and coworkers [64] reported a series of solution-processable phosphorescent iridium complexes with iridium units being covalently attached to and in conjugation with oligo(9,9-dioctylfluorenyl-2,7-diyl) [(FO)$_n$] to form complexes [Ir(ppy-(FO)$_n$)$_2$(acac)] or [Ir(btp-(FO)$_n$)$_2$(acac)]. Based on ITO/PEDOT:PSS/emitting polymer/Ca/Al, [Ir(ppy-(FO)$_n$)$_2$(acac)] with $n = 10$ (**40**) and 30 (**41**) emitted mostly triplet emission at 570 and 612 nm but with a small peak (444 nm) due to PF singlet emission. [Ir(btp-(FO)$_n$)$_2$(acac)] with $n = 5$ (**42**), 10 (**43**), 20 (**44**), and 40 (**45**) all showed exclusive triplet emissions in their EL spectra (main peaks around 665 nm) pointing to charge trapping at the iridium complex being the dominant process under EL excitation. EL efficiencies are larger for [Ir(ppy-(FO)$_n$)$_2$(acac)] with longer FO segments (0.045% for **40** and 0.07% for **41**), and this increasing trend of EL efficiencies with longer FO segments also occurs for [Ir(btp-(FO)$_n$)$_2$(acac)] (0.15, 0.2, 0.45, and 1.5% for **42–45**, respectively). These results might be due to the reduced self-quenching for polymer with longer FO segments (namely, lower concentration of Ir complex).

40: n = 10, **41:** n = 30   **42:** n = 5, **43:** n = 10, **44:** n = 20, **45:** n = 40

Cao and coworkers synthesized copolymers **46** (feed ratio of Ir complex is 3 mol %) with 9,9-dihexylfluorene-*alt*-N-2-ethylhexyl-carbazole segments and backbone β-diketonate moieties coordinating to iridium [65]. A saturated red-emitting PLED with the emission peak at 628 nm, maximum external quantum efficiency of 0.6% at 38.5 mA cm^{-2}, and maximum luminance of 541 cd m^{-2} at 15 V were achieved from the device ITO/PEDOT:PSS/**46** blended with 2-(4-biphenylyl)-5-(4-*tert*butylphenyl)-1,3,4-oxadiazole (40%)/Ba/Al. For **46**, the incorporation of Cz and iridium complex units in the backbone can reduce its barriers for both hole and electron injections, as compared with PFO, because its HOMO and LUMO levels (5.49 and 2.47 eV, respectively) are closer to the work functions of PEDO:PSST (4.9 eV) and Ba (2.2 eV) than those of PFO (5.77 and 2.91 eV, respectively).

3.2
White Emitters

In addition to green- and red-light emitters, white PLEDs have also received great attention due to their potential applications in backlight for full-color displays and lighting.

3.2.1
Physical Blending System

Due to the many reported examples utilizing physical blending systems, we have only selected a few important examples for this section. Shu and coworkers used the blend of the blue-emitting PF, **21**, with two fluorene-derived fluorescent dyes [66], FFBFF (green emitter) and FTBTF (red emitter), to realize white light emission. The resulting EL device (ITO/PEDOT:PSS/**21** with 0.18 wt. % FFBFF and 0.11 wt. % FTBTF/Ca/Ag) exhibited maximum external quantum efficiency of 0.82% and maximum luminance of 12 900 cd m^{-2}; while its EL spectra changed with the applied voltages (from CIE coordinates (0.36, 0.37) at 6 V to (0.34, 0.34) at 12 V).

We prepared two PFs, **47** and **48**, which exhibited rather high bipolar charge mobilities in the order of 10^{-3} to 10^{-4} cm^2 V^{-1} s^{-1} [67]. The light-

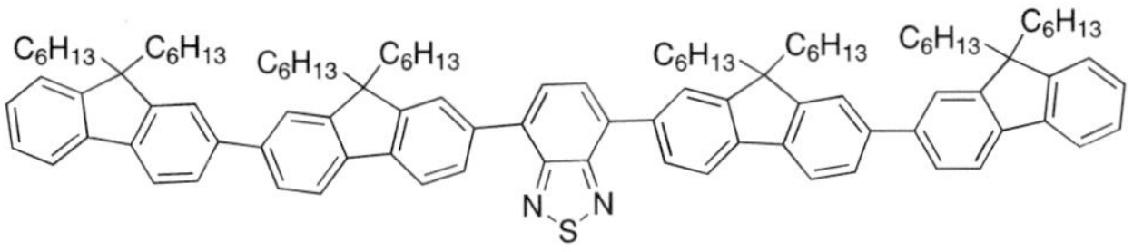

FFBFF

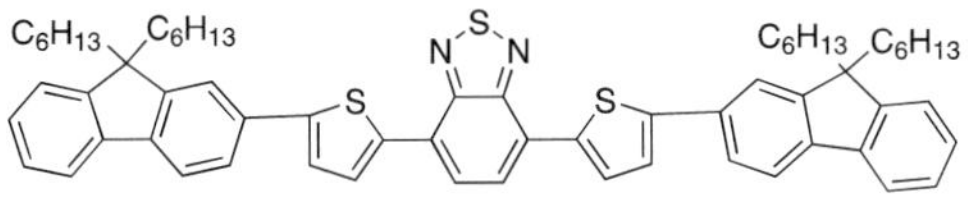

FTBTF

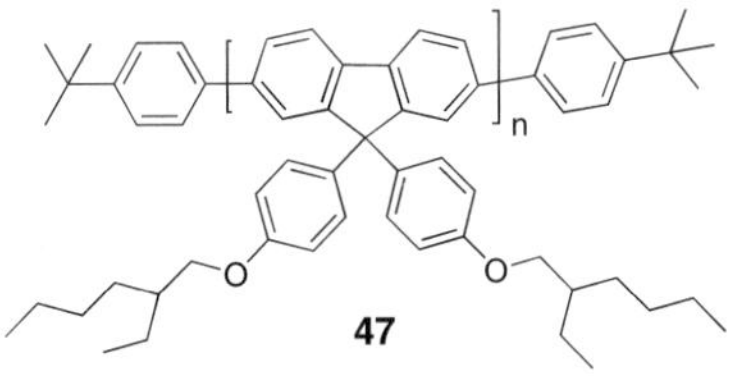

47

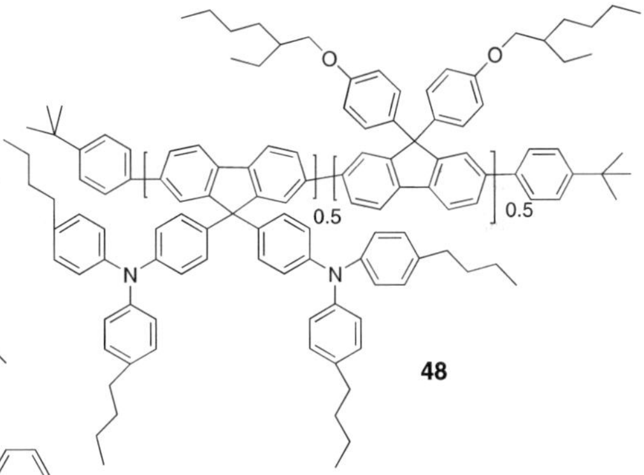

48

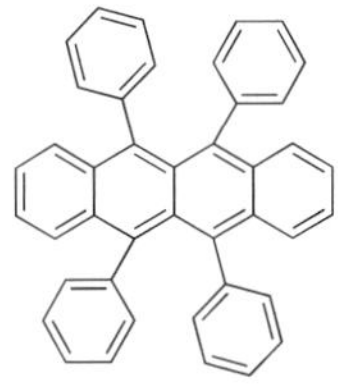

Rubrene

emitting devices (ITO/PEDOT:PSS/emitting layer/CsF/Ca/Al) with rubrene-doped **47** showed the maximal luminances 36 000 and 70 000 cd m^{-2} with the maximal efficiencies 3.5 and 9 cd A^{-1} for the white (doped with 0.25 wt. % rubrene) and yellow emissions (doped with 0.5 wt. % rubrene), respectively. For rubrene doped **48**, the maximal luminance and efficiency are 56 000 cd m^{-2} and 9 cd A^{-1} for white-emitting devices, and 72 000 cd m^{-2} and 14 cd A^{-1} for yellow emission, respectively. Furthermore, the EL profiles of the investigated devices are nearly independent of the applied voltages.

3.2.2
Chemical Strategy: Single-Polymer Approach

To yield white emission, it is difficult to generate emissions covering the whole visible region via a single polymer. Moreover, blending systems (polymer/polymer or polymer/molecular material) to realize white-light emission usually suffer from intrinsic phase separation and bias-dependent spectra. Therefore, creation of a strategy to realize white light from a single polymer is necessary.

The first example of a white PLED achieved by a single polymer with SPF backbone was realized by Covion Corporation [68]. The white-light emitting polymers (the detailed chemical structures were not disclosed) were synthesized by copolymerization of PPV monomer unit (emits green) and bisthiophenylbenzothiadiazole (emits red) derivatives onto the SPF backbone. The device (ITO/PEDOT:PSS/polymer/Ba/Al) emitted white light with CIE coordinates (0.36, 0.41) and exhibited a maximum efficiency of 7.8 cd A^{-1} and a maximum luminance of ca. 10 000 cd m^{-2}.

Shu and coworkers developed white-light emitting polymers **49** through incorporating benzothiadiazole (emits green) and bisthiophenylbenzothiadiazole (emits red) moieties into the backbone of a blue bipolar PF [69]. White electroluminescence can be achieved via incomplete energy transfer. The white [CIE coordinates (0.37, 0.36)] EL device based on this PF (ITO/PEDOT:PSS/ **49**/TPBI/Mg:Ag/Ag) exhibits a maximum external quantum efficiency of 2.22%, maximum luminance of 5000 cd m^{-2}, and high turn-on voltage of 5.5 V.

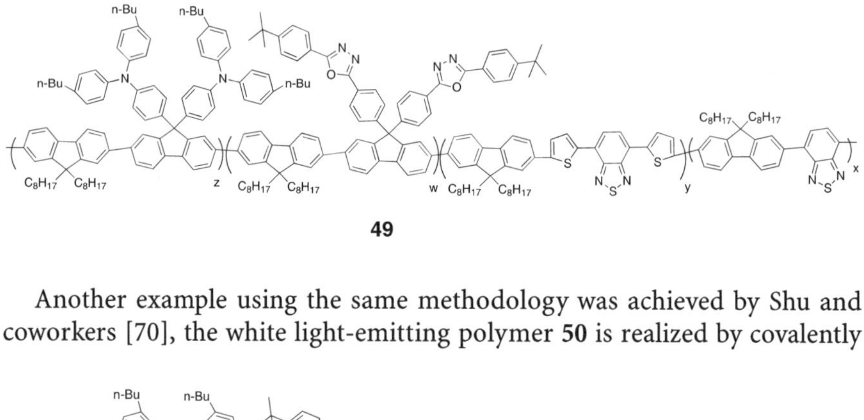

49

Another example using the same methodology was achieved by Shu and coworkers [70], the white light-emitting polymer **50** is realized by covalently

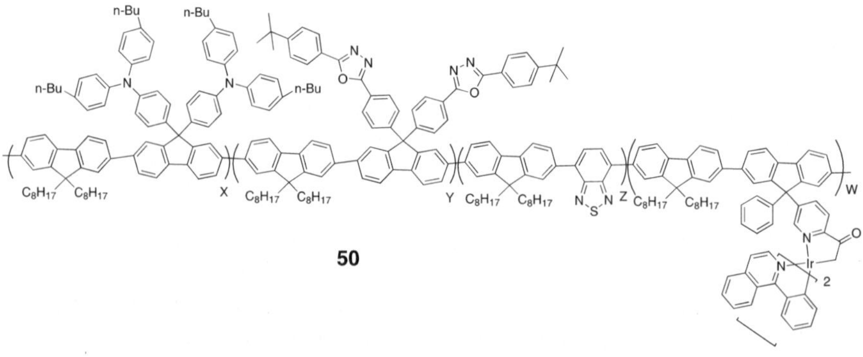

50

attaching a green fluorophore as well as a red phosphor into the backbone of bipolar PF. The EL device (ITO/PEDOT:PSS/50/CsF/Al) exhibited a maximum efficiency of 8.2 cd A^{-1}, maximum luminance of ca. 4000 cd m^{-2}, and low turn-on voltage of 2 V, but with drifted white emissions from CIE coordinates (0.35, 0.38) to (0.33, 0.36).

4
Alteration of Interfaces Between PFs and Electrodes

In order to achieve better hole or electron injection for enhancement of device performance, introduction of appropriate hole-transport and electron-injection materials into the interfaces of anode/PFs and PFs/cathode, respectively, is another useful approach in addition to the chemical and physical modifications of PF chain structure. Besides, choosing a cathode material with a work function close to LUMO levels of PFs has been proven to be a practical method for performance improvement.

4.1
Hole-Transporting Layer

He et al. reported a high-performance electroluminescence device based on bilayer conjugated polymer structures consisting of a hole-transport polymer **51** and a green-emissive polymer **52** prepared by the spin-coating technique on a glass substrate [71]. With this insertion of hole-transporting polymer, the device showed a high brightness ($\sim$10 000 cd m^{-2} at 0.84 mA mm^{-2}), good emission efficiency ($\sim$14.5 cd A^{-1}) and luminous efficiency (2.26 lm W^{-1}), and high external quantum efficiency (3.8%).

Marks and coworkers used a siloxane-derived hole-transport material, 4,4′-bis[(p-trichlorosilylpropylphenyl)phenylamino]biphenyl (TPD-Si$_2$), to modify the ITO anode surface via formation of a self-assembly monolayer (SAM) [72]. Due to the close Ip of TPD-Si$_2$ (5.5 eV) to the HOMO level of PFO (5.9 eV), the PFO-based device with ITO anode treated with TPD-Si$_2$ SAM exhibited a maximum brightness and external quantum efficiency of 7000 cd m^{-2} and 0.35%, respectively, which were about two orders of magnitude larger than those for the bare ITO device (work function of ITO 4.7 eV).

Even compared to a device with PEDOT:PSS (HOMO 5.2 eV) as HTL, the SAM-modified device exhibited higher maximum brightness by a factor of three, and a comparable efficiency.

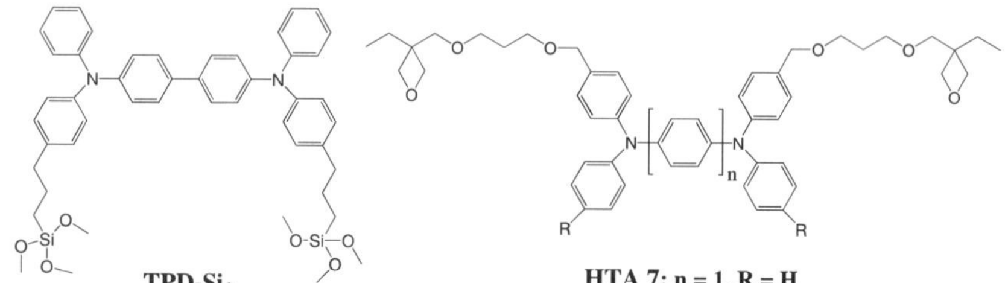

TPD-Si$_2$

HTA 7: n = 1, R = H
HTA 8: n = 2, R = OMe
HTA 9: n = 2, R = H

In order to achieve a better hole injection via the concept of graded HTL, Müller et al. synthesized a series of cross-linkable oxetane-functionalized hole-transporting TAA (HTA 7, 8, 9) with different HOMO levels (5.29, 5.41, and 5.48 eV, respectively) [73]. They reported that the device based on graded HTL (spin-coating and then cross-linking in the order HTA 7, 8, and 9) and polymer **16** (end-capper feed ratio 0.3%) exhibited a better efficiency of 2.7 cd A^{-1} than the 0.64 cd A^{-1} of the device with only HTA 9, demonstrating that the graded HTL concept is useful in improving hole-injection ability.

Further, Friend and coworkers [74] demonstrated molecular-scale interface engineering, also based on the graded HTL concept, for an improvement of hole injection from ITO anode to emitting layer consisting of PFO blended with 5 wt. % **52**. The PEDOT:PSS materials with different doping levels from saturated–doped (charge-per-ring ratio $y = 0.3$, designated as p^{++}), to the intermediate doping levels of $y = 0.25$ and 0.2 (designated as p^+ and p, respectively), and to the lowest doping level of $y = 0.02$ (designated as i) can be acquired via reducing as-received PEDOT:PSS (saturated–doped) by various equivalent moles of hydrazine hydrate. The device with the graded (PEDOT:PSS/**53**)$_5$(PSS/**53**) ($2p^{++}$-p^+-p-i) interlayer between ITO and emitting layer exhibited a current density more than twice as large as the p^{++}-PEDOT:PSS device, and four times that of the i-PEDOT:PSS device (here, calcium was used as cathode for all devices). This result further demonstrates that appropriate grade HTL can improve hole injection into the emitting layer.

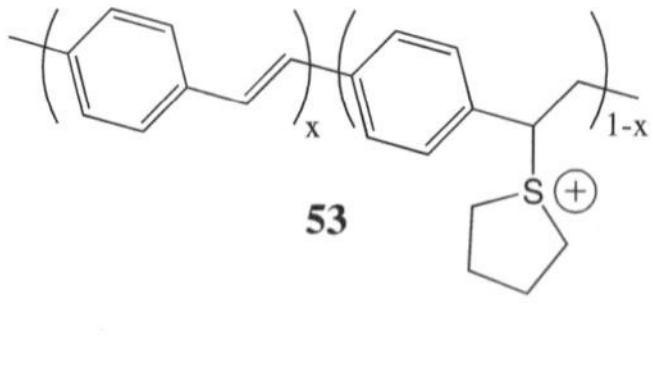

53

Although PEDOT:PSS is one of the most often adopted hole-transport materials, fundamental knowledge about how the process parameters (such as film thickness, backing conditions after coating, etc.) of this material affect PLED behaviors is still inadequate. Our group has demonstrated that migration of PSS from PEDOT:PSS into emitting layer may exert significant effects on the effective hole-injection barrier in blue-light emitting devices based on PFO [75]. In comparison to the hole-injection barrier 0.9 eV for bare ITO, the introduction of a PEDOT:PSS layer actually decreases the hole-injection barrier because the hole-injection barrier from PEDOT:PSS to PFO increases from 0.5 to 0.85 eV while PEDOT:PSS film thickness increases from 10 to 50 nm. This result was attributed to p-doping of PFO at the interface by free PSS chains. The thinner PEDOT:PSS film provides less p-doping and thus a lower hole-injection barrier. Therefore, the PFO-based PLED (ITO/PEDOT:PSS/PFO/CsF/Ca/Al) with lower PEDOT:PSS thickness (15 nm) can exhibit the highest device brightness and efficiency (8374 cd m^{-2} and 0.75 cd A^{-1}, respectively) among those devices with thicker PEDOT:PSS films (3000 cd m^{-2} and 0.5 cd A^{-1} for 50 nm thick PEDOT:PSS).

In addition to the effect of PEDOT:PSS film thickness on the hole-injection barrier, our group [76] found that the Ip value of PEDOT:PSS spin-coated onto the ITO anode decreases as the delay time to baking after spin-coating increases, from 5.45 eV (0 h) to 5.32 eV (48 h) as shown in Fig. 6. The trend moves toward the Ip of the neutral PEDOT, around 3.93–4.0 eV [77, 78]. This results in an increase of the hole-injection barrier to emitting polymer (PFO). The IP change was attributed to dedoping of PEDOT in PEDOT:PSS due to a reaction of ITO with protons in PSS and with those in doped PEDOT in the presence of water. Based on the device configuration of ITO/PEDOT:PSS/PFO/CsF/Ca/Al with PEDOT:PSS film varying with delay time to baking, the maximum brightness and efficiency are 15 600 cd m^{-2} (9 V) and 3.3 cd A^{-1} (3 V), respectively, for the device without delay. These are better than those for the device with 6 h delay (corresponding

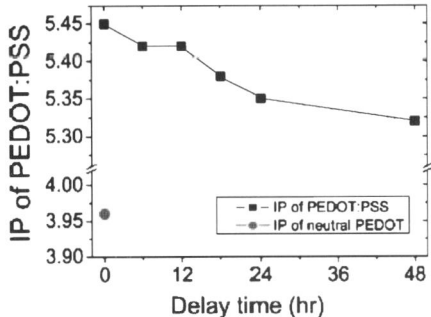

Fig. 6 IP values of 20 nm PEDOT:PSS films at different delay times to baking after spin-coating on ITO, and IP of neutral PEDOT (taken from [76])

Table 2 Ionization potential and the F/C atom ratio of the three CFx layers (taken from [79])

CFx layer	Ionization potential [eV]	F/C ratio
CFx (20 W)	5.7	1.51
CFx (35 W)	5.6	1.47
CFx (50 W)	5.3	1.33

performance: 1.9 cd A^{-1} at 3 V and 10 000 cd m^{-2} at 9 V). In other words, PEDOT:PSS film should be baked right after spin-coating in order to get better performance of the bipolar device.

Introduction of CFx thin film on top of the ITO anode as HTL via plasma polymerization of CHF$_3$ can also enhance device performance of PFO-based PLED, as reported by us [79]. At the optimal C/F atom ratio using the radio frequency power 35 W (see Table 2) as determined by X-ray photoelectron spectrometer, the device performance based on the ITO/CFx(35 W)/PFO/CsF/Ca/Al configuration is optimal having maximum current efficiency of 3.1 cd A^{-1} and maximum brightness of 8400 cd m^{-2}; much better than 1.3 cd A^{-1} and 1800 cd m^{-2} for the device with PEDOT:PSS as HTL. The improved device performance was attributed to a better balance between hole and electron fluxes because the CFx (35 W) layer possesses an Ip value of 5.6 eV (see Table 2), as determined by ultraviolet photoelectron spectroscopy data, and therefore causes a lower hole-injection barrier to the PFO layer (0.2 eV) than that of 0.7 eV for PEDOT:PSS.

4.2
Cathode Materials and Electron-Injection/Hole-Blocking Layer

The most straightforward approach to facilitate electron injection from cathode to emitting layer is to choose the cathode material with work function close to the LUMO level of the emitting layer for a reduction of the electron-injection barrier [56]. As shown in Table 3, metals with various work function values can be chosen as optimum cathode for efficient electron injection. However, a minimization of the electron-injection barrier is difficult to achieve since low work-function metals (such as Ca, Mg, Li, and Cs) are highly reactive and can easily interact with the organic layer onto which they are evaporated to quench excitons [80]. In addition, these metals are sensitive to oxygen and moisture in the environment, causing a quick degradation of these metals if lacking proper encapsulation.

Recently, alkali-metal fluoride/metal bilayer cathodes, especially for LiF and CsF, have shown substantial electron-injection ability [80–82]. Friend and coworkers demonstrated the minimization of the electron-injection barrier with the use of LiF as bilayer cathode (LiF/Ca) via electroabsorption

Table 3 Electronic properties of common electrode metals (taken from [56])

Element	Ionization potential [eV]	Preferred work function [eV]
Cs	3.89	2.14
K	4.34	2.30
Ba	5.21	2.70
Na	5.14	2.75
Ca	6.11	2.87
Li	5.39	2.90
Mg	7.65	3.66
In	5.79	4.12
Ag	7.58	4.26
Al	5.99	4.28
Nb	6.88	4.30
Cr	6.77	4.50
Cu	7.73	4.65
Si	8.15	4.85
Au	9.23	5.10

measurements [80]. They found that the electron-injection barrier for the LiF/Ca cathode is the lowest among all the cathode materials (including Al, LiF/Al, Ca, CsF/Al, and LiF/Ca). The result is that the PFO-based device with LiF/Ca cathode (ITO/PEDOT:PSS/PFO/LiF/Ca/Al) exhibits the largest current density (45 mA/cm^2) and highest luminance (1600 cd m^{-2}) at the driving voltage of 5 V. However, there is still no consensus over the mechanism behind the enhancement of electron injection for the use of this bilayer cathode and several interpretations have been proposed, as summarized by this group [80].

In addition to the appropriate choice of cathode material for better electron injection, the insertion of an electron-injection layer between cathode and emitting polymer has been reported. Yang and coworkers introduced the nanoscaled interfacial layer, calcium acetylacetonate [Ca(acac)$_2$], between the aluminum cathode and green-emitting PF blend (5BTF8, consisting of 5 wt. % 52 and 95 wt. % PFO), which led to improved device efficiency (28 cd A^{-1} at 2650 cd m^{-2}), higher than the device using calcium/aluminum as the cathode by a factor of three [83]. This result was attributed to the multiple functions of Ca(acac)$_2$:

1. It enhanced the injection of electrons due to electrons being the minor charge carriers in 5BTF8
2. It provided a buffer layer to prevent the quenching of luminescence from the aluminum electrode
3. It behaved as a hole-blocking layer for increasing hole–electron recombination probability

Ca(acac)$_2$

PEO

Another material, calcium carbonate (Cs$_2$CO$_3$), was applied to PFO-based PLED by Yang and coworkers and the luminous efficiency of a device with Cs$_2$CO$_3$/Al is higher than the device with only Ca as the cathode by a factor of 1.3, due to the better electron-injection and hole-blocking ability [84]. Recently, they also demonstrated that the improved electron injection is a result of the reaction of Cs$_2$CO$_3$ with thermally evaporated Al, which can reduce its work function to 2.1 eV by forming the Al–O–Cs complex. The optimum Cs$_2$CO$_3$ thickness is 15 Å with maximum power efficiency of 15 lm W^{-1} [85].

Recently, our group demonstrated a novel dual-functional composite layer having a superior hole-blocking effect along with promising electron-transport capability for PFO [86]. The dual-functional composite layer is composed of a non-conjugated polymer (polyethyleneoxide, PEO) as well as an inorganic salt (Cs$_2$CO$_3$), which allows an effective recombination of

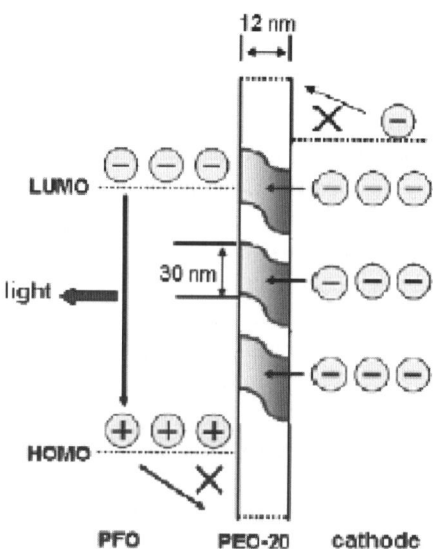

Fig. 7 Schematic diagram of charge-transport mechanism in PFO/PEO20/cathode. The *gray region* of the PEO20 layer is the Cs$_2$CO$_3$-rich region (taken from [86])

electrons and holes and results in a high device performance. In particular, a PFO-based device with this composite layer (PEO doped with 20 wt. % Cs_2CO_3, designated as PEO20) and a composite bilayer cathode Cs_2CO_3 (2 nm)/Ca (3 nm) (which can enhance the electron injection dramatically), ITO/PEDOT:PSS/PFO/PEO20/Cs_2CO_3/Ca/Al, gives a maximum brightness of 27 000 cd m^{-2} and current efficiency of 3.5 cd A^{-1} with pure blue emission at CIE coordinates (0.16, 0.07). This performance is better than that of the device without PEO20 [corresponding performance: 12 000 cd m^{-2}, 1.4 cd A^{-1}, and (0.16, 0.15)]. The hole-blocking effect of the composite layer is mainly because the incorporation of non-conjugated polymer PEO with high HOMO level can provide a large barrier for hole transport out of PFO (HOMO 5.8 eV). On the other hand, the presence of PEO film also reduces the electron current density (due to the large barrier for electron injection from cathode to PEO) but the addition of Cs_2CO_3 in PEO film can enhance the electron flux, indicating that the presence of Cs_2CO_3 in PEO acts as a channel for electron transport through the PEO film. The charge-transport mechanism in a PFO/PEO20/cathode is clearly illustrated in Fig. 7. In Fig. 7, electron carriers can smoothly transport through the PEO20 layer by passing through the Cs_2CO_3-rich region (electron-transport channel), but electron injection to the LUMO level of PEO is difficult. Furthermore, a majority of the hole carriers are blocked at the PFO/PEO20 interface and can effectively recombine with electron carriers.

5
Summary

The methods to improve performance of PF-based PLEDs in terms of emission efficiency, brightness, and operating voltage described in this review essentially promote a balance between hole and electron fluxes in the emitting layer and an enhancement of charge fluxes. The methods include modifying PF chain structures through chemical or physical methods, choosing a cathode material to minimize the electron-injection barrier, and inserting hole-transport and electron-injection layers to reduce hole and electron-injection barriers, respectively. The chemical method, the incorporation of HTM or ETM into PF chains, can tune HOMO and LUMO levels of PFs to match the work functions of anode and cathode, respectively, and thus to enhance hole and electron fluxes. Physical manipulation on PF chain structures can achieve balanced hole and electron fluxes; for example, the presence of β-phase can provide electron trapping and promoted hole mobility and result in the highest device performance (3.85 cd A^{-1} and 34 300 cd m^{-2}) among the reported pure-blue PLEDs. Color tuning to cover all visible regions can be achieved via physical blending or chemical incorporation of PFs with fluorescent or phosphorescent chromophores.

References

1. Burroughes JH, Bradley DDC, Brown AR, Marks RN, Mackay K, Friend RH, Burns PL, Holmes AB (1990) Nature 347:539
2. Kraft A, Grimsdale AC, Holmes AB (1998) Angew Chem Int Edit 37:402
3. Horowitz G (1998) Adv Mater 10:365
4. Ling MM, Bao ZN (2004) Chem Mater 16:4824
5. Svensson M, Zhang FL, Veenstra SC, Verhees WJH, Hummelen JC, Kroon JM, Inganäs O, Andersson MR (2003) Adv Mater 15:988
6. Coakley KM, McGehee MD (2004) Chem Mater 16:4533
7. Ariu M, Lidzey DG, Lavrentiev M, Bradley DDC, Jandke M, Strohriegl P (2001) Synth Met 116:217
8. Bernius MT, Inbasekaran M, O'Brien J, Wu WS (2000) Adv Mater 12:1737
9. Neher D (2001) Macromol Rapid Commun 22:1365
10. Scherf U, List EJW (2002) Adv Mater 14:477
11. Baldo MA, O'Brien DF, You Y, Shoustikov A, Sibley S, Thompson ME, Forrest SR (1998) Nature 395:151
12. Janietz S, Bradley DDC, Grell M, Giebeler C, Inbasekaran M, Woo EP (1998) Appl Phys Lett 73:2453
13. Woudenbergh TV, Wildeman J, Blom PWM, Bastiaansen JJAM, Langeveld-Voss BMW (2004) Adv Funct Mater 14:677
14. Redecker M, Bradley DDC, Inbasekaran M, Woo EP (1998) Appl Phys Lett 73:1565
15. Redecker M, Bradley DDC, Inbasekaran M, Woo EP (1999) Appl Phys Lett 74:1400
16. Chua LL, Zaumseil J, Chang JF, Ou ECW, Ho PKH, Sirringhaus H, Friend RH (2005) Nature 434:194
17. Ohmori Y, Uchida M, Muro K, Yoshino K (1991) Japan J Appl Phys 30:1941
18. Grice AW, Bradley DDC, Bernius MT, Inbasekaran M, Wu WW, Woo EP (1998) Appl Phys Lett 73:629
19. Nakazawa YK, Carter SA, Nothofer HG, Scherf U, Lee VY, Miller RD, Scott JC (2002) Appl Phys Lett 80:3832
20. Lee JH, Hwang DH (2003) Chem Commun, p 2836
21. Yu WL, Pei J, Huang W, Heeger AJ (2000) Adv Mater 12:828
22. Wu Y, Li J, Fu Y, Bo Z (2004) Org Lett 6:3485
23. Vak D, Chun C, Lee CL, Kim JJ, Kim DY (2004) J Mater Chem 14:1342
24. Muller CD, Falcou A, Reckefuss N, Rojahn M, Wiederhirn V, Rudati P, Frohne H, Nuyen O, Becker H, Meerholz K (2003) Nature 421:829
25. Li Y, Ding J, Day M, Tao Y, Lu J, D'iorio M (2004) Chem Mater 16:2165
26. Ego C, Grimsdale AC, Uckert F, Yu G, Srdanov G, Müllen K (2002) Adv Mater 14:809
27. Chen XW, Liao JL, Liang YM, Ahmed MO, Tseng HE, Chen SA (2003) J Am Chem Soc 125:636
28. Liao JL, Chen X, Liu CY, Chen SA, Su CH, Su AC (2007) J Phys Chem B 111:10379
29. Miteva T, Meisel A, Knoll W, Nothofer HG, Scherf U, Müller DC, Meerholz K, Yasuda A, Neher D (2001) Adv Mater 13:565
30. Liu MS, Jiang X, Herguth P, Jen AKY (2001) Chem Mater 13:3820
31. Wu FI, Reddy DS, Shu CF, Liu MS, Jen AKY (2003) Chem Mater 15:269
32. Su HJ, Wu FI, Su CF, Tung YL, Chi Y, Lee GH (2005) J Polym Sci A 43:859
33. Hung MC, Liao JL, Chen SA, Chen SH, Su AC (2005) J Am Chem Soc 127:14576
34. Shu CF, Dodda R, Wu FI, Liu MS, Jen AKY (2003) Macromolecules 36:6698
35. Wu FI, Shih PI, Shu CF, Tung YL, Chi Y (2005) Macromolecules 38:9028
36. Chen YC, Huang GS, Hsiao CC, Chen SA (2006) J Am Chem Soc 128:8549

37. Yuan MC, Shih PI, Chien CH, Shu CF (2007) J Polym Sci A 45:2925
38. Chien CH, Shih PI, Wu FI, Shu CF, Chi YJ (2007) Polym Sci A 45:2073
39. Chen SH, Chou HL, Su AC, Chen SA (2004) Macromolecules 37:6833
40. Chen SH, Su AC, Chu SU, Chen SA (2005) Macromolecules 38:379
41. Chen SH, Su AC, Chen SA (2005) J Phys Chem B 109:10067
42. Ariu M, Lidzey DG, Bradley DDC (2000) Synth Met 111/112:607
43. Apperloo JJ, Janssen RAJ, Malenfant PRL, Frechet JMJ (2000) Macromolecules 33:7038
44. Grell M, Bradley DDC, Ungar G, Hill J, Whitehead KS (1999) Macromolecules 32:5810
45. Lu HH, Liu CY, Chang CH Chen SA (2007) Adv Mater 19:2574
46. Chunwaschirasiri W, Tanto B, Huber DL, Winokur MJ (2005) Phys Rev Lett 94:107402
47. Lu HH, Liu CY, Jen TH, Liao JL, Tseng HE, Huang CW, Hung MC, Chen SA (2005) Macromolecules 38:10829
48. Prieto I, Teetsov J, Fox MA, Bout DAV, Bard AJ (2001) J Phys Chem A 105:520
49. Weinfurtner KH, Fujikawa H, Tokito S, Taga Y (2000) Appl Phys Lett 76:2502
50. Chen X, Tseng HE, Liao JL, Chen SA (2005) J Phys Chem B 109:17496
51. Bliznyuk VN, Carter SA, Scott JC, Klärner G, Miller RD, Miller DC (1999) Macromolecules 32:361
52. Lee JI, Klaerner G, Miller RD (1999) Chem Mater 11:1083
53. List EJW, Guentner R, Freitas PSD, Scherf U (2002) Adv Mater 14:374
54. Gaal M, List EJW, Scherf U (2003) Macromolecules 36:4236
55. Gamerith S, Gaal M, Romaner L, Nothofer HG, Güntner R, Freitas PSD, Scherf U, List EJW (2003) Synth Met 139:855
56. Bernius MT, Inbasekaran M, O'Brien J, Wu W (2000) Adv Mater 12:1737
57. Pei J, Yu WL, Huang W, Heeger AJ (2000) Chem Commun, p 1631
58. Hou Q, Xu Y, Yang W, Yuan M, Peng J, Cao Y (2002) J Mater Chem 12:2887
59. Lim E, Jung BJ, Shim HK (2003) Macromolecules 36:4288
60. Liu MS, Luo J, Jen AKY (2003) Chem Mater 15:3496
61. Ego C, Marsitzky D, Becker S, Zhang J, Grimsdale AC, Müllen K, MacKenzie JD, Silva C, Friend RH (2003) J Am Chem Soc 125:437
62. Cho NS, Hwang DH, Jung BJ, Lim E, Lee J, Shim HK (2004) Macromolecules 37:5265
63. Wu W, Inbasekaran M, Hudack M, Welsh D, Yu W, Cheng Y, Wang C, Kram S, Tacey M, Bernius M, Fletcher R, Kiszka K, Munger S, O'Brien J (2004) Microelectron J 35:343
64. Sandee AJ, Williams CK, Evans NR, Davies JE, Boothby CE, Köhler A, Friend RH, Holmes AB (2004) J Am Chem Soc 126:7041
65. Zhang K, Chen Z, Yang C, Gong S, Qin J, Cao Y (2006) Macromol Rapid Commun 27:1926
66. Kim JH, Herguth P, Kang MS, Jen AKY, Tseng YH, Shu CF (2004) Appl Phys Lett 85:1116
67. Peng KY, Huang CW, Liu CY, Chen SA (2007) Appl Phys Lett 91:093502
68. Buchhauser D, Scheffel M, Rogler W, Tschamber C, Heuser K, Hunze A, Gieres G, Henseler D, Jakowetz W, Diekmann K, Winnacjer A, Becjer H, Busing A, Falcou A, Rau L, Vogele S, Gottling S (2004) Proc SPIE 5519:70
69. Chuang CY, Shih PI, Chien CH, Wu FI, Shu CF (2007) Macromolecules 40:247
70. Wu FI, Yang XH, Neher D, Dodda R, Tseng YH, Shu CF (2007) Adv Funct Mater 17:1085
71. He Y, Gong S, Hattori R, Kanicki J (1999) Appl Phys Lett 74:2265
72. Yan H, Huang Q, Cui J, Veinot JGC, Kern MM, Marks TJ (2003) Adv Mater 15:835
73. Müller CD, Braig T, Nothofer H, Arnoldi M, Grall M, Scherf U, Nuyken O, Meerholz K (2000) Chem Phys Chem 1:207

74. Ho PKH, Kim JS, Burroughes JH, Becker H, Li SFY, Brown TM, Cacialli F, Friend RH (2000) Nature 404:481
75. Chang CH, Liao JL, Hung MC, Chen SA (2007) Appl Phys Lett 90:063506
76. Chang CH, Chen SA (2007) Appl Phys Lett 91:103514
77. Sotzing GA, Reynolds JR, Steel PJ (1997) Adv Mater 9:795
78. Xing KZ, Fahlman M, Chen XW, Inganäs O, Salancek WR (1997) Synth Met 89:161
79. Hsiao CC, Chang CH, Lu HH, Chen SA (2007) Org Electron 8:343
80. Brown TM, Friend RH, Millard IS, Lacey DJ, Butler T, Burroughes JH, Cacialli F (2003) J Appl Phys 93:6159
81. Brown TM, Friend RH, Millard IS, Lacey DJ, Burroughes JH, Cacialli F (2001) Appl Phys Lett 79:174
82. Chan MY, Lai SL, Fung MK, Lee CS, Lee ST (2004) J Appl Phys 95:5397
83. Xu Q, Ouyang J, Yang Y, Ito T, Kido J (2003) Appl Phys Lett 83:4695
84. Huang J, Li G, Wu E, Xu Q, Yang Y (2006) Adv Mater 18:114
85. Huang J, Xu Z, Yang Y (2007) Adv Funct Mater 17:1966
86. Hsiao CC, Hsiao AE, Chen SA (2008) Adv Mater DOI: 10.1002/adma.200702150

Poly(dibenzosilole)s

Wallace W. H. Wong[1,2] (✉) · Andrew B. Holmes[1,2]

[1] School of Chemistry, University of Melbourne, 3010 Victoria, Australia
wwhwong@unimelb.edu.au
[2] Bio21 Institute, University of Melbourne, 3010 Victoria, Australia

1	Introduction	86
2	Synthesis	87
3	Chemical and Physical Properties	91
4	Organic Electronic Devices	93
5	Conclusion	96
References		97

Abstract Poly(dibenzosilole)s are an emerging class of polymers with similar optoelectronic properties to polyfluorenes. With increased stability towards oxidation, several poly(dibenzosilole)-based devices, such as light emitting diodes, have shown improved performance over their polyfluorene counterparts. As a consequence of reduced conjugation in the polymer chain, some poly(dibenzosilole)s have high triplet excited state energies, which make them suitable hosts for blue triplet emitters in electrophosphorescent devices.

Keywords Conjugated polymer · Dibenzosilole · Light emitting diode · Organic electronic materials

Abbreviations

CIE	Commission internationale de l'éclairage
CV	Cyclic voltammetry
EL	Electroluminescence
GPC	Gel permeation chromatography
HOMO	Highest occupied molecular orbital
ITO	Indium tin oxide
LUMO	Lowest unoccupied molecular orbital
OFET	Organic field effect transistor
OLED	Organic light emitting diode
OSC	Organic solar cell
PCBM	[6,6]-Phenyl-C_{61} butyric acid methyl ester
PEDOT:PSS	Poly(3,4-ethylenedioxythiophene) poly(styrenesulfonate)
PF8	Poly(9,9-dioctylfluorene)
PL	Photoluminescence

1
Introduction

The area of conjugated polymers has grown into a major field of research since the discovery of highly conductive polyacetylenes in 1977 [1, 2]. Other properties of conjugated polymers have drawn increasing attention including electroluminescence for organic light emitting diodes (OLED) [3], high charge carrier mobility for organic field effect transistors (OFET) [4], tunable electronic properties for application in organic solar cells (OSC) [5] and response to physical and chemical stimuli for sensory applications [6]. Over the past decade, significant efforts have been devoted to developing conjugated polymers for application in multicolour OLED displays [7–9]. Polyfluorene and derivatives have emerged as the dominant class of polymers for commercial application [10–13]. They have been shown to exhibit high luminescence quantum efficiencies, thermal stability and good solubility. However, blue polyfluorene-based OLEDs suffer from poor spectral stability, which is characterized by the emergence of a broad and featureless low energy green emission band in the luminescence spectrum. This spectral instability has been linked to excimer formation as well as keto defects in the polymer chain [14–16]. Excimer formation can be suppressed by the introduction of bulky substituents to the polymer chain to increase interchain distances. Keto defects can be eliminated by careful synthetic preparation [17] or by replacing the vulnerable C-9 bridgehead atom with a heteroatom that is not easily oxidised. Although nitrogen [18–20], phosphorus [21] and sulfur [22] analogues of polyfluorene have all been prepared and some have shown stable blue emission, their low solubility in various organic solvents has limited processibility. Figure 1 shows the structures of some fluorene analogues.

Silicon and germanium have emerged as promising candidates as polyfluorene analogues (Fig. 1) [23]. They belong to the same periodic group as car-

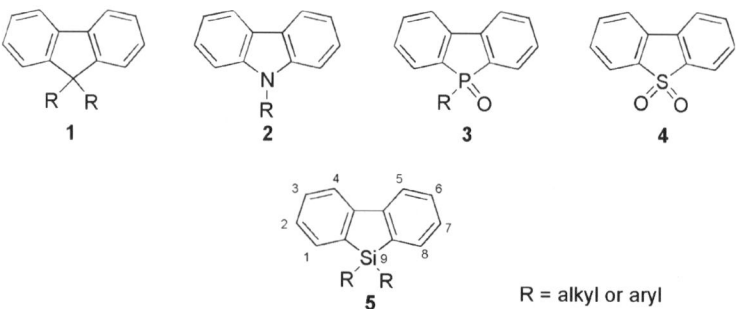

Fig. 1 Structural illustrations of fluorene 1, carbazole 2, dibenzophosphole oxide 3, dibenzothiophene dioxide 4 and dibenzosilole 5 with ring numbering

bon, and have been extensively exploited in both organic and inorganic semiconductors. Aside from the potential of eliminating low wavelength emission, their solubility and processibility are close to that of polyfluorenes A further attraction of such a polymer is the expected enhancement of electron affinity owing to the $\sigma^*-\pi^*$ conjugation, as observed in analogous molecular and polymeric siloles [24, 25]. The silicon analogue, poly(dibenzosilole), has attracted some attention in recent years as a polymer for blue OLEDs as well as in other organic electronic applications. This article will survey the current literature concerning poly(dibenzosilole)s. A more general review of silicon-containing polymers has recently appeared [26].

2
Synthesis

Early examples of the dibenzosilole unit were reported by Gilman and Gorsich in the 1950s [27–29]. The silicon analogue of 9,9-diphenylfluorene, 9,9-diphenyldibenzosilole, was obtained from the dilithiation of 2,2′-dibromobiphenyl followed by treatment with diphenyldichlorosilane [29]. 9,9-Dimethyldibenzosilole has also been prepared by reacting phenyl radicals with diphenyldimethylsilane [30]. The first report of incorporating the dibenzosilole unit in a polymer came in 1983 [31], but it was not until the 1990s that several patents covering dibenzosiloles were filed. Their applications include electroluminescence [32, 33], semiconducting material for dye sensitized solar cells [34] and photoresist applications [35]. However, full synthetic procedures for dibenzosilole monomers and their corresponding polymers were only reported in recent years. The synthesis and electroluminescent properties of poly(2,7-dibenzosilole) 12 were reported by Holmes and co-workers in 2005 [23]. 2,7-Dibromodibenzosilole 10 was obtained from the selective lithiation of 4,4′-dibromo-2,2′-diiodobiphenyl 9 and treatment with dichlorodihexylsilane (Scheme 1). 4,4′-Dibromo-2,2′-diiodobiphenyl 9 was synthesised from 2,5-dibromonitrobenzene 1 following an Ullmann coupling, reduction, diazotisation/iodination sequence. The boronic acid pinacol ester 11 was obtained from 2,7-dibromodibenzosilole by lithiation and treatment with 2-isopropoxy-4,4′,5,5′-tetramethyl-1,3,2-dioxaboralane. Suzuki cross-coupling polymerisation of the dibromo and diboronic acid pinacol ester derivatives gave poly(2,7-dibenzosilole) in excellent yield. Huang and coworkers have also synthesised copolymers containing the dibenzosilole unit from 2,7-dibromodibenzosilole monomer 16 [36]. This monomer was obtained from o-dianisidine 13 by diazotization, iodination followed by lithiation and quenching with dichlorodimethylsilane (Scheme 2). Suzuki coupling has also been used to make copolymers of 2,7-dibenzosilole unit with thiophene [37] and dithienyl-benzothiadiazole [38] with application in thin-film transistors and polymer solar cells respectively.

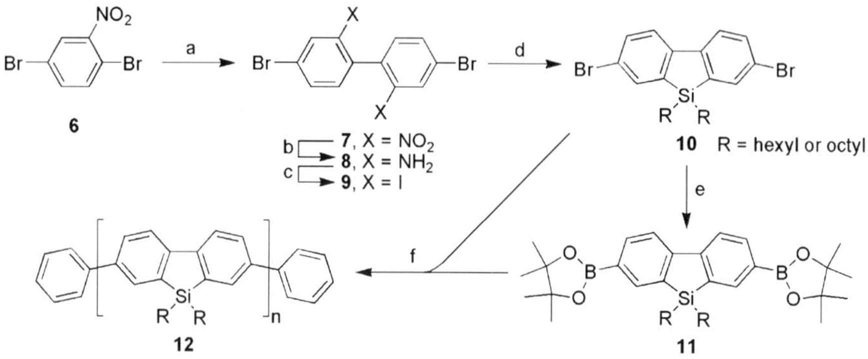

Scheme 1 Synthesis of poly(2,7-dibenzosilole) [23]. Reagents and conditions: (a) Cu, DMF, 125 °C, 88%; (b) Sn, HCl, EtOH, 110 °C (bath temp), 72%; (c) nitrosylsulfuric acid, concentrated H_2SO_4, 0 °C, then aq. KI, –10 to 50 °C, 30%; (d) t-BuLi (4 equiv), THF, –90 to –78 °C, then dichlorodihexylsilane or dichlorodioctylsilane, 25 °C, 52%; (e) t-BuLi, diethyl ether, –78 °C, then 2-isopropoxy-4,4′,5,5′-tetramethyl-1,3,2-dioxaboralane, 25 °C, 86%

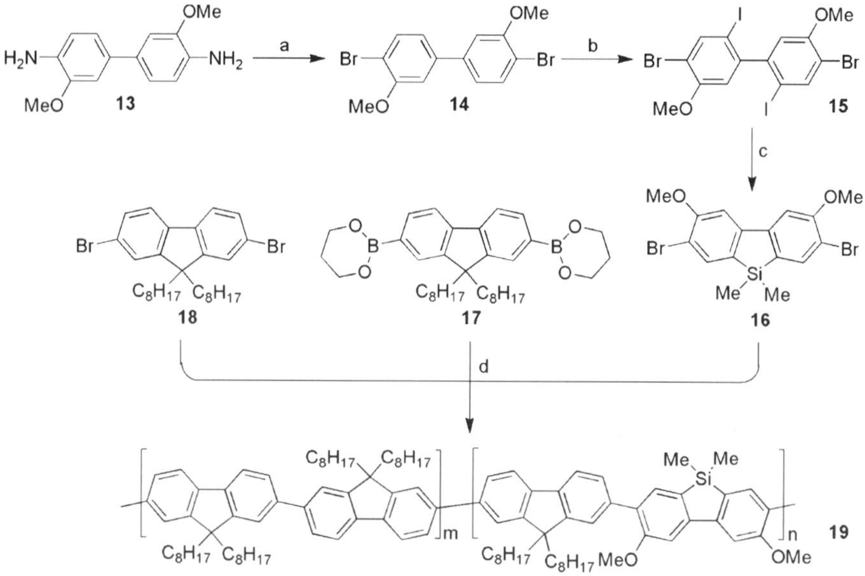

Scheme 2 Synthesis of poly(2,7-fluorene-co-2,7-dibenzosilole) [36]. Reagents and conditions: (a) $NaNO_2$, CuBr, HBr (40%), 0 °C; (b) I_2, KIO_3, H_2SO_4, AcOH, 90 °C; (c) 2 equiv n-BuLi, 2 equiv dichlorodimethylsilane, THF, –100 °C; (d) $Pd(PPh_3)_4$ (2 mol %), toluene/2 M K_2CO_3, 90 °C

3,6-Dibenzosilole monomers have also been reported in the literature [39–41]. Two routes were reported by Holmes and coworkers starting from 2,2′-dibromobiphenyl (Scheme 3) [41]. In Route 1, 2,2′-dibromobiphenyl was converted to 2,2′-diiodobiphenyl by treatment with n-BuLi followed by iodine. Regioselective bromination followed by selective silacyclisation afforded dibenzosilole 23. Double lithiation of 23 and boronation gave dibenzosilole boron acid ester 24. Cao and coworkers obtained the 3,6-dibenzosilole monomers through diamine intermediates using the diazoti-

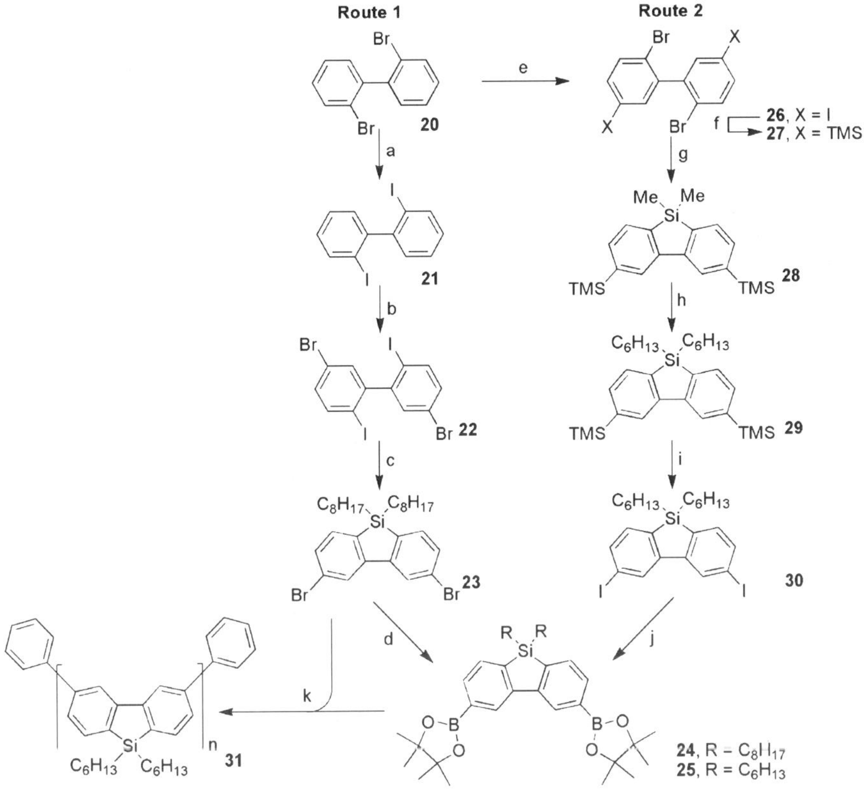

Scheme 3 Synthesis of poly(3,6-dibenzosilole) reported by Holmes and coworkers [41]. Reagents and conditions: (a) n-BuLi, THF, −78 to 25 °C, 24 h, then I$_2$ in Et$_2$O, 0 °C, 2 h, 85%; (b) Br$_2$, Fe, CHCl$_3$, 50 °C, 24 h, 50%; (c) t-BuLi (4.2 equiv), THF, −90 °C, (C$_8$H$_{17}$)$_2$SiCl$_2$, 25 °C, 64%; (d) t-BuLi, THF, −78 °C, then 2-isopropoxy-4,4′,5,5′-tetramethyl-1,3,2-dioxaborolane, 25 °C, 75%; (e) I$_2$, NaIO$_4$, conc. H$_2$SO$_4$, AcOH, Ac$_2$O, 24 h, 40%; (f) n-BuLi, −78 °C, 24 h, then TMSCl, −78 to 25 °C, 24 h, 84%; (g) t-BuLi, −78 to 25 °C, 24 h, SiMe$_2$Cl$_2$, −78 to 25 °C, 24 h, 81%; (h) n-HexLi, −78 °C, 15 min, 95%; (i) ICl, CH$_2$Cl$_2$, 25 °C, 1 h, 90%; (j) t-BuLi, THF, −78 °C, then 2-isopropoxy-4,4′,5,5′-tetramethyl-1,3,2-dioxaborolane, −78 to 25 °C, 24 h, 42%; (k) 23, 24, Pd(OAc)$_2$, tricyclohexylphosphine, Et$_4$NOH, toluene, 90 °C, then PhB(OH)$_2$, 2 h, then PhBr, 2 h, 93%

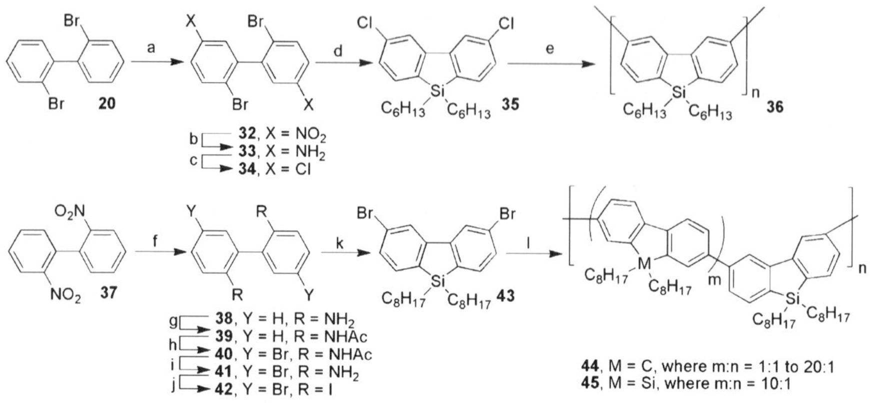

Scheme 4 Synthesis of poly(3,6-dibenzosilole) reported by Cao and coworkers [39, 40, 42]. Reagents and conditions: (a) HNO_3/H_2SO_4; (b) HCl/Fe, EtOH; (c) $NaNO_2$/HCl, CuCl; (d) BuLi, –65 °C, $(C_6H_{13})_2SiCl_2$; (e) $NiCl_2$, triphenylphosphine, 2,2′-bipyridine, Zinc, DMF; (f) Fe, HCl; (g) acetic anhydride, Na_2CO_3; (h) NBS, DMF; (i) HCl, EtOH; (j) $NaNO_2$/HCl, KI; (k) n-BuLi, –65 °C, $(C_8H_{17})_2SiCl_2$; (l) $Pd(OAc)_2/Et_4NOH$, tricyclohexylphosphine, **17** and **18** or **10** and **11**

zation strategy (Scheme 4) [39, 40]. Using these monomers, homopolymers **31** and **36**, as well as various high energy copolymers [39, 42] **44** and **45**, have been synthesised by Yamamoto [40] and Suzuki [41] coupling methods (Schemes 3 and 4).

In a recent communication, an asymmetric dibenzosilole moiety was reported [43]. Using an iridium-catalysed [2 + 2 + 2] cycloaddition of silicon-bridged diynes and alkynes, various asymmetric dibenzosilole units were obtained in yields of up to 93% (Scheme 5). These novel dibenzosilole compounds can potentially lead to AB monomers for polymerisation. Ladder-

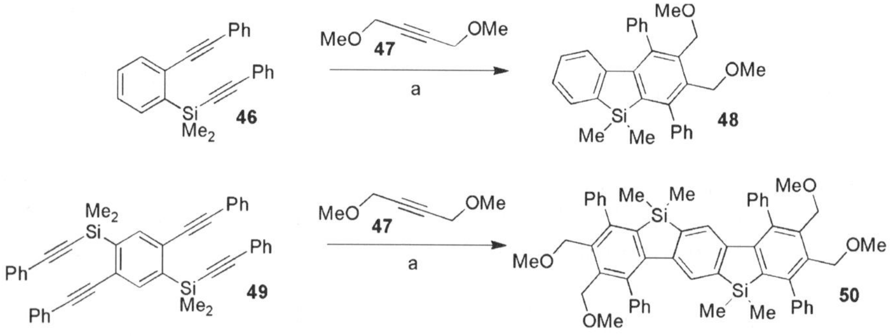

Scheme 5 Synthesis of dibenzosilole using iridium-catalyzed [2 + 2 + 2] cycloaddition [43]. Reagents and conditions: (a) $[IrCl(COD)]_2$ 2.5 mol %, PPh_3 10 mol %, Bu_2O, 110 °C, 24 h

type dibenzosilole compounds, such as **50**, were also synthesised using the same method opening an opportunity for the synthesis of dibenzosilole-based rigid planar polymers (Scheme 5).

Non-conjugatively linked 9,9'-dibenzosilole polymers have also been reported [44, 45]. In an early paper, a dehydrogenation catalyst, bis(1,5-cyclo-octadiene)palladium, was used in the synthesis of poly(9,9-dibenzosilole) **52** (Scheme 6) [45]. Recently, poly(dibenzosilole-vinylene)s **54** were obtained from the platinum-catalysed hydrosilylation of 9,9-dihydrodibenzosilole **51** with 9,9-diethynyldibenzosilole **53** (Scheme 6) [44].

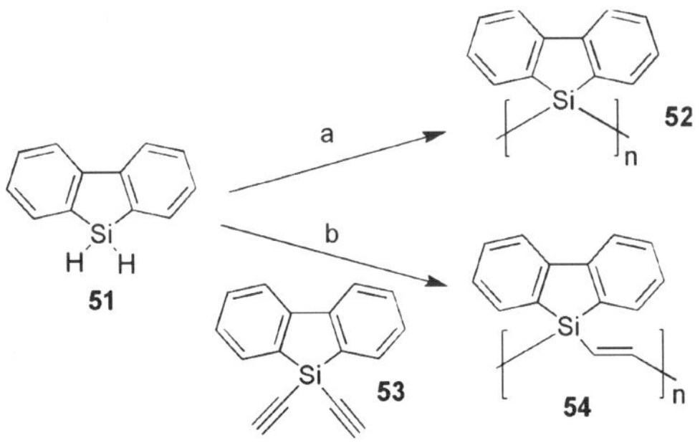

Scheme 6 Synthesis of poly(dibenzosilole-vinylene) [44] and poly(9,9-dibenzosilole) [45]. Reagents and conditions: (a) Pd(COD)$_2$, toluene, 80 °C, 24 h; (b) H$_2$PtCl$_6$, toluene, 70 °C, 24 h

3
Chemical and Physical Properties

The optical properties of poly(9,9-dihexyl-2,7-dibenzosilole) **12** are remarkably similar to those of poly(9,9-dioctyl-2,7-fluorene) (PF8) [23]. The absorption maximum in the UV-vis spectrum of a thin film of 7 (λ_{max} = 390 nm) is comparable with that of PF8 (λ_{max} = 389 nm), and the optical band gaps, determined from the λ_{0-0} band edges, are 293 eV for both polymers. The photoluminescence (PL) emission maximum (excitation at 325 nm) of a film of polymer **12** at 425 nm and its two vibronic sidebands at 449 and 482 (CIE coordinates x = 0.15, y = 0.11, PL efficiency = 62%) are all within 3 nm of the corresponding bands for PF8 (PL efficiency = 60%). However polymer **12**, with a glass transition temperature of 149 °C, is thermally more stable than PF8. Moreover, on annealing at 250 °C for 16 h under air and ambient light, the PL spectrum of polymer **12** remains unchanged whereas a broad green band centered at 535 nm emerged in the PL spectrum of PF8 after annealing (Fig. 2). The HOMO and LUMO levels of polymer **12** in thin films were determined by cyclic voltammetry (CV) to be – 5.77 and – 2.18 eV respec-

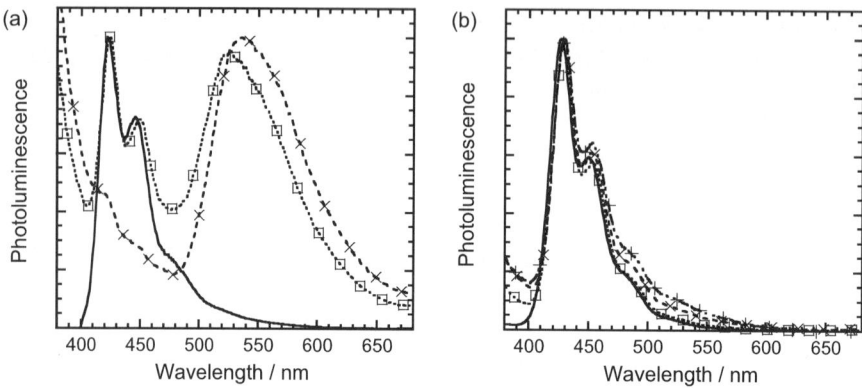

Fig. 2 Normalized photoluminescence spectra of thin films of polymers before annealing (○), and after annealing for 2 h at 150 °C (□), 4 h at 200 °C (×) and 24 h at 250 °C (+) under ambient light and atmosphere. **a** Polyfluorene PF8. **b** Poly(dibenzosilole) **12**

tively, which is about 0.1 eV lower than that of PF8 measured under the same conditions (Fig. 3) [23].

The PL spectrum of a thin film of poly(3,6-dibenzosilole) **31** at 77 K exhibited a 0–0 transition at 3.5 eV and a second maximum at 3.3 eV (excitation at 4.4 eV) [41]. The phosphorescence emission spectrum at 77 K consists of a broad band exhibiting vibronic structure (excitation at 3.9 eV). The polymer triplet energy level was taken to be the onset of triplet emission at 2.55 eV. This is considerably higher than the triplet energy of commonly used polyfluorenes (2.1 eV) [10, 46] making it a host for phosphorescent emitters without the risk of energy back-transfer onto the polymer.

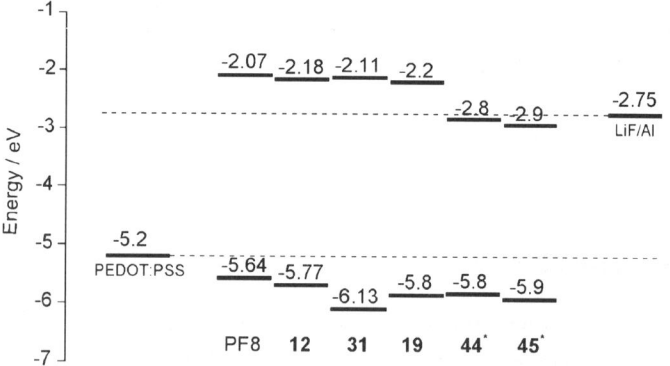

Fig. 3 A schematic diagram showing the HOMO and LUMO energy levels of various polymers of fluorene and dibenzosilole, as well as the energy levels of PEDOT:PSS and LiF/Al electrode. * The energy levels of polymers **44** and **45** are reported in comparison to differing HOMO and LUMO levels for PF8 at – 5.80 and – 2.87 respectively [39, 42]

Cao and coworkers reported almost identical figures for the same polymer with no endcaps **36** [40]. Thermal gravimetric analysis showed excellent thermal stability with decomposition temperature at 442 °C, while its glass transition temperature is similar to that of poly(9,9-dioctyl-2,7-fluorene) at 83 °C. It is worth noting that molecular weights of poly(3,6-dibenzosilole)s (23 000/11 000 M_w/M_n and polydispersity of 2.1) [41] are much lower than that of poly(2,7-dibenzosilole)s (425 000/109 000 M_w/M_n and polydispersity of 3.9) [23], measured using gel permeation chromatography (GPC), referenced against polystyrene standards. This is also observed in polyfluorenes and other bridged phenylene polymers in general. From CV measurements, the HOMO level was determined to be – 5.65 eV for polymer **31** [41] (Fig. 3) and – 6.1 eV for polymer **36** [40], while the LUMO energy level was estimated to be approximately – 2.1 eV for both polymers.

The properties of copolymers involving dibenzosilole monomers have been reported by several groups [23, 36–39, 42]. Huang and coworkers studied a series of poly(9,9-dioctyl-2,7-fluorene-co-2,7-dibenzosilole)s **19** with varying ratios of fluorene and dibenzosilole units [36]. With increasing ratio of dibenzosilole to fluorene, the absorption ($\lambda_{max} \sim 380$ nm) and emission spectra (~ 420 nm) are slightly blue shifted. This change in absorption and emission spectra was attributed to the shortening of average conjugation length in the copolymer with increasing incorporation of the dibenzosilole unit. Similar to PF8, a green emission at around 530 nm was observed for the copolymers after annealing in air at 200 °C. The HOMO and LUMO energy levels of this series of copolymers are very similar to that of PF8 at – 5.8 and – 2.2 eV respectively (Fig. 3). The increased proportion of the dibenzosilole unit in the polymer slightly raised the HOMO and lowered the LUMO giving a smaller band gap [36]. The reported LUMO of copolymers **44** and **45** is significantly lower than that of other dibenzosilole polymers shown in Fig. 3. This is most likely an artifact of the method of energy level measurement used by the investigators as the LUMO and HOMO levels of poly(9,9-dioctylfluorene) are reported to be – 2.87 and – 5.80, respectively [39, 42]. Other copolymers designed for specific device applications will be discussed in the next section [37, 38].

4
Organic Electronic Devices

Although polyfluorene has been a popular active layer material for OLEDs, it suffers from oxidative degradation characterized by the emergence of a green emission band. Poly(2,7-dibenzosilole), with its higher thermal stability and resistance to oxidation, is a potential alternative to polyfluorene [23]. OLEDs fabricated from poly(2,7-dibenzosilole), with a configuration of ITO/PEDOT:PSS/polymer(65 nm)/Ba/Al, showed emission maxima at 431 and 451 nm and operated more efficiently than a device fabricated from PF8.

With its high triplet energy, poly(3,6-dibenzosilole) is a suitable host for triplet emitters. Electrophosphorescent devices containing poly(3,6-dibenzosilole) as polymer host and iridium complexes as triplet emitters have been reported [41, 47]. In one study, a green phosphorescent dopant, *fac*-tris[2-(2-pyridyl-κN)-5-methylphenyl]iridium(III), was blended into a toluene solution of polymer **31** (8 wt. %) and deposited in a device with configuration ITO/PEDOT:PSS/polymer blend/LiF/Al [41]. The electroluminescence (EL) spectrum of this device showed an emission maximum at around 540 nm. No emission was detected from the polymer indicating complete energy transfer from the polymer host to the green dopant. Cao and coworkers reported a similar system with a blue phosphorescent iridium dopant [47]. In comparison with the use of polyvinylcarbazole (PVK) as the polymer host, poly(3,6-dibenzosilole) **36** devices gave much higher external quantum and luminous efficiencies. Cao and coworkers also observed an interesting phenomenon whereby the PL spectrum of the polymer/iridium blend showed emission from the polymer only while the EL spectrum showed emission from the iridium complex alone (Fig. 4) [47]. This difference in PL and EL spectra was attributed to a Förster transfer mechanism in the PL process in contrast to a charge trapping mechanism in the EL process. The

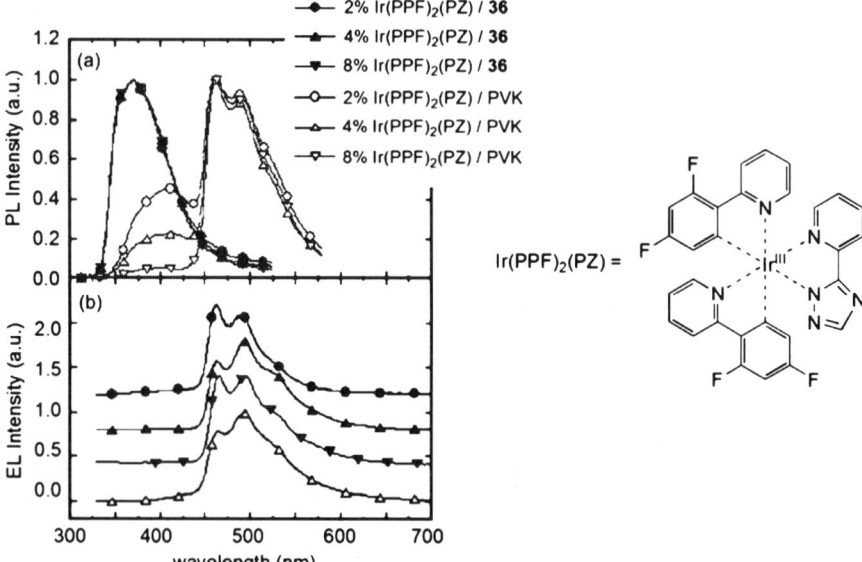

Fig. 4 a PL spectra of the iridium complex with different concentration doped into **36** and PVK, respectively. **b** EL spectra of devices Ir complex doped into **36** and PVK. Reprint with permission from [47]. © 2006, American Institute of Physics

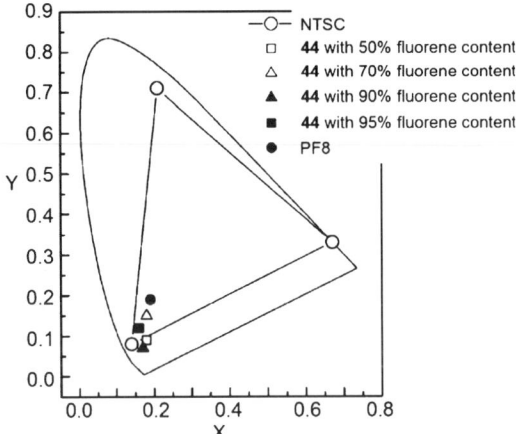

Fig. 5 CIE coordinates of OLED devices fabricated from copolymers **31** [39]. Reproduced by permission of the Royal Society of Chemistry

authors speculate that the triplet energy level of poly(3,6-dibenzosilole) **36** is lower than that of the iridium complex, allowing the back transfer of energy in PL. Using PVK as the polymer host, which has no low energy triplet states, both EL and PL spectra showed emission from the iridium complex only [48].

Polymer LEDs with copolymers containing dibenzosiloles have been reported in the literature [39, 42]. By varying the 3,6-dibenzosilole content in a fluorene-based polymer, superior colour purity and optimum external quantum and luminous efficiencies were obtained (Fig. 5) [39]. Compared to polyfluorene, these copolymers are also stable to thermal annealing. Similar results were reported for a 2,7-dibenzosilole-*co*-3,6-dibenzosilole polymer [42].

A copolymer of 2,7-dibenzosilole and dithienylbenzothiadiazole **55** has been employed as the hole transport material in organic solar cells [38]. In CV measurements, the HOMO and LUMO energy of the polymer were determined to be 5.70 and 3.81 eV, respectively, making it a suitable partner for the commonly used electron transport material, [6,6]-phenyl-C_{61} butyric acid methyl ester (PCBM), in bulk heterojunction solar cells. Devices, with structure ITO/PEDOT:PSS/polymer-PCBM 1 : 4/Al, tested under AM 1.5 G illumination showed a power conversion efficiency of 16% with open circuit voltage of 0.97 V and a fill factor of 55% (Fig. 6). These figures are comparable to other polymer/PCBM bulk heterojunction solar cells reported to date. A very similar polymer to **55** has been reported recently in a red light-emitting device [49]. Along with a green light-emitting poly(2,7-dibenzosilole-*co*-benzothiadiazole), it was found that their device efficiencies are higher than that of the fluorene analogues.

Copolymers of dibenzosilole and thiophene have also been used as semiconductors for OFET [37]. Carrier mobilities approaching 0.1 cm^2/Vs have

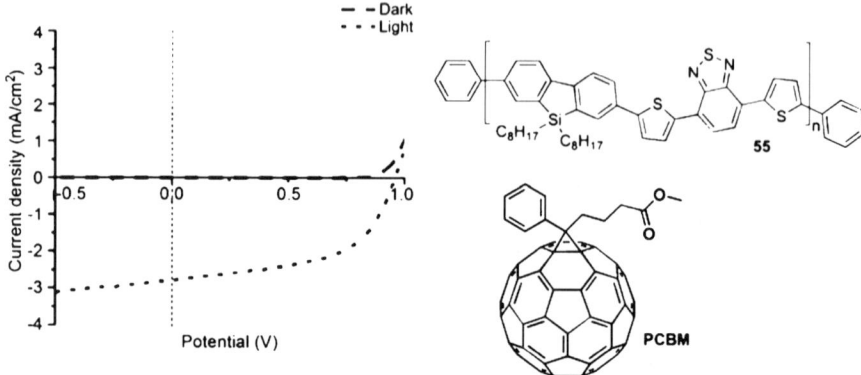

Fig. 6 Current density-potential characteristic of the PCBM/copolymer 55 4 : 1 bulk heterojunction organic solar cell under illumination with AM 1.5 G solar simulated light (*dotted line*) and in the dark (*dashed line*) [38]. © Wiley. Reproduced with permission

been measured in these OFETs. In a different area of application for dibenzosilole polymers, a non-conjugated 9,9-dibenzosilole polymer was used in the detection of various explosive chemicals [44]. Many high explosives contain multiple nitro substituents making them great electron acceptors. Explosive detection was achieved by fluorescence quenching of the polymer by electron transfer to the explosive compound. A 0.3 ng/cm^2 solid state detection limit was achieved for TNT with the 9,9-dibenzosilole polymer.

5
Conclusion

Poly(dibenzosilole)s are a relatively new class of compounds for the area of organic electronic materials. Poly(dibenzosilole)s are not easily oxidised and have advantages of solubility and processability over other polyfluorene analogues, such as carbazoles, dibenzophosphole oxides and dibenzothiophene dioxides. Several groups have already incorporated poly(dibenzosilole)s in the latest organic electronic devices such as OLEDs, OFETs and OSCs and have found improved performance over similar polyfluorene-based devices. With continual advancement in their synthesis, dibenzosilole-based polymers are set to match the popularity of polyfluorenes in organic electronic materials.

Acknowledgements We thank the Australian Research Council (ARC), Commonwealth Scientific and Industrial Research Organisation (CSIRO), Victorian Endowment for Science, Knowledge and Innovation (VESKI), University of Melbourne and DAAD/Go8 exchange scheme for generous financial support.

References

1. Shirakawa H, Louis EJ, MacDiarmid AG, Chiang CK, Heeger AJ (1977) J Chem Soc Chem Commun 578–580
2. Chiang CK, Fincher CR Jr, Park YW, Heeger AJ, Shirakawa H, Louis EJ, Gau SC, MacDiarmid AG (1977) Phys Rev Lett 39:1098–1101
3. Shim H-K, Jin J-I (2002) Adv Polym Sci 158:193–243
4. Zaumseil J, Sirringhaus H (2007) Chem Rev 107:1296–1323
5. Guenes S, Neugebauer H, Sariciftci NS (2007) Chem Rev 107:1324–1338
6. Thomas SW III, Joly GD, Swager TM (2007) Chem Rev 107:1339–1386
7. Friend RH, Gymer RW, Holmes AB, Burroughes JH, Marks RN, Taliani C, Bradley DDC, dos Santos DA, Bredas JL, Logdlund M, Salaneck WR (1999) Nature 397:121–128
8. Kraft A, Grimsdale AC, Holmes AB (1998) Angew Chem Int Edit 37:403–428
9. Burroughes JH, Bradley DDC, Brown AR, Marks RN, Mackay K, Friend RH, Burn PL, Holmes AB (1990) Nature 347:539–541
10. Scherf U, List EJW (2002) Adv Mater 14:477–487
11. Rees ID, Robinson KL, Holmes AB, Towns CR, O'Dell R (2002) MRS Bull 27:451–455
12. Leclerc M (2001) J Polym Sci A 39:2867–2873
13. Bernius MT, Inbasekaran M, O'Brien J, Wu W (2000) Adv Mater 12:1737–1750
14. Gaal M, List EJW, Scherf U (2003) Macromolecules 36:4236–4237
15. Lupton JM, Craig MR, Meijer EW (2002) Appl Phys Lett 80:4489–4491
16. List EJW, Guentner R, Scanducci de Freitas P, Scherf U (2002) Adv Mater 14:374–378
17. Cho SY, Grimsdale AC, Jones DJ, Watkins SE, Holmes AB (2007) J Am Chem Soc 129:11910–11911
18. Sigwalt P, Wegner G, Morin J-F, Leclerc M, Ades D, Siove A (2005) Macromol Rapid Commun 26:761–778
19. Kim Y-G, Thompson BC, Ananthakrishnan N, Padmanaban G, Ramakrishnan S, Reynolds JR (2005) J Mater Res 20:3188–3198
20. Morin J-F, Leclerc M (2001) Macromolecules 34:4680–4682
21. Makioka Y, Hayashi T, Tanaka M (2003) JP 2 004 256 718, 20 030 227, 16 pp
22. Perepichka II, Perepichka IF, Bryce MR, Palsson L-O (2005) Chem Commun, pp 3397–3399
23. Chan KL, McKiernan MJ, Towns CR, Holmes AB (2005) J Am Chem Soc 127:7662–7663
24. Palilis LC, Makinen AJ, Uchida M, Kafafi ZH (2003) Appl Phys Lett 82:2209–2211
25. Yamaguchi S, Endo T, Uchida M, Izumizawa T, Furukawa K, Tamao K (2000) Chem Eur J 6:1683–1692
26. Chen J, Cao Y (2007) Macromol Rapid Commun 28:1714–1742
27. Gilman H, Gorsich RD (1958) J Am Chem Soc 80:3243–3246
28. Gilman H, Gorisch RD (1958) J Am Chem Soc 80:1883–1886
29. Gilman H, Gorsich RD (1955) J Am Chem Soc 77:6380–6381
30. Coutant RW, Levy A (1967) J Organomet Chem 10:175–176
31. Gverdtsiteli I, Diberidze DE, Chernyshev E (1983) Chem Abs 101:73765
32. Suzuki K, Hashimoto Y, Senoo A, Ueno K (2001) EP 1 120 839, 20 010 126, 47 pp
33. Kreuder W, Lupo D, Salbeck J, Schenk H, Stehlin T (1994) DE 4 442 052, 19 941 125, 18 pp
34. Bach U, Graetzel M, Salbeck J, Weissoertel F, Lupo D (1997) DE 19 711 713, 19 970 320, 46 pp
35. Sasaki A (1999) JP 2 000 284 480, 19 990 330, 18 pp
36. Chen R-F, Fan Q-L, Liu S-J, Zhu R, Pu K-Y, Huang W (2006) Synth Met 156:1161–1167

37. Usta H, Lu G, Facchetti A, Marks TJ (2006) J Am Chem Soc 128:9034–9035
38. Boudreault P-LT, Michaud A, Leclerc M (2007) Macromol Rapid Commun 28:2176–2179
39. Wang E, Li C, Mo Y, Zhang Y, Ma G, Shi W, Peng J, Yang W, Cao Y (2006) J Mater Chem 16:4133–4140
40. Mo Y, Tian R, Shi W, Cao Y (2005) Chem Commun, pp 4925–4926
41. Chan KL, Watkins SE, Mak CSK, McKiernan MJ, Towns CR, Pascu SI, Holmes AB (2005) Chem Commun, pp 5766–5768
42. Wang E, Li C, Peng J, Cao Y (2007) J Polym Sci A 45:4941–4949
43. Matsuda T, Kadowaki S, Goya T, Murakami M (2007) Org Lett 9:133–136
44. Sanchez JC, DiPasquale AG, Rheingold AL, Trogler WC (2007) Chem Mater 19:6459–6470
45. Chauhan BPS, Shimizu T, Tanaka M (1997) Chem Lett, pp 785–786
46. Hertel D, Setayesh S, Nothofer H-G, Scherf U, Mullen K, Bassler H (2001) Adv Mater 13:65–70
47. Zhang X, Jiang C, Mo Y, Xu Y, Shi H, Cao Y (2006) Appl Phys Lett 88:051116/1–051116/3
48. Sudhakar M, Djurovich PI, Hogen-Esch TE, Thompson ME (2003) J Am Chem Soc 125:7796–7797
49. Wang E, Li C, Zhuang W, Peng J, Cao Y (2008) J Mater Chem 18:797–801

›
Poly(2,7-carbazole)s and Related Polymers

Pierre-Luc T. Boudreault · Nicolas Blouin · Mario Leclerc (✉)

Canada Research Chair on Electroactive and Photoactive Polymers,
Département de Chimie, Université Laval, Quebec City, Quebec G1K 7P4, Canada
mario.leclerc@chm.ulaval.ca

1	Introduction	100
2	Carbazoles	101
2.1	Synthesis of Monomers	101
2.2	Synthesis and Properties of Poly(2,7-carbazole)s	103
3	Indolo[3,2-b]carbazoles	108
3.1	Synthesis and Properties	108
3.2	Synthesis and Properties of Related Polymers	114
4	Diindolo[3,2-b:2′,3′-h]carbazoles and Bisindenocarbazoles	115
4.1	Synthesis and Properties	115
4.2	Synthesis and Properties of Related Polymers	118
5	Conclusion	120
	References	120

Abstract This chapter reviews the recent progress made in the synthesis and characterization of carbazole-based oligomers and polymers. In fact, four classes of those p-type semiconducting materials will be presented: oligomers and polymers derived from carbazoles, indolo[3,2-b]carbazoles, diindolo[3,2-b : 2′,3′-h]carbazoles, and bisindenocarbazoles. Synthetic pathways used to obtain ladder-type polymers with rod-like structure will also be reported. The characterization of the optical and electrical properties will be discussed with a strong emphasis on the structure–property relationships. Great attention will also be paid to the performances obtained with these materials used as active layers in different types of devices such as field-effect transistors, light-emitting diodes, photovoltaic cells, and thermoelectric devices.

Keywords Bisindenocarbazoles · Carbazoles · Conjugated polymers ·
Diindolocarbazoles · Indolocarbazoles

Abbreviations
BHJ   Bulk heterojunction
BHT   Butylated hydroxytoluene
BIC   Bisindenocarbazole
CB   Chlorobenzene
DIC   Diindolo[3,2-b:2′,3′-h]carbazole
EDOT 3,4-Ethylenedioxythiophene
HOMO Highest occupied molecular orbital

IC	Indolo[3,2-b]carbazole
LUMO	Lowest unoccupied molecular orbital
M_n	Number-average molecular weight
ODCB	*ortho*-Dichlorobenzene
OFET	Organic field-effect transistor
OLED	Organic light-emitting diode
PC	Photovoltaic cell
PCE	Power conversion efficiency
Pd(PPh$_3$)$_4$	Tetrakis(triphenylphosphine) palladium(0)
Pd$_2$(dba)$_3$	Tris(dibenzylideneacetone)dipalladium(0)
PDIC	Polydiindolo[3,2-b:2′,3′-h]carbazole
PIC	Polyindolo[3,2-b]carbazole
PLED	Polymer light-emitting diode
PPh$_3$	Triphenylphosphine
TBAH	Tetrabutylammonium hydroxide
TD	Thermoelectric device
THF	Tetrahydrofuran
UV-Vis	Ultraviolet-visible
V_{OC}	Open circuit potential

1
Introduction

Conjugated oligomers and polymers have drawn much attention over the past decades due to their great potential for numerous electrical and optical applications [1–3]. They have been used as active layers in many devices such as non-linear optical transducers [4], sensors [5, 6], light-emitting diodes [7], field-effect transistors [8, 9], and photovoltaic cells [10–12]. Many conjugated polymers have been extensively studied over the years, among them: polyacetylenes [13], polyanilines [14], poly(2,5-thiophene)s [15], poly(1,4-phenylene)s [16], and poly(2,7-fluorene)s [17, 18]. In particular, poly(2,7-fluorene) derivatives have revealed very promising electrical and optical properties and are the subject of many reviews in this special issue.

Poly(2,7-carbazole)s have also been studied over the past few years [19]. In fact, the presence of a nitrogen atom at the 9-position of this fluorene-like molecule leads to several chemical and physical changes. For instance, because of the unpaired electrons on the nitrogen atom, this monomer is now fully aromatic and that results in oligomers and polymers with good chemical and environmental stabilities. Moreover, other carbazole derivatives such as indolo[3,2-b]carbazoles (ICs) and diindolo[3,2-b:2′-3′-h]carbazoles (DICs) (see Scheme 1) exhibit also interesting properties. For instance, the electron-rich IC unit shows very promising properties as p-type semiconducting materials because of its compact organization in the solid state [20]. However, because ICs and DICs have a tendency to form rigid-rod structures, that makes them harder to solubilize.

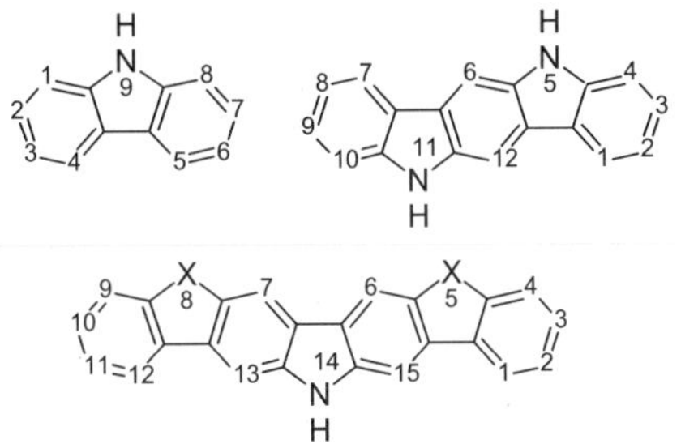

Scheme 1 Chemical structure of carbazole, indolo[3,2-*b*]carbazoles, diindolo[3,2-*b*:2′,3′-*h*]-carbazoles (X = N), and bisindenocarbazoles (X = C)

Along these lines, we want to highlight the recent advances that have been made in the synthesis and characterization of poly(2,7-carbazole)s and related polymers. The synthesis of ladder-type derivatives with extended π systems like indolo[3,2-*b*]carbazoles, diindolo[3,2-*b*:2′,3′-*h*]carbazoles, and bisindenocarbazoles will also be presented. The structure–property relationships of all these materials will be discussed as well as their utilization as active layers in several organic electronic devices, such as light-emitting diodes (LEDs), field-effect transistors (FETs), photovoltaic cells (PCs), and thermoelectric devices (TDs).

2
Carbazoles

2.1
Synthesis of Monomers

Although *N*-vinylcarbazoles and 3,6-functionalized carbazoles can be readily synthesized from 9*H*-carbazole, the preparation of 2,7-functionalized carbazoles is not as straightforward. In contrast to the fluorene unit, the 3,6- (*para* from the nitrogen atom) and 1,8-positions (*ortho* from the nitrogen atom) are highly activated in the electron-rich carbazole unit. In fact, both 2- and 7-positions are less reactive, prohibiting any direct and specific aromatic substitution reactions at these positions. Therefore, alternative methodologies have been developed to obtain 2,7-functionalized carbazoles. Those synthetic strategies usually require functionalized 4,4′-biphenyl precursors with an additional activated group at the 2(-2′) position(s) for subsequent ring-closure reaction (Scheme 2).

The first efficient synthetic strategy for 2,7-functionalized carbazoles was reported by Smith and Brown in 1951 [21, 22]. The crucial step in this synthesis is the formation of the azide compound; generally starting from an

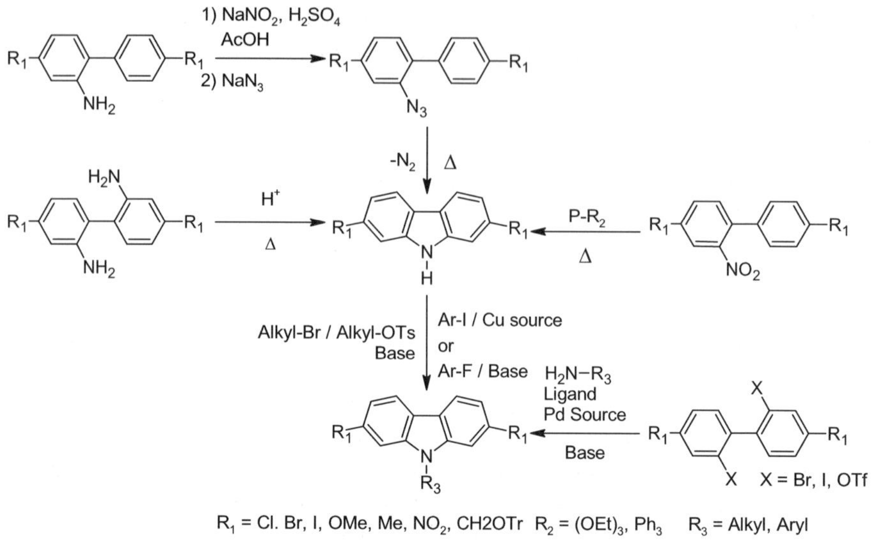

Scheme 2 Synthetic pathways for 2,7-functionalized carbazoles

amino group converted into a diazonium salt. Then, the ring-closure reaction is carried out from the azide group. In 1953, Heinrich reported the synthesis of 2,7-functionalized carbazoles directly from 2,2′-diamino-biphenyl using phosphoric acid at 200 °C [23]. Later on, other proton sources, such as Nafion H membrane, have been used, improving the conversion yield up to 87–93% [24]. In 1965, Cadogan reported a more useful ring-closure reaction (50–60% yield) [25, 26], directly from 2-nitro-biphenyl derivatives through a reflux in P(OEt)$_3$. Recently, the Cadogan ring-closure reaction was optimized, mainly to eliminate the N-ethylcarbazole by-product, through microwave synthesis [27] or substituting P(OEt)$_3$ by PPh$_3$ [28]. The yields are then within 70–90%. In general, the Cadogan ring-closure reaction affords 2,7-functionalized carbazoles in only two or three steps from commercially available starting compounds [29–31].

To enhance processability and to modulate their properties (and those of the resulting polymers), the synthesis of the monomers is generally completed by N-alkylation or N-arylation of the 2,7-functionalized carbazole (Scheme 2). Once synthesized, if useful, additional groups can be then easily added at the 3,6- and 1,8-positions. Indeed, various groups such as halogen, alkyl, nitrile, nitro, ketone, or silane groups have been introduced [31–35]. Finally, in the last few years, novel methodologies [36–38] based on Buchwald and Hartwig palladium cross-coupling reactions have also been developed (Scheme 2). These methods afford milder reaction conditions, tolerant to a wide variety of functional groups. Furthermore, the resulting 2,7-functionalized carbazoles are directly obtained in an N-alkylated or N-arylated form. However, additional steps are generally necessary to prepare the required 4,4′-biphenyl precursor with the proper functional groups (iodide, bromide, or triflate) in 2,2′-positions, limiting the scope of this synthetic pathway.

2.2
Synthesis and Properties of Poly(2,7-carbazole)s

Most of the initial work on the preparation of poly(2,7-carbazole)s [29, 39, 40] was adapted from poly(1,4-phenylene)'s [41–44] and poly(2,7-fluorene)'s [18, 45] chemistry. Indeed, as reported by Morin and Leclerc in 2001 [29], Suzuki [44], Stille [46, 47], and Yamamoto [48, 49] coupling polymerization reactions have led to the first poly(2,7-carbazole) derivatives. Starting from N-octyl-2,7-dichlorocarbazole, the corresponding homopolymer was prepared from Yamamoto coupling. Alternating copolymers were also obtained from N-octyl-2,7-diiodocarbazole using both Suzuki and Stille coupling reactions. The resulting homopolymers and copolymers had limited molecular weights (ca. 5 kDa). Additional key steps for the development of poly(2,7-carbazole) derivatives have then been performed, including the efficient synthesis of 2,7-dibromocarbazole [31] and 2,7-diboronic ester derivatives [34, 39, 50], leading to a wide variety of copolymers.

Initially, poly(2,7-carbazole)s have been developed for PLED applications. For instance, to obtain blue-light-emitting diodes, these semiconducting materials need to have a large band-gap, good and balanced charge injection and transport properties while being air-stable. Preliminary data on homopolymers suggested a good potential for such applications. In fact, poly(2,7-carbazole)s exhibit blue fluorescence at 415–440 nm with relatively high quantum yields [29, 51]. Furthermore, as previously mentioned, the carbazole unit is fully aromatic, providing a good air-stability [17, 52]. Therefore, several poly(2,7-carbazole) derivatives have been developed to maximize the electroluminescence efficiency for blue PLED applications (Scheme 3). To facilitate device fabrication, different linear and bulky alkyl side chains have been placed on the nitrogen atom (P1-B to E). Bulky alkyl side chains (P1-C, P1-E) [40, 51, 53, 54] show an important increase of the polymer solubility and molecular weight. Aryl substitution (P1-F to K) at the nitrogen position has a similar impact on the polymer solubility and molecular weight [53, 55–57]. Moreover, this strategy also increases the solid-state fluorescence quantum yield up to 0.40 [56]. Indeed, homopolymers based on 2,7-carbazole with proper side chains exhibit relatively high number-average molecular weights (M_n) (30–93 kDa) and a strong blue emission. Alternatively, some 2,7-carbazole-based copolymers (P2 and P3) also lead to blue-light-emitting materials [58, 59]. It has been revealed that poly(2,7-carbazole)s show some cross-linking at the 3,6-positions [39] though it does not seem to significantly alter their overall device performances in blue PLEDs [60, 61]. Recently, solutions to this potential problem have been de-

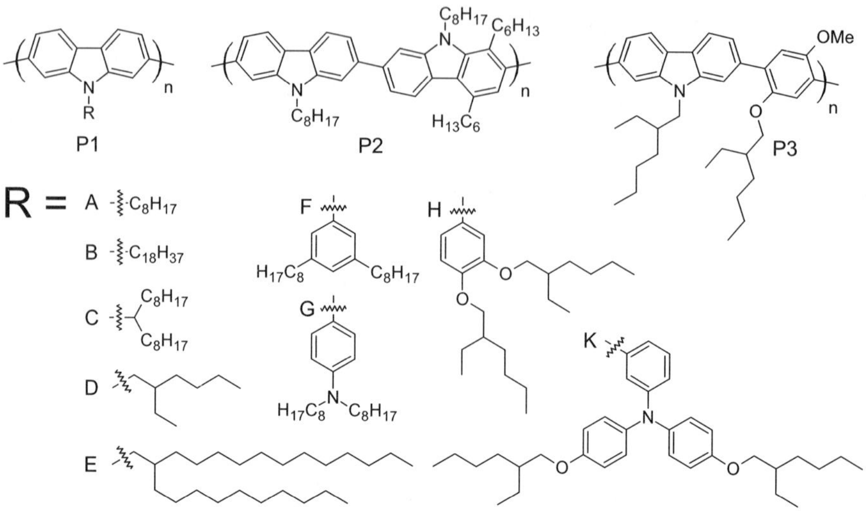

Scheme 3 Blue-light-emitting poly(2,7-carbazole) derivatives

veloped. The protection of the 3,6-positions with functional groups such as methyl or nitrile moieties afford more stable materials [62, 63]. In PLEDs, P3, P1-K, and P1-H show blue electroluminescence up to 1500 cd m^{-2} at 10 V [64, 65].

Through the preparation of new alternating copolymers, low band-gap poly(2,7-carbazole) derivatives have also been successfully obtained (Scheme 4) [40]. The internal charge transfer (ICT) from an electron-rich unit to an electron-deficient moiety usually results in conjugated polymers with lower band-gaps [66–70]. Taking advantage of this strategy, green-light-emitting polymer was obtained with quinoxaline (P4), whereas red-light-emitting polymer was obtained with thiophene dioxide-based pentamer (P5). Poly(2,7-carbazolenevinylene) derivatives (P6 to P10) show similar behaviors with band-gaps varying from 2.37 to 2.15 eV [30]. Only P7 exhibits orange-red (656 nm) luminescence in the solid state. Strong solid-state interactions may quench the fluorescence of the other polymers (P6, P8, P9) [30, 71]. Horner–Emmons and Knoevenagel polymerization reactions were used to synthesize these polymers. Since these two methods are metal-free, they should lead to high purity materials which are required for micro-electronic applications [72].

In contrast, as a charge transport-based application, OFETs need good structural ordering and film-forming properties [73–78]. Different side chains can be incorporated on the carbazole unit to specifically address these issues. Indeed, different alkyl and aryl substituents have been tested to obtain processable and high molecular weight poly(2,7-carbazole) derivatives. The best results have been obtained with a secondary alkyl side chain and

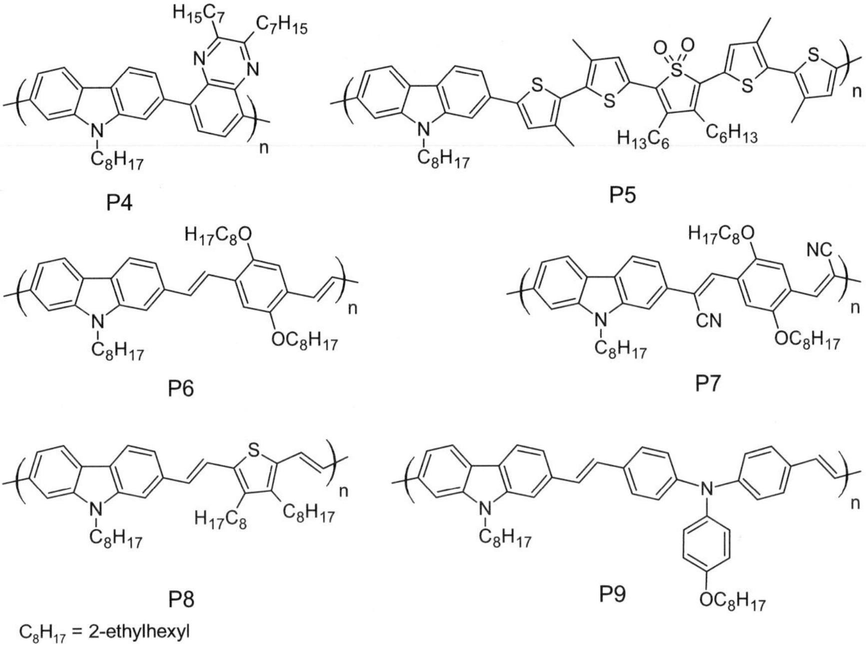

Scheme 4 Low band-gap poly(2,7-carbazole) derivatives

a 3,5-dialkyl-phenyl side chain on the nitrogen atom (a structure that looks like that of 9,9-dialkylfluorene [79, 80]) (Scheme 5). Furthermore, optimized Stille copolymerization reactions have led to relatively high molecular weight materials (M_n between 20–30 kDa). P10 and P11 exhibit modest hole mobilities (10^{-4}–10^{-3} cm^2 V^{-1} s^{-1}) for OFET applications but stable performances, attributed to the good air-stability of the polymers [53].

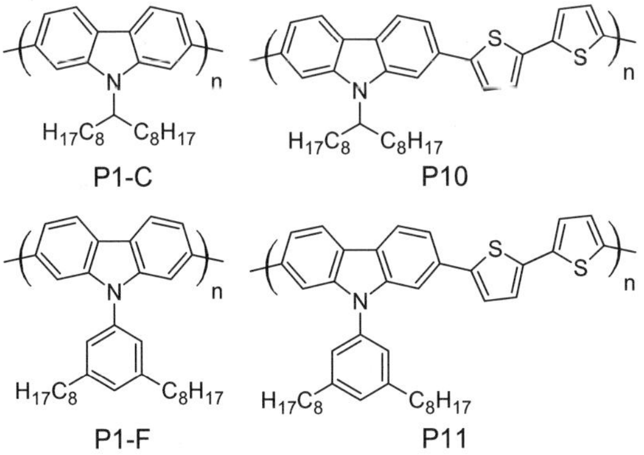

Scheme 5 Poly(2,7-carbazole) derivatives for OFETs

On the other hand, these hole mobility values and their relatively low HOMO energy levels make some poly(2,7-carbazole)s derivatives excellent candidates for applications in bulk heterojunction (BHJ) PCs [10–12, 81]. The first PC based on a homopolymer was reported by Müllen et al. [54]. The limited device performances are probably related to the poor solar spectral match of this polymer [39, 53]. Taking advantage of the easy band-gap modulation of poly(2,7-carbazole) derivatives, several low band-gap polymers have been designed. For instance, poly(2,7-carbazolenevinylene) derivatives (Scheme 6) were studied in a conventional device architecture, using PCBM-C_{61} as the electron-transporting unit [82]. The resulting polymers have optical band-gaps ranging from 2.3 to 1.7 eV while keeping the HOMO energy level below the air oxidation threshold [83]. Although low band-gap polymers were obtained, the reported modest performances (power conversion efficiencies (PCE) up to 0.8%) are most probably limited by the poor solubility of the polymers and their low molecular weight. Furthermore, the vinylene bonds are now known to degrade as a function of time due to photo-oxidation reactions [84].

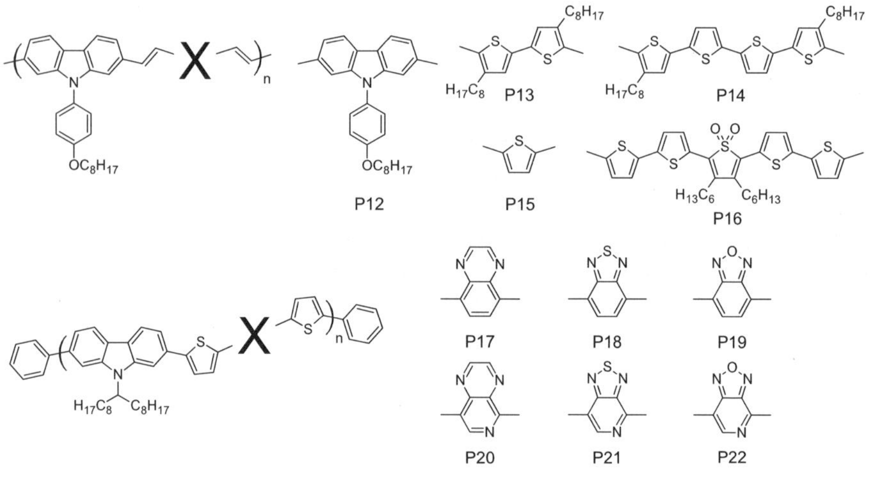

Scheme 6 Poly(2,7-carbazole) derivatives for PCs

A second generation of poly(2,7-carbazole) derivatives was designed to solve the previously observed problems. Selected alkyl and aryl chains developed for OFET applications were tested (Scheme 5), but in this case, alkyl side chains were clearly the best solution for PCs [85]. Using optimized Suzuki coupling polymerization, these polymers combine good solubility, high molecular weight, good film-forming properties, and good hole mobility [85, 86]. Interestingly, the utilization of poly(2,7-carbazole) derivative P18, with a hole mobility up to 3×10^{-3} cm^2 V^{-1} s^{-1} and an optical band-gap of 1.9 eV, leads to a PCE value of 3.6% [85, 86].

To help the design of optimized polymeric materials for BHJ solar cells, several models have been recently proposed [87–89]. The combination of these models and DFT calculations has recently led to the synthesis of several other poly(2,7-carbazole) derivatives (P17, P19–P22). Symmetric polymers (P17–P19) show better structural organization than asymmetric polymers (P20–P22), resulting in higher hole mobilities and power conversion efficiencies. Moreover, their low HOMO energy levels (ca. (−5.6)–(−5.4)eV) provide an excellent air stability and relatively high V_{OC} values (between 0.71–0.96 V).

Another charge transport-based application (i.e. thermoelectric devices) revealed some interesting results. The two important parameters to characterize these devices are the Seebeck coefficient and the electrical conductivity. Theoretical calculations demonstrated potentially high Seebeck coefficients for poly(2,7-carbazole) derivatives [33, 35]. Moreover, the good hole mobilities observed in OFET and PC applications may lead to high electrical conductivities in the doped form. These two reasons motivated the development of poly(2,7-carbazole) derivatives for thermoelectric applications. Taking advantage of the possibility of additional functionalizations, hexyl chains were added at the 3,6-positions, providing better solubility, higher molecular weights (M_n from 7 to 52 kDa), and, more importantly, better film-forming properties (Scheme 7) than their non-alkylated counterparts [33, 35]. These polymers (P23–P26) exhibit a relatively high Seebeck coefficient (55 to 600 μV/K). However, low electrical conductivities (ca. 5×10^{-3} S cm^{-1}) limit the applications of these polymers. To improve the electrical conductivity, the alkyl chain on the nitrogen atom was substituted with a benzoyl moiety. Such poly(2,7-carbazole) derivatives had previously demonstrated better electrical conductivities [39]. Even without the presence

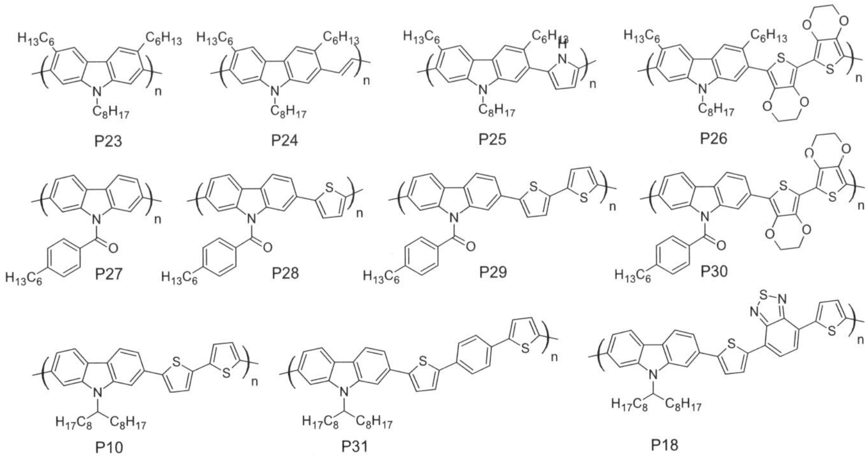

Scheme 7 Poly(2,7-carbazole) derivatives as thermoelectric materials

of additional hexyl side chains at the 3,6-positions, these polymers (P27–P30) exhibit high M_n (22–34 kDa). As expected, they show higher electrical conductivities (from 1×10^{-2} to 3×10^{-1} S cm^{-1}) while keeping a relatively high Seebeck coefficient (ca. 60–70 μV/K). However, the lack of structural ordering in the polymers still limits the overall electrical conductivity. When the linear alkyl chain on the nitrogen atom is substituted by a secondary alkyl chain (P10, P18, P30), the resulting materials are more ordered, resulting in much higher electrical conductivities (up to 200 S cm^{-1}) while keeping a good Seebeck coefficient (ca. 30 μV/K) [90].

3
Indolo[3,2-b]carbazoles

After those extensive studies performed on poly(2,7-carbazole) derivatives, another very promising class of materials emerged, namely indolocarbazoles. Basically, it is a ladder-type extension of the carbazole unit. Bergman et al. recently published a review on the synthesis of all indolocarbazole regioisomers [91] but we will focus here on indolo[3,2-b]carbazole (IC) derivatives. First, it was thought that their pentacene-like structure should allow charge-transport properties close to those of this organic semiconductor, which is among the best candidates for the active layer in p-type OFETs [92]. Moreover, since it is known that pentacene is very sensitive to light, atmospheric oxygen, and practically insoluble in most common organic solvents, it was anticipated that IC derivatives should solve these major problems. Indeed, the two nitrogen atoms in the IC molecule create a completely aromatic molecule which lowers the HOMO energy level. Consequently, this should increase the photostability of the material. Secondly, long chains can easily be added as side chains on the nitrogen atoms which should lead to highly soluble and processable materials. Aromatic substituents can also be added at both ends of the IC moiety such as phenyl, thiophene, diphenyl-4-aminophenyl, fluorene, or carbazole units. Those functional groups were added to enhance the molecular packing of the materials [20, 93]. In addition, the presence of reactive groups at the 2,8- or 3,9-positions should lead to the synthesis of numerous homopolymers and copolymers [94–96]. Comonomers, such as thiophene, bithiophene, and ethylenedioxythiophene (EDOT), have been already copolymerized with the IC unit, providing copolymers with different and interesting properties.

3.1
Synthesis and Properties

Since the synthesis of heteroacenes usually requires multiple steps; there are only a few synthetic routes that have been developed to obtain the IC unit. As shown in Scheme 8, Wakim et al. used a nine-step synthetic pathway [97], the final IC framework being formed via a Cadogan ring-closure reaction. Since this reaction is not regioselective, methyl protecting groups were essential to obtain the desired regioisomer. Moreover, the Cadogan ring closure is performed in P(OEt)$_3$ and is refluxed overnight to afford the IC unit in only a 40% yield. Afterward, the compound was alkylated with n-bromooctane under reflux in NaOH, TBAH, and acetone with 90% isolated yield. Blouin et al. developed a double Cadogan ring closure without methyl protecting groups [95]. The double Cadogan is performed on a dinitrotriphenyl precursor but the yields are very low (6–8%). This synthetic pathway is, however, much quicker (only two steps from commercially available product). Afterward, the compound was alkylated with bromoalkyl with high isolated yields.

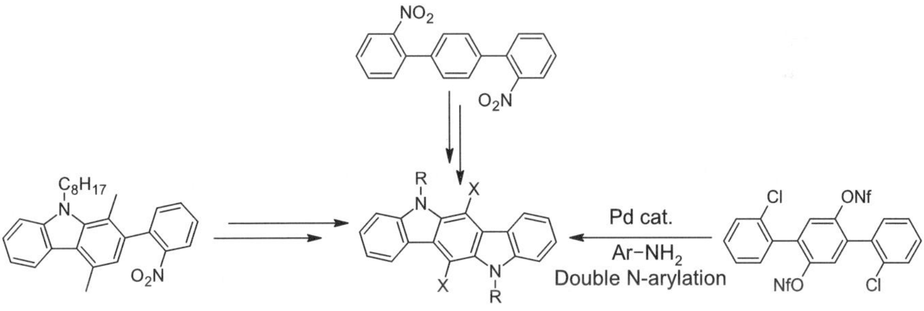

Scheme 8 Synthesis of ICs via Cadogan ring closure and double N-arylation

Recently, another synthetic pathway has been developed by Kawaguchi et al. [98], involving a palladium-catalyzed double N-arylation reaction (Scheme 8). Before the synthesis of the IC, they proved that it was possible to synthesize the carbazole unit starting from various 2,2'-dihalobiphenyls and aminoaryls using Pd$_2$(dba)$_3$, a phosphine ligand, and a base. Good yields (70–85%) were obtained using this double N-arylation reaction [36, 37, 99]. Although this method is very efficient, it is still limited regarding the functional groups that can be added on the IC unit. It is worth noting that to synthesize oligomers or polymers from those compounds, they have to contain reactive substituents at both ends.

Dehaen et al. [100, 101] found another interesting way to synthesize the IC backbone. They first used a condensation of indole and aldehyde in the presence of iodine in acetonitrile to afford 3,3'-bis(indolyl)methane. The IC moiety was obtained by treating the previous compound with triethyl orthoformate and a catalytic amount of a strong acid (sulfuric acid or methanesulfonic acid) in methanol. 6,12-Disubstituted-5,11-dihydroindolo[3,2-b]carbazoles can be obtained in moderate yields (23–50%). However, this pathway to synthesize ICs is not the most efficient to afford materials for applications in electronic devices because of its limitation to 6,12-disubstituted compounds.

The most common way to afford efficiently ICs is via the double Fischer indolization of cyclohexane-1,4-dione bis(phenylhydrazone) in strong acidic conditions (H_2SO_4 in AcOH), which was first developed by Robinson in 1963 [102]. As reported by Yudina and Bergman [103], the substituted double hydrazone can be obtained in 50–60% yields with a double condensation of substituted phenylhydrazines on cyclohexane-1,4-dione in ethanol. Next, Fischer cyclization is performed by adding the double hydrazone to a cold mixture of H_2SO_4 and AcOH. After the addition is completed, the mixture is heated up to 110 °C for 15 min and then cooled overnight. The indolization reaction allows the synthesis of the IC unit with various functional groups [103–105]. For example, 2,8-dimethoxy-5,11-diBOC-indolo[3,2-b]carbazole was obtained in a 20% yield and 2,8-difluoro-5,11-dihydroindolo[3,2-b]carbazole was synthesized with yields from 26 to 48%. Scheme 9 shows the general reaction and different products that can be obtained from this pathway. Fischer cyclization affords indolo[2,3-b]carbazole, indolo[2,3-c]carbazole and also the same two isomers with the non-aromatic central ring. Those isomers can be readily oxidized to obtain the fully aromatic compound by heating up to 175 °C in ethylene glycol.

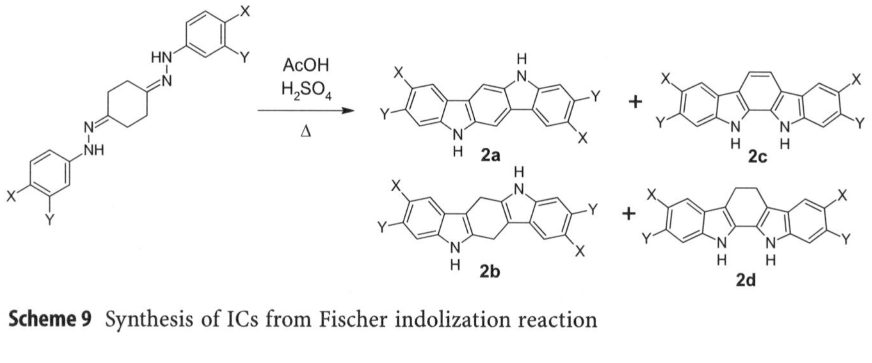

Scheme 9 Synthesis of ICs from Fischer indolization reaction

Even though the yields are usually low (20–50%) for the desired regioisomer (separated from recrystallization), this synthetic pathway is simple and versatile. For instance, numerous substituents can be added on the phenylhydrazine starting material such as methoxy, fluoride, chloride, and bromide either at the 3- or 4-positions to give the 2,8- or 3,9-disubstituted indolocarbazole derivatives. These reactive groups are important to synthesize different IC-based materials afterward.

The first utilization of an IC unit in organic electronics (e.g. a hole transporting molecule in OLEDs) was published by Ong et al. in 1999 [106]. They were able to add naphtyl substituents on the nitrogen atoms (**3a**) of the 5,11-dihydroindolo[3,2-*b*]carbazole with an Ullmann reaction. The yields go from 33 to 77% [107]. Alkylated ICs have also been reported, including compounds **3b–e**. All these materials show very interesting hole mobilities [108].

Some other examples were made by the same group [108, 109], including ICs with chlorine substituents at the 2,8- and 3,9-positions. The IC unit was synthesized using the Fischer indolization mentioned above. *n*-Dodecyl substituents were added on the nitrogen atoms by solubilizing the dichloro-5,11-dihydroindolocarbarbazole in DMSO in the presence of *n*-dodecyl bromide, sodium hydroxide, and a catalytic amount of a phase transfer agent, benzyltriethylammonium chloride. The yields for this reaction average 90%. The chemical structures of all those derivatives are presented in Scheme 10.

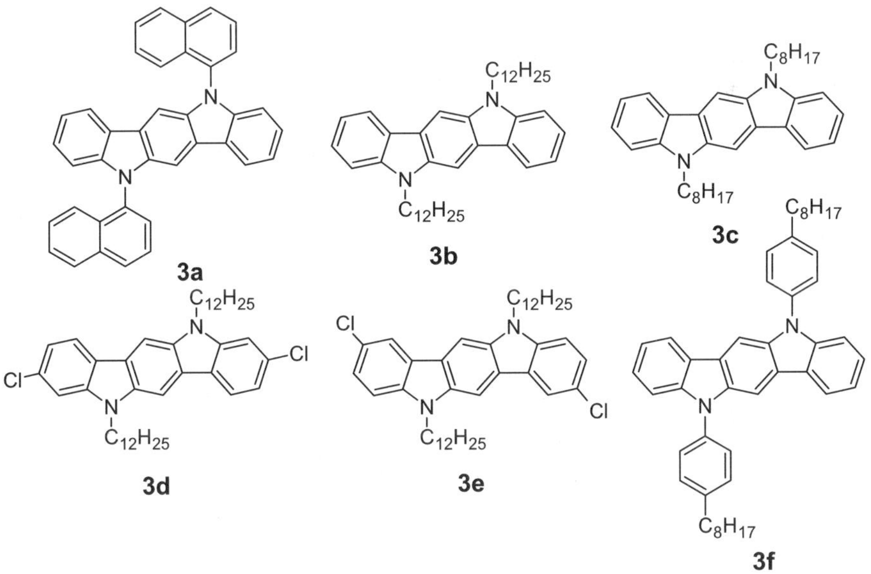

Scheme 10 Chemical structure of different IC derivatives

Recently, another pathway was used to synthesize indolo[3,2-*b*]carbazole-based molecules containing aromatic substituents on both ends of the material. First, octyl side chains are added on the nitrogen atoms; then, aromatic substituents can be added by using either Suzuki or Stille cross-coupling starting with 2,8- or 3,9-dibromo-5,11-dioctylindolo[3.2-b]carbazoles. As shown in Scheme 11, the addition of phenyl, thiophene, biphenylamino-phenyl, fluorene, and even carbazole units has recently been reported in the literature [20, 93].

The spectroscopic properties of the synthesized ICs are quite interesting. When there is no substituent on both ends of the IC unit, the maximum

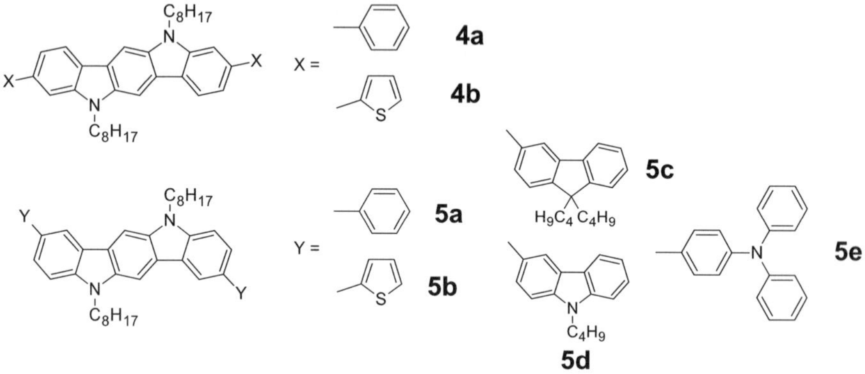

Scheme 11 Chemical structure of 3,9- and 2,8-disubstituted ICs

absorption wavelength is located between 336 and 342 nm. The addition of aromatic units on both ends of the IC unit leads to a red shift of the absorption going from 349 to 383 nm [20, 93]. Moreover, all these molecules have a HOMO energy level lower than 5.12 eV. This value is lower than most of those reported for regioregular polythiophenes, which are usually between 4.9 and 5.0 eV [110, 111]. Those values should lead to a better stability against oxidation. For the materials presented in Scheme 10, which do not contain any aromatic group at both extremities, the conjugation length is shorter and consequently, the HOMO energy levels are lower, varying between 5.5 to 5.7 eV [108, 109]. In short, all those materials should have a better air stability than acenes and thiophene-based materials [73, 110–118]. It has to be noted that theoretical calculations fit reasonably well with the experimental data obtained for the IC-based materials [107, 119].

In parallel, single crystals were grown from some of the above-mentioned compounds. In fact, Boudreault et al. [20] have been able to grow crystals from compounds **3c–f** in hexanes and dichloromethane. Interestingly, the addition of aromatic or halogen units is an efficient way to obtain short contacts, not only in one dimension but also in two or even three dimensions (see Fig. 1). Indeed, when it is substituted with alkyl side chains (Fig. 1a), the molecular packing is less efficient with a distance of 3.49 Å along the π-stacking direction and of 3.75 Å along the herringbone direction. Despite a good short contact distance along the π-stacking direction, relatively low hole mobilities (around 10^{-3} cm^2 V^{-1} s^{-1}) were obtained. This is probably due to the fact that the contact distance along the herringbone direction is much higher. This could force the charges to circulate only in one dimension. On the other hand, by adding phenyl side chains on the nitrogen atoms of the IC (Fig. 1b), the hole mobility increases up to 0.12 cm^2 V^{-1} s^{-1}. This significant enhancement can be explained by the creation of numerous short contacts along each axis of the lattice.

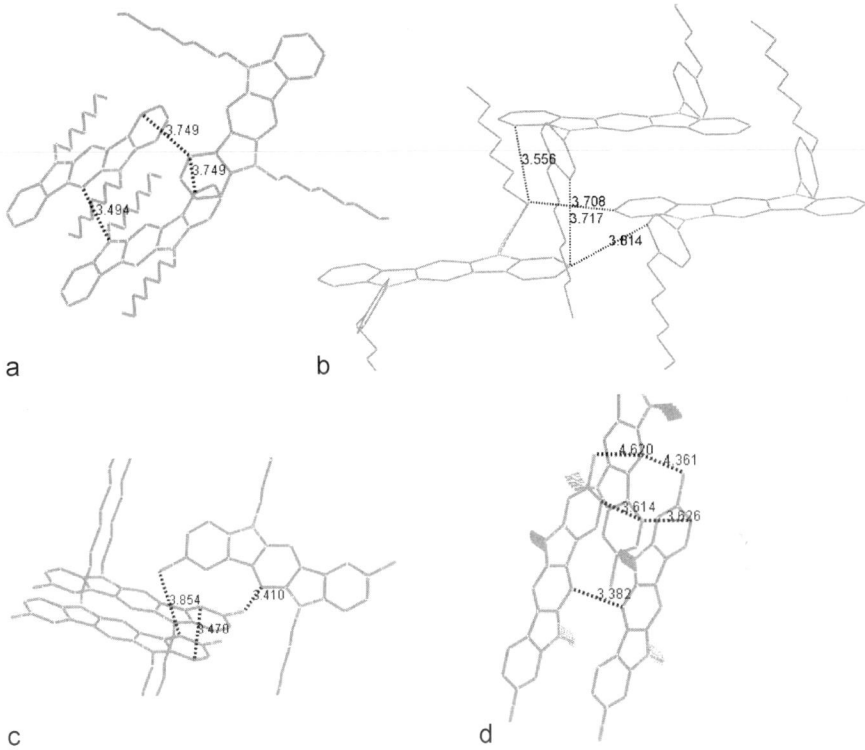

Fig. 1 Crystal structure of 5,11-dioctylindolo[3,2-b]carbazole (**a**), 5,11-bis(p-octylphenyl)-indolo[3,2-b]carbazole (**b**), 2,8-dichloro-5,11-didodecylindolo[3,2-b]carbazole (**c**) and 3,9-dichloro-5,11-didodecylindolo[3,2-b]carbazole (**d**). Hydrogen atoms have been removed to clarify the view

Furthermore, as shown in Fig. 1c, the resulting molecular packing for 2,8-dichloro-5,11-didodecylindolo[3,2-b]carbazole is quite compact with short contact distances of 3.47 Å along the π-stack and of 3.41 Å along the herringbone direction. This efficient organization leads to field-effect mobilities as high as $0.14\,\text{cm}^2\,\text{V}^{-1}\,\text{s}^{-1}$. The packing is not as efficient when the chlorine atoms are located at the 3- and 9-positions (Fig. 1d). Perhaps this is because the C to Cl short contact distance is significantly higher when the chlorine atoms are located at such positions. This C to Cl distance goes from 3.41 to 4.36 Å which should have an important impact on the OFET performances. In fact, the hole mobility decreases by at least one order of magnitude ($0.01\,\text{cm}^2\,\text{V}^{-1}\,\text{s}^{-1}$) for 3,9-dichloro-5,11-didodecyl indolo[3,2-b]-carbazole. Currently, for OFET applications, the best IC-based material is 3,9-diphenyl-5,11-dioctylindolo[3,2-b]carbazole (compound **4a**) with a hole mobility reaching $0.2\,\text{cm}^2\,\text{V}^{-1}\,\text{s}^{-1}$, obtained with vacuum-sublimated thin-films [20].

3.2
Synthesis and Properties of Related Polymers

As shown in Scheme 12, there are a few possibilities for the synthesis of IC-based polymers. Homopolymers were synthesized starting from 2,8- or 3,9-dichloroindolo[3,2-b]carbazole derivatives using Yamamoto nickel(0) polymerization [49, 95] and Zn or FeCl$_3$-mediated dehalogenative coupling polymerization [96]. The alkyl side chains length is of great importance to obtain high molecular weight materials. The PICs (P32-A, P33-A) synthesized by Blouin et al. exhibited low M_n (ranging from 3 to 6.3 kDa) due to the very poor solubility of the materials [95]. Ong et al. were able to raise the M_n values up to 11 kDa by adding long n-dodecyl (P32-B, P33-B) or p-dodecylphenyl side chains (P32-B, P33-D) [96]. However, these polymers are poorly soluble, probably due to the rigid rod-like structure of the IC unit. As observed for poly(2,7-carbazole)s, bulky side chains result in more soluble and higher molecular weight polymers. For instance, PIC derivatives (P32 and P33) with side-chains C or E lead to higher M_n (ranging from 12 to 27 kDa) [120]. Alternatively, Lévesque et al. were able to obtain significantly higher M_n (up to 26 kDa) by using N-benzoyl side chains where the benzoyl group is substituted by two octyloxy side chains (P32-G, P33-G) [35]. For IC-based polymers, hole mobilities as high as 0.02 cm^2 V^{-1} s^{-1} have been obtained by Ong et al. [96] for 2-coupled polymers (P32-B). The 3-coupled PICs showed mobility values about one order of magnitude lower [120].

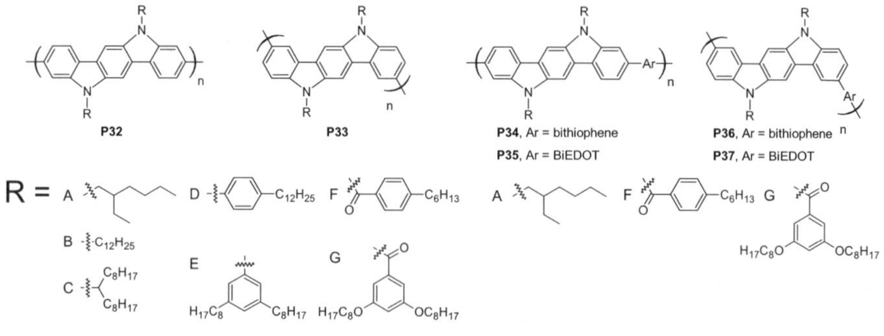

Scheme 12 Chemical structure of some IC-based polymers

The rigid-rod nature of these polymers was also proven by their UV-Vis absorption spectra. Indeed, very similar UV-Vis absorption spectra were obtained in the solid state and in solution [95]. Moreover, as observed with different IC derivatives, the conjugation through the 3- and 9-positions is more efficient than through the 2- and 8-positions. It has been shown that the conjugation through the 2- and 8-positions should be associated with the nitrogen atoms while when the conjugation goes through the 3- and 9-positions, it rather adopts a p-phenylene-like pathway.

In order to co-polymerize the IC unit, Suzuki and Stille polymerizations have been used. First, Blouin et al. [94] were able to obtain polyindolo[3,2-b]-carbazole derivatives with bithiophene or biEDOT as co-monomers. Unfortunately, these studies demonstrated a relatively low oxidation potential for these polymers (especially for P35 and P37), limiting their applications in OFETs and PCs. However, for doped state applications, these polymers may exhibit interesting properties [35]. For instance, when copolymerized with bithiophene, the resulting copolymer shows a good electrical conductivity (as high as $0.7\,\mathrm{S\,cm^{-1}}$) but a low Seebeck coefficient of $4.3\,\mu\mathrm{V\,K^{-1}}$ [35]. Finally, the UV-Vis absorption maxima are similar for poly(2,8-indolocarbazole-alt-bithiophene) and poly(2,8-indolocarbazole-alt-bis(3,4-ethylenedioxythiophene)). A broad absorption band is centered at 430 nm whereas, for the 3- and 9-substituted copolymers, the broad band is centered around 490–500 nm [94].

4
Diindolo[3,2-b:2′,3′-h]carbazoles and Bisindenocarbazoles

Once again, to obtain processable pentacene-like molecules (with even longer ladder-type π systems) [2, 121], different diindolo[3,2-b:2′,3′-h]carbazole (DIC) and bisindenocarbazole (BIC) derivatives have been synthesized. Since halogen atoms can be quite easily added at both ends of those molecules, several polymers have been synthesized from DIC and BIC monomers. The syntheses and the properties of the related oligomers and polymers will be described in this section.

4.1
Synthesis and Properties

The synthesis of all precursors leading to DICs and BICs begins with 2,7-disubstituted carbazoles. The synthesis of DICs have been first reported by Bouchard et al. [34] and involves seven steps from 1,3,6,8-tetrabromo-2,7-dimethoxy-N-octylcarbazole (Scheme 13). This methodology takes advantage of the higher selectivity at the 3,6- and 1,8-positions on the carbazole rings (see Sect. 2.1). In this pathway, the 1- and 8-positions of the carbazole unit have to be protected with methyl groups to afford a selective Cadogan ring-closure reaction. Compound **6** is methylated by the double lithiation of the bromine atoms at the 1- and 8-positions of the carbazole moiety in a 64% yield. The two remaining bromine atoms at the 3- and 6-positions are removed with a double lithiation, which is then hydrolyzed with water to afford compound **8** in 83% yield. Deprotection of the methoxy groups is achieved by using BBr$_3$ in anhydrous dichloromethane. Afterward, the diol was treated with Tf$_2$O to afford compound **10** in a 59% yield. The triflate groups allow one to introduce boronic esters via a Masuda reaction [122]. Suzuki cross-coupling was then performed in the presence of commercially available 1-bromo-2-nitrobenzene over 3 days to afford compound **12** in a 73% yield. The DIC unit is then obtained by Cadogan ring closure which is achieved by treating with P(OEt)$_3$, the yield of this final step is 30%. In the second pathway, the final step utilized to obtain the DIC unit involves a copper-catalyzed Ullmann reaction which is completely regioselective [95, 123]. Consequently, the carbazole precursor does not need to be protected with methyl groups at the 1- and 8-positions. The final reaction yields are around 60%, independent of the nature of the groups at both ends of the DIC. Obviously, the synthesis of DICs is still very difficult and the overall yields are relatively low.

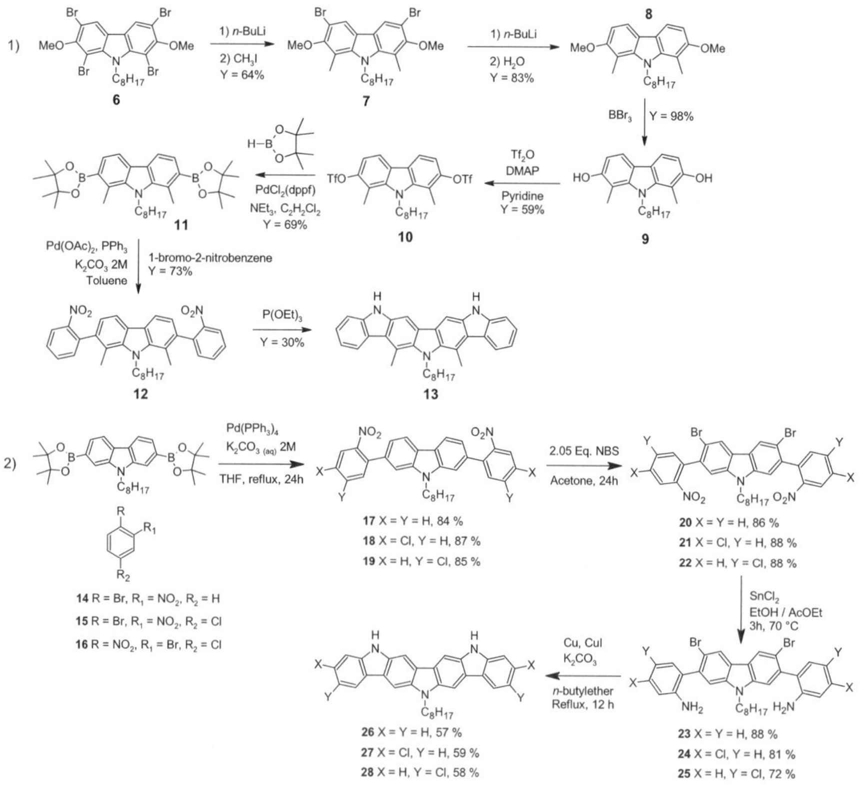

Scheme 13 Synthesis of DIC units

On the other hand, substituted and unsubstituted BICs were synthesized using two relatively efficient and easy procedures (Scheme 14). In the first method, the starting compound is 2,7-dibromocarbazole on which is added benzoyl groups at the 3- and 6-positions by a Friedel–Craft acylation with benzoyl anhydride to give 76% of isolated yield [124]. The reaction has also been carried out with 4-octylbenzoyl chloride with a similar yield. A Suzuki cross-coupling is then performed with 4-chlorophenylboronic acid to synthesize a tetraphenylene diketone in a good yield (82%). This diketone is treated with a slight excess of 4-octylphenyllithium to produce the corresponding diol and a ring-closure reaction afforded compound 32 as the new ladder-type BIC monomer [125].

The second pathway used has been described by Sonntag and Strohriegl [126] and it basically leads to small oligomeric materials. The synthetic method is quite similar to that described above. The starting materials are also 2,7-dibromocarbazoles with different alkyl chains on the nitrogen atoms such as *sec*-butyl, 2-ethylhexyl, and methyl groups. First, 2-acetylphenyl groups are added by Suzuki cross-coupling with excellent yields (around 90%). The resulting diketone is reduced quantitatively to a diol by treating it with lithium triethylborohydride. The ring-closure reaction of the diol was carried out with boron trifluoride etherate as a Lewis-acid catalyst. Since the 3- and 6-positions of the carbazole are highly activated, only one regioisomer is obtained. The final step is the alkylation of the BIC core with different side chains. They are introduced by treatment with *n*-BuLi and the corresponding alkyl halide. The fact that the last step is the alkylation reaction is a great advantage here since the target molecule can be tailored by changing the length of the alkyl chain.

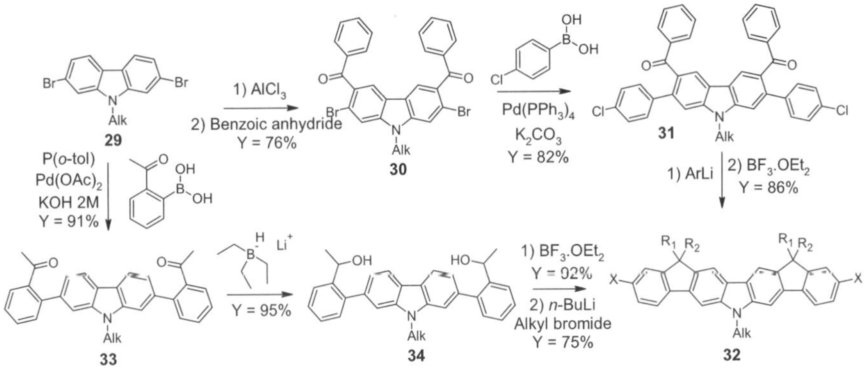

Scheme 14 Two pathways used to synthesize BICs

Finally, Sonntag and Strohriegl [127] recently described a new BIC derivative containing p-hexylphenyl on both ends. p-Hexylphenyl units were added by Suzuki cross-coupling using P(*o*-tol)$_3$ and Pd(OAc)$_2$ as the catalyst to afford 79% of isolated yield. The compound was found to be thermally stable and also shows excellent electrochemical stability.

4.2
Synthesis and Properties of Related Polymers

A great part of the work on the BIC unit has been done for developing precursors to ladder or semi-ladder polymers. Two different ways have been used so far, the direct synthesis of the BIC unit and the polymerization of this unit or the polymerization of 3,6-dibenzoyl-substituted carbazole and a ring-closure afterward. This last pathway affords a completely ladder-type polymer which is of great interest for various applications.

To the best of our knowledge, there is only one paper by Müllen et al. [125] that presents the polymerization of the BIC unit. The semi-ladder polymer was synthesized by a nickel(0)-mediated Yamamoto-type polymerization using dichlorobisindenocarbazole or dibromobisindenocarbazole as precursors. The two polymers were synthesized in good yields (77 and 70%). They both show a strong blue emission in solution with a maximum centered at 445 nm.

Furthermore, it was possible to obtain carbazole-based ladder polymers via the polymerization of two different carbazole units, one containing boronic esters and the other bearing benzoyl groups at the 3- and 6-positions in order to close the rings afterward (P39) (Scheme 15) [50]. It is also pos-

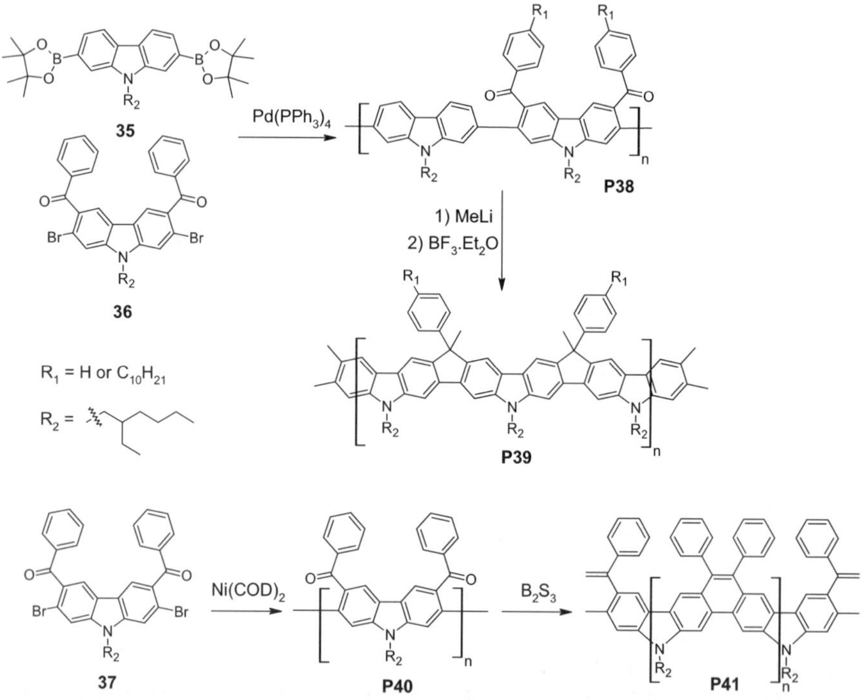

Scheme 15 Synthesis of carbazole-based ladder polymers

sible to polymerize a 3,6-dibenzoyl-2,7-dibromo-N-(2-ethylhexyl) carbazole unit by using a Yamamoto-type polymerization to afford a precursor polymer (P40), as also shown in Scheme 15. Then, the precursor is treated with boron sulfide to perform the ring-closure reaction, affording the desired ladder-type polymer (P41).

In parallel, Scherf et al. [128] have also reported a ladder-type polymer containing carbazole moieties. Their strategy is quite similar to those presented above; the authors copolymerized the carbazole moiety with a phenyl bearing two benzoyl side chains. After the polymerization is completed, the ketone moieties located on the phenyl are reduced into alcohol groups by a reaction with MeLi. The ring-closure reaction of the polyalcohol was carried out by treatment with boron trifluoride etherate, affording the ladder polymer.

Moreover, semi-ladder polymers based on DIC units were obtained in much more classical ways. 2,11- and 3,10-dichloro DIC derivatives were polymerized through a nickel(0)-mediated Yamamoto polymerization reaction [95]. The resulting polymers are soluble in ODCB and partially soluble in CB. The absorption maxima of the BIC-based polymers clearly indicate a planar structure. The absorption maxima of the polymeric precursors show a primary maximum around 350 nm compared to the ladder-type polymers where the maximum is centered around 470 nm [50]. This significant red shift clearly indicates the higher degree of conjugation in the planar rigid polymers. The fluorescence spectra of those polymers also show the same trend.

In addition, the ladder polymers, shown in Scheme 15, exhibit oxidation potentials between 0.71 and 0.84 V (vs. Ag/AgCl). The HOMO energy levels are similar to MEH-PPV [10], and significantly higher than poly[2,7-(9,9-dioctylfluorene)] [129], which can be attributed to the electron-donating effect of the nitrogen atom. However, the electrochemical properties are quite different for semi-ladder polymers, e.g. PDIC [95] or poly(ladder-type tetraphenylene) [125]. For instance, both DIC-based polymers show two different oxidation processes centered at 0.43 and 1.02 V vs. Ag/AgCl. As for IC copolymers, these studies reveal the relatively low oxidation potential for these polymers resulting in possible air oxidation in ambient environment, limiting the possibilities of applications for such polymers. The absorption spectra of those polymers are slightly red-shifted compared to the PIC homologues. The absorption maxima in solution are centered at 431 nm for P3DIC and 386 nm for P2DIC. This important difference between the two polymers is explained by the conjugation going through the nitrogen atoms when it is substituted at the 2- and 11-positions compared to a p-phenylene-like conjugation pathway for the 3- and 10-substituted polymer. Moreover, the BIC-based polymers absorption spectra with maxima at 434 and 427 nm show a high conjugation length [125]. The oxidation potential was found to be at 1.20 and 1.23 V vs. Ag/AgCl, ensuring good air stability.

These polymers have been tested in OLEDs and as thermoelectric materials. As mentioned above, Müllen et al. [125] reported two polymers with strong blue emission in solution. The absorption maximum is centered at 445 nm. They also exhibit high luminescence typically over 700–900 cd m^{-2}. Furthermore, PDICs were tested in thermoelectric devices. Unfortunately, the power factor is only of 10^{-9} W K^{-2} m^{-1}, because oxidation with ferric chloride can only reach the first oxidation process of those polymers which leads to conductivity of about 10^{-2} S cm^{-1} [35].

5
Conclusion

As shown in this review, many different carbazole-based oligomers and polymers can now be synthesized quite efficiently. These materials exhibit interesting and tunable properties which enable them to be used in many different applications such as light-emitting diodes, field-effect transistors, photovoltaic cells, and thermoelectric devices. In particular, processable and air-stable carbazole-based materials have shown excellent hole transport properties which make them among the most promising materials for applications in solar cells and transistors. Although some structural optimizations (molecular weight, solubility, band-gap, etc.) still need to be done, it is believed that they should lead to commercial products in the near future.

References

1. Skotheim TA, Reynolds JR (2007) Handbook of Conducting Polymers. Taylor & Francis Group, Boca Raton, Fl
2. Klauk H (2006) Organic Electronics. Wiley-VCH, Weinheim
3. Hadziioannou G, Malliaras GG (2007) Semiconducting Polymers. Wiley-VCH, Weinheim
4. Yu L, Chen M, Dalton LR (1990) Chem Mater 2:649–659
5. Leclerc M, Faid K (1997) Adv Mater 9:1087–1094
6. McQuade DT, Pullen AE, Swager TM (2000) Chem Rev 100:2537–2574
7. Burroughes JH, Bradley DDC, Brown AR, Marks RN, Mackay K, Friend RH, Burns PL, Holmes AB (1990) Nature 347:539–541
8. Dimitrakopoulos CD, Malenfant PRL (2002) Adv Mater 14:99–117
9. Katz HE (2004) Chem Mater 16:4748–4756
10. Brabec CJ, Sariciftci NS, Hummelen JC (2001) Adv Funct Mater 11:15–26
11. Hoppe H, Sariciftci NS (2004) J Mater Res 19:1924–1945
12. Coakley KM, McGehee MD (2004) Chem Mater 16:4533–4542
13. Lam JWY, Tang BZ (2005) Acc Chem Res 38:745–754
14. Kang ET, Neoh KG, Tan KL (1998) Prog Polym Sci 23:277–324
15. Roncali J (1992) Chem Rev 92:711–738
16. Kim DY, Cho HN, Kim CY (2000) Prog Polym Sci 25:1089–1139
17. Scherf U, List EJW (2002) Adv Mater 14:477–487
18. Leclerc M (2001) J Polym Sci A Polym Chem 39:2867–2873
19. Morin J-F, Leclerc M, Adès D, Siove A (2005) Macromol Rapid Commun 26:761–778

20. Boudreault PLT, Wakim S, Blouin N, Simard M, Tessier C, Tao Y, Leclerc M (2007) J Am Chem Soc 129:9125–9136
21. Smith PAS, Brown BB (1951) J Am Chem Soc 73:2435–2437
22. Smith PAS, Brown BB (1951) J Am Chem Soc 73:2438–2441
23. Heinrich L (1953) Chem Ber 86:522–524
24. Yamato T, Hideshima C, Suehiro K, Tashiro M, Prakash GKS, Olah GA (1991) J Org Chem 56:6248–6250
25. Cadogan JIG, Carmeron-Wood M, Makie RK, Searle RJG (1965) J Chem Soc, pp 4831–4837
26. Cadogan JIG (1969) Synthesis, pp 11–17
27. Appukkuttan P, Van der Eycken E, Dehaen W (2005) Synlett, pp 127–133
28. Freeman AW, Urvoy M, Criswell ME (2005) J Org Chem 70:5014–5019
29. Morin JF, Leclerc M (2001) Macromolecules 34:4680–4682
30. Morin JF, Drolet N, Tao Y, Leclerc M (2004) Chem Mater 16:4619–4626
31. Dierschke F, Grimsdale AC, Müllen K (2003) Synthesis, pp 2470–2472
32. Bouchard J, Wakim S, Leclerc M (2004) Synth Commun 34:2737–2742
33. Levesque I, Gao X, Klug DD, Tse JS, Ratcliffe CI, Leclerc M (2005) React Funct Polym 65:23–36
34. Bouchard J, Wakim S, Leclerc M (2004) J Org Chem 69:5705–5711
35. Levesque I, Bertrand PO, Blouin N, Leclerc M, Zecchin S, Zotti G, Ratcliffe CI, Klug DD, Gao X, Gao F, Tse JS (2007) Chem Mater 19:2128–2138
36. Nozaki K, Takahashi K, Nakano K, Hiyama T, Tang H-Z, Fujiki M, Yamaguchi S, Tamao K (2003) Angew Chem Int Edit 42:2051–2053
37. Kuwahara A, Nakano K, Nozaki K (2005) J Org Chem 70:413–419
38. Kitawaki T, Hayashi Y, Ueno A, Chida N (2006) Tetrahedron 62:6792–6801
39. Zotti G, Schiavon G, Zecchin S, Morin JF, Leclerc M (2002) Macromolecules 35:2122–2128
40. Morin JF, Leclerc M (2002) Macromolecules 35:8413–8417
41. Rehahn M, Schluter A-D, Wegner G, Feast WJ (1989) Polymer 30:1060–1062
42. Vahlenkamp T, Wegner G (1994) Macromol Chem Phys 195:1933–1952
43. Remmers M, Schulze M, Wegner G (1996) Macromol Rapid Commun 17:239–252
44. Schlüter AD (2001) J Polym Sci A Polym Chem 39:1533–1556
45. Ranger M, Rondeau D, Leclerc M (1997) Macromolecules 30:7686–7691
46. Babudri F, Cicco SR, Farinola GM, Naso F, Bolognesi A, Porzio W (1996) Macromol Rapid Commun 17:905–911
47. Bao Z, Chan WK, Yu L (1995) J Am Chem Soc 117:12426–12435
48. Yamamoto T, Morita A, Miyazaki Y, Maruyama T, Wakayama H, Zhou ZH, Nakamura Y, Kanbara T, Sasaki S, Kubota K (1992) Macromolecules 25:1214–1223
49. Yamamoto T (2003) Synlett, pp 425–450
50. Dierschke F, Grimsdale AC, Müllen K (2004) Macromol Chem Phys 205:1147–1154
51. Iraqi A, Wataru I (2004) Chem Mater 16:442–448
52. List EJW, Guentner R, Freitas PS, Scherf U (2002) Adv Mater 14:374–378
53. Wakim S, Blouin N, Gingras E, Tao Y, Leclerc M (2007) Macromol Rapid Commun 28:1798–1803
54. Li J, Dierschke F, Wu J, Grimsdale AC, Müllen K (2006) J Mater Chem 16:96–100
55. Iraqi A, Simmance TG, Yi H, Stevenson M, Lidzey DG (2006) Chem Mater 18:5789–5797
56. Kobayashi N, Koguchi R, Kijima M (2006) Macromolecules 39:9102–9111
57. Yi H, Iraqi A, Stevenson M, Elliott CJ, Lidzey DG (2007) Macromol Rapid Commun 28:1155–1160

58. Morin J-F, Boudreault P-L, Leclerc M (2002) Macromol Rapid Commun 23:1032–1036
59. Fu Y, Bo Z (2005) Macromol Rapid Commun 26:1704–1710
60. Morin J-F, Beaupre S, Leclerc M, Levesque I, D'Iorio M (2002) Appl Phys Lett 80:341–343
61. Drolet N, Beaupre S, Morin J-F, Tao Y, Leclerc M (2002) J Opt A Pure Appl Opt 4:S252–S257
62. Iraqi A, Pickup DF, Yi H (2006) Chem Mater 18:1007–1015
63. Iraqi A, Pegington RC, Simmance TG (2006) J Polym Sci A Polym Chem 44:3336–3342
64. Kobayashi N, Kijima M (2007) Appl Phys Lett 91:081113–081113
65. Drolet N (2006) PhD Thesis. Université Laval, Quebec City
66. Havinga EE, ten Hoeve W, Wynberg H (1993) Synth Met 55:299–306
67. Zhang QT, Tour JM (1998) J Am Chem Soc 120:5355–5362
68. Mullekom HAM, Vekemans JAJM, Havinga EE, Meijer EW (2001) Mater Sci Eng R32:1–40
69. Ajayaghosh A (2003) Chem Soc Rev 32:181–191
70. Roncali J (2007) Macromol Rapid Commun 28:1761–1775
71. Belletête M, Durocher G, Hamel S, Côté M, Wakim S, Leclerc M (2005) J Chem Phys 122:104303–104309
72. Krebs FC, Nyberg RB, Jorgensen M (2004) Chem Mater 16:1313–1318
73. Sirringhaus H, Brown PJ, Friend RH, Nielsen MM, Bechgaard K, Langeveld-Voss BMW, Spiering AJH, Janssen RAJ, Meijer EW, Herwig P, de Leeuw DM (1999) Nature 401:685–688
74. Sirringhaus H, Wilson RJ, Friend RH, Inbasekaran M, Wu W, Woo EP, Grell M, Bradley DDC (2000) Appl Phys Lett 77:406–408
75. Salleo A (2007) Mater Today 10:38–45
76. McCulloch I, Heeney M, Bailey C, Genevicius K, MacDonald I, Shkunov M, Sparrowe D, Tierney S, Wagner R, Zhang W, Chabinyc ML, Kline RJ, McGehee MD, Toney MF (2006) Nat Mater 5:328–333
77. Kline RJ, McGehee MD, Kadnikova EN, Liu J, Frechet JMJ, Toney MF (2005) Macromolecules 38:3312–3319
78. Kline RJ, McGehee MD, Liu ENKJ, Fréchet JMJ (2003) Adv Mater 15:1519–1522
79. Leclerc M, Ranger M, Bélanger-Gariépy F (1998) Acta Crystallogr C: Cryst Struct Commun C54:799–801
80. Ranger M, Leclerc M (1999) Macromolecules 32:3306–3313
81. Gunes S, Neugebauer H, Sariciftci NS (2007) Chem Rev 107:1324–1338
82. Leclerc N, Michaud A, Sirois K, Morin JF, Leclerc M (2006) Adv Funct Mater 16:1694–1704
83. de Leeuw DM, Simenon MMJ, Brown AR, Einhard REF (1997) Synth Met 87:53–59
84. Alstrup J, Norrman K, Jorgensen M, Krebs FC (2006) Sol Energ Mater 90:2777–2792
85. Blouin N, Michaud A, Leclerc M (2007) Adv Mater 19:2295–2300
86. Blouin N, Michaud A, Gendron D, Wakim S, Blair E, Neagu-Plesu R, Belletete M, Durocher G, Tao Y, Leclerc M (2008) J Am Chem Soc 130:732–742
87. Thompson BC, Kim YG, Reynolds JR (2005) Macromolecules 38:5359–5362
88. Koster LJA, Mihailetchi VD, Blom PWM (2006) Appl Phys Lett 88:093511–093513
89. Scharber MC, Mühlbacher D, Koppe M, Denk P, Waldauf C, Heeger AJ, Brabec CJ (2006) Adv Mater 18:789–794
90. Aïch RB, Blouin N, Bouchard A, Leclerc M (2008) Chem Mater (submitted)

91. Bergman J, Janosik T, Wahlstrom N (2001) Indolocarbazoles. In: Katritzky AR (ed) Advances in Heterocyclic Chemistry. Academic Press, New York, pp 1–71
92. Klauk H, Halik M, Zschieschang U, Schmid G, Radlik W (2002) J Appl Phys 92:5259–5263
93. Zhao H-P, Tao X-T, Wang P, Ren Y, Yang J-X, Yan Y-X, Yuan C-X, Liu H-J, Zou D-C, Jiang M-H (2007) Org Electron 8:673–682
94. Blouin N, Leclerc M, Vercelli B, Zecchin S, Zotti G (2006) Macromol Chem Phys 207:175–182
95. Blouin N, Michaud A, Wakim S, Boudreault PLT, Leclerc M, Vercelli B, Zecchin S, Zotti G (2006) Macromol Chem Phys 207:166–174
96. Li Y, Wu Y, Ong BS (2006) Macromolecules 39:6521–6527
97. Wakim S, Bouchard J, Simard M, Drolet N, Tao Y, Leclerc M (2004) Chem Mater 16:4386–4388
98. Kawaguchi K, Nakano K, Nozaki K (2007) J Org Chem 72:5119–5128
99. Nakano K, Hidehira Y, Takahashi K, Hiyama T, Nozaki K (2005) Angew Chem Int Edit 44:7136–7138
100. Gu R, Hameurlaine A, Dehaen W (2006) Synlett, pp 1535–1538
101. Gu R, Hameurlaine A, Dehaen W (2007) J Org Chem 72:7207–7213
102. Robinson B (1963) J Chem Soc, p 3097
103. Yudina LN, Bergman J (2003) Tetrahedron 59:1265–1275
104. Bergman J (1970) Tetrahedron 26:3353–3355
105. Tholander J, Bergman J (1999) Tetrahedron 55:12595–12602
106. Hu N-X, Xie S, Popovic Z, Ong B, Hor A-M, Wang S (1999) J Am Chem Soc 121:5097–5098
107. Belletete M, Blouin N, Boudreault PLT, Leclerc M, Durocher G (2006) J Phys Chem A 110:13696–13704
108. Wu Y, Li Y, Gardner S, Ong BS (2005) J Am Chem Soc 127:614–618
109. Li Y, Wu Y, Gardner S, Ong BS (2005) Adv Mater 17:849–853
110. Bao Z, Dodabalapur A, Lovinger AJ (1996) Appl Phys Lett 69:4108–4110
111. Ong BS, Wu Y, Liu P, Gardner S (2004) J Am Chem Soc 126:3378–3379
112. Gundlach DJ, Lin YY, Jackson TN, Nelson SF, Schlom DG (1997) IEEE Electr Dev Lett 18:87–89
113. Sundar V, Zaumseil J, Podzorov V, Menard E, Willett R, Someya T, Gershenson M, Rogers JA (2004) Science 303:1644–1646
114. Afzali A, Dimitrakopoulos CD, Breen TL (2002) J Am Chem Soc 124:8812–8813
115. Garnier F, Yassa A, Hajlaoui R, Horowitz G, Deloffre F, Servet B, Ries S, Alnot P (1993) J Am Chem Soc 115:8716–8721
116. Katz HE, Torsi L, Dodabalapur A (1995) Chem Mater 7:2235–2237
117. Laquindanum JG, Katz HE, Lovinger AJ (1998) J Am Chem Soc 120:664–672
118. Li XC, Sirringhaus H, Garnier F, Holmes AB, Moratti SC, Feeder N, Clegg W, Teat SJ, Friend RH (1998) J Am Chem Soc 120:2206–2207
119. Belletete M, Boudreault PLT, Durocher G, Leclerc M (2007) Theochem 824:15–22
120. Blouin N, Wakim S, Tao Y, Leclerc M (2007) Polym Prepr 48:292–293 (Am Chem Soc, Div Polym Chem)
121. Grimsdale A, Müllen K (2006) Adv Polym Sci 199:1–82
122. Murata M, Oyama T, Watanabe S, Masuda Y (2000) J Org Chem 65:164–168
123. Wakim S, Bouchard J, Blouin N, Michaud A, Leclerc M (2004) Org Lett 6:3413–3416
124. Dierschke F, Grimsdale A, Müllen K (2003) Synthesis, pp 2470–2472
125. Mishra AK, Graf M, Grasse F, Jacob J, List EJW, Mullen K (2006) Chem Mater 18:2879–2885

126. Sonntag M, Strohriegl P (2006) Tetrahedron 62:8103–8108
127. Sonntag M, Strohriegl P (2006) Tetrahedron Lett 47:8313–8317
128. Patil SA, Scherf U, Kadashchuk A (2003) Adv Funct Mater 13:609–614
129. Janietz S, Bradley DDC, Grell M, Giebeler C, Inbasekaran M, Woo EP (1998) Appl Phys Lett 73:2453–2455

Polyfluorenes with On-Chain Metal Centers

Shu-Juan Liu · Qiang Zhao · Bao-Xiu Mi · Wei Huang (✉)

Jiangsu Key Lab of Organic Electronics
& Information Displays and Institute of Advanced Materials (IAM),
Nanjing University of Posts & Telecommunications (NUPT),
66 XinMoFan Road, 210003 Nanjing, China
wei-huang@njupt.edu.cn

1	Introduction	126
2	Polyfluorenes Containing Iridium(III) Complexes	127
2.1	Polyfluorenes with Neutral Iridium(III) Complexes	127
2.1.1	Polyfluorenes with Neutral Iridium(III) Complexes in the Side-Chain	127
2.1.2	Polyfluorenes with Neutral Iridium(III) Complexes in the Main-Chain	130
2.2	Polyfluorenes with Charged Iridium(III) Complexes	133
3	Polyfluorenes Containing Platinum(II) Complexes	135
4	Polyfluorenes Containing Europium(III) Complexes	138
5	Polyfluorenes Containing Rhenium(I) or Ruthenium(II) Complexes	140
6	Conclusions and Outlook	142
	References	142

Abstract Polyfluorenes containing metal complexes, especially phosphorescent heavy-metal complexes, are a type of important optoelectronic materials. The present review summarizes the synthesis and optoelectronic properties of polyfluorenes with phosphorescent heavy-metal complexes on the chain. Efficient energy transfer from polymer main-chain to metal centers occurs in these host–guest systems. The promise of strong emission and the easiness for processing in organic electronic devices provide incentives to develop these materials intensively. The range of applications of these materials spans the whole field of interaction between light and electricity. Especially, attention is given to the interesting optoelectronic properties and promising applications in organic light-emitting diodes (OLEDs), memory devices, and sensors.

Keywords Energy transfer · Metal complexes · Optoelectronic devices · Polyfluorenes

Abbreviations
Alq3 Al(III) triquinolinolate
bpy 2,2′-Bipyridine
BT Benzothiadiazole
Cz Carbazole
EL Electroluminescence
LEC Light-emitting electrochemical cell

MLCT	Metal-to-ligand charge-transfer
OLED	Organic light-emitting diode
PBD	2-(4-Biphenyl)-5-(t-butylphenyl)-1,3,4-oxadiazole
PEDOT	Poly(3,4-ethylenedioxythiophene)
PF	Polyfluorene
PL	Photoluminescence
PLED	Polymer light-emitting diode
PPV	Poly(phenylvinylene)
PT	Polythiophene
PtTPP	Pt(II) tetraphenylporphyrin
PVK	Polyvinyl carbazole
WORM	Write-once read-many-times
WPLED	White polymer light-emitting diode

1
Introduction

In recent years, conjugated polymers containing metal complexes, especially phosphorescent heavy-metal complexes, have attracted increasing attention, since the utility of phosphorescent heavy-metal complexes can increase the luminescence efficiency in optoelectronic applications by harvesting the large percentage of triplet excitons created upon electron–hole recombination [1–5]. The energy can be transferred efficiently from conjugated main-chain (as host materials) to metal complexes (as guest) in the main- or side-chain of polymers. In addition, polymers with on-chain metal centers can efficiently avoid the intrinsic defects in corresponding blend system, such as phase-separation [6].

Current widely studied conjugated polymers include poly(phenylvinylene) (PPV), polythiophene (PT), and polyfluorene (PF). Although PPV and its derivatives have high photoluminescence (PL) and electroluminescence (EL) efficiencies, they usually suffer from photooxidative degradation [7]. Compared with PPV, PT and its derivatives have low PL and EL efficiencies [8]. So far, polyfluorenes have emerged as the most attractive blue-emitting materials due to their high efficiency, charge transport properties, and good chemical and thermal stability [9–15]. Therefore, they are usually selected as host materials for phosphorescent heavy-metal complexes [16–18]. The most important criterion for serving as phosphorescent host is that triplet energy level of host materials should be above that of dopant in order to satisfy exothermic energy transfer. This is a crucial factor for achieving high phosphorescence efficiency in the host–guest system [19]. The triplet energy level of polyfluorene derivatives is about 2.15 eV [20], which renders them act as the efficient host for red-light emitting phosphors, but the photoluminescence efficiencies will be reduced when used as the host for green-light and yellow-light emitting phosphors. Nevertheless, the triplet energy level of polyfluorenes can be tuned through modifying their chemical structure.

To date, there have been numerous examples of polyfluorenes incorporating phosphorescent heavy-metal complexes, including iridium(III) (Ir), platinum(II) (Pt), europium(III) (Eu), rhenium(I) (Re), and ruthenium(II) (Ru) complexes. The polymers are usually prepared by Suzuki-type or Yamamoto-type polymerization method of fluorene monomers and complex monomers containing active groups on ligands. In this review, we will concentrate on their preparation, optical properties, as well as applications in optoelectronic devices. Meanwhile, the feature of intramolecular interaction between polymer and metal complex is elucidated; the advantages of these polymeric phosphorescent materials serving as organic light-emitting diodes (OLEDs) component are illustrated.

2
Polyfluorenes Containing Iridium(III) Complexes

Among phosphorescent metal complexes, Ir(III) complexes have been considered as one of the best phosphorescent material candidates because they show intense phosphorescence at room temperature and significantly shorter emission lifetime compared with other heavy-metal complexes, which are crucial for utility of phosphorescent materials. Moreover, the emission colour can be tuned easily over the entire visible region by simply modifying the chemical structures of ligands [21–23].

Ir(III) complexes can be divided into two classes. One class includes the neutral complexes containing cyclometalated ligands and other anionic ligands. The other class includes charged Ir(III) complexes containing neutral bidentate ligands (e.g., bipyridine and phenanthroline derivatives). Correspondingly, the subsequent discussion of polyfluorenes containing Ir(III) complexes can be also divided into two parts, namely, polyfluorenes with neutral Ir(III) complexes and with charged complexes in the chain. Furthermore, a major subdivision arises from a consideration of the location of the Ir(III) centres; these can be either in the main-chain or in the side-chain.

2.1
Polyfluorenes with Neutral Iridium(III) Complexes

Polyfluorenes with neutral iridium(III) complexes in the side-chain and main-chain of polymers are discussed, respectively.

2.1.1
Polyfluorenes with Neutral Iridium(III) Complexes in the Side-Chain

Introducing Ir(III) complexes into the side-chain using a chemical bond is an effective way to produce polyfluorenes with on-chain iridium centres. Chen et al. [4] first reported high-efficiency red-light emission from polyfluorenes grafted with cyclometalated iridium complexes (polymer 1–7). These polymers were prepared by Suzuki-type or Yamamoto-type polymerization method. Polymer 7 based device showed red emission with highest efficiency of 2.8 cd/A at 7 V. The incorporation of carbazole (Cz) can significantly increase the efficiency and lower the turn-on voltage. They demon-

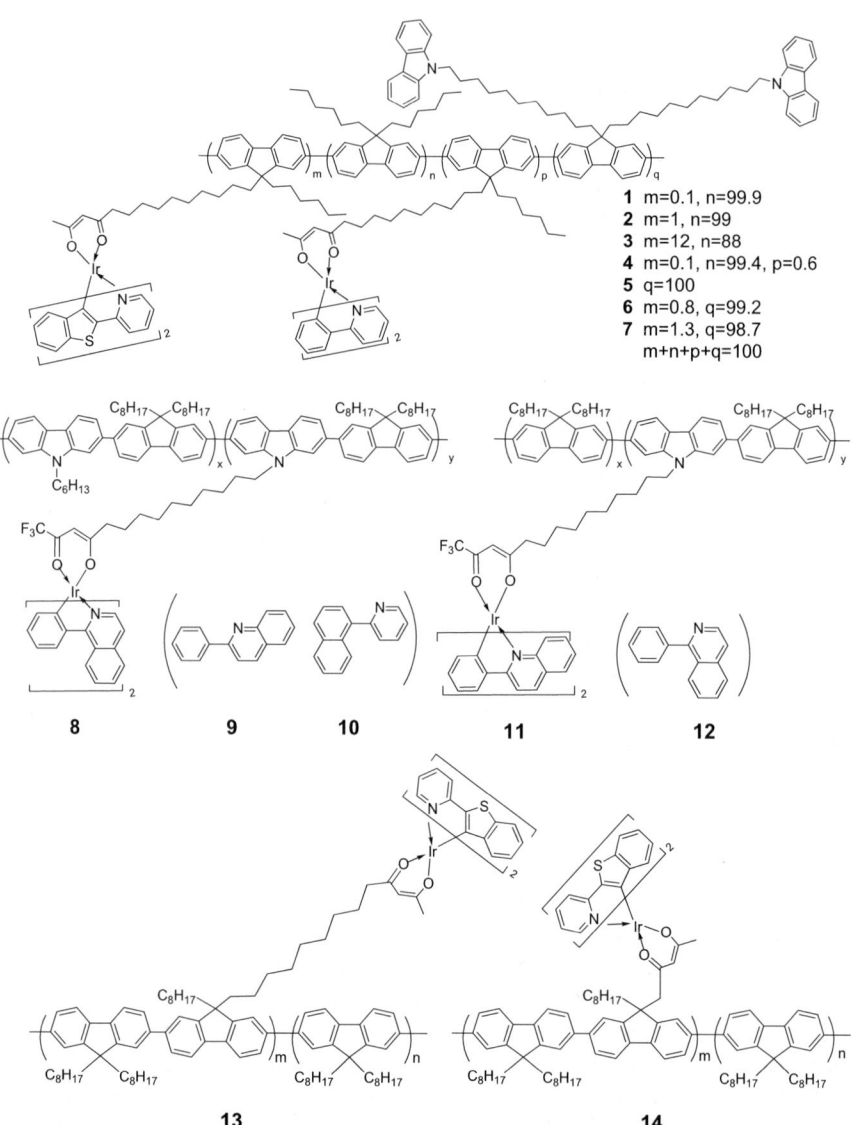

Scheme 1 Polyfluorene derivatives with Ir(III) complex on the side-chain

strated that both the energy transfer from PF main chain and from an electroplex formed between main-chain fluorene and side-chain carbazole moieties to the red Ir(III) complex can significantly enhance the device performance.

Subsequently, Yang et al. also prepared a series of fluorene-alt-carbazole copolymers grafted with cyclometalated Ir(III) complexes (polymer 8–12) [24]. The red-light emitting devices were realized and showed the highest external quantum efficiency of 4.9% ph/el, the luminous efficiency of 4.0 cd/A, and a peak emission of 610 nm.

Proper conformation between the Ir(III) complex guest and the fluorene segment host will play an important role in the energy transfer between the two parts [25]. The face-to-face conformation is beneficial to π–π interactions between host and guest [25], which are known to be the key requirement for Dexter triplet energy transfer [26]. The conformation may be determined by the length of side-chain. Thus, the incorporation of different alkyl between the polymer host and phosphorescent guest is an important design principle for achieving high efficiencies in those electrophosphorescent organic light-emitting diodes for which the triplet energy levels of the host and guest are similar [25]. The investigation of the property–structure relationship of polymer 13 and 14 confirmed that long alkyl between the polymer host and phosphorescent guest might be helpful in suppression of the back transfer of triplets from the red phosphorescent iridium complex to the polyfluorene backbone, which could result in higher photo- and electroluminescence efficiencies.

White polymer light-emitting diodes (WPLEDs) have been attracting increasing attention because they can be easily fabricated using wet processes, including the spin-coating, screen printing, and ink-jet printing techniques, which are expected to be lower in cost in mass production, especially in the production of large-area panel displays. Therefore, it is desirable to obtain white-light emission from a single-component material, so as to avoid drawbacks of polymer-blend or small-molecule-doped polymer systems. White-light emission could be obtained from a single polymer with three individual emission species by introducing a small number of Ir(III) complexes into the side chain [27]. In polymer 15, fluorene is used as the blue-emissive component, benzothiadiazole (BT) and the Ir(III) complex act as green-emissive and red-emissive chromophores, respectively. By changing the contents of BT and Ir(III) complex in the polymer, the electroluminescence spectrum from a single polymer can be adjusted to achieve white-light emission. The devices exhibit a maximum luminance efficiency of 6.1 cd/A at a current density of 2.2 mA/cm^2. A maximum luminance of $10\,110 \text{ cd/m}^2$ was achieved at a current density of 345 mA/cm^2. Introducing orange phosphorescent iridium complex into polyfluorene can also obtain white-light-emitting polymers (polymer 16 and 17) [28]. Tuning the content of Ir(III) complex in the polymer 16, WPLEDs with a maximum luminous efficiency

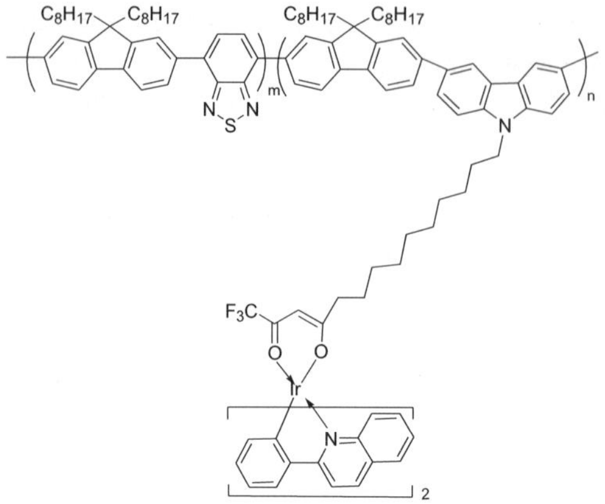

Scheme 2 White light-emitting polymers based on polyfluorene derivatives with Ir complex on the side-chain

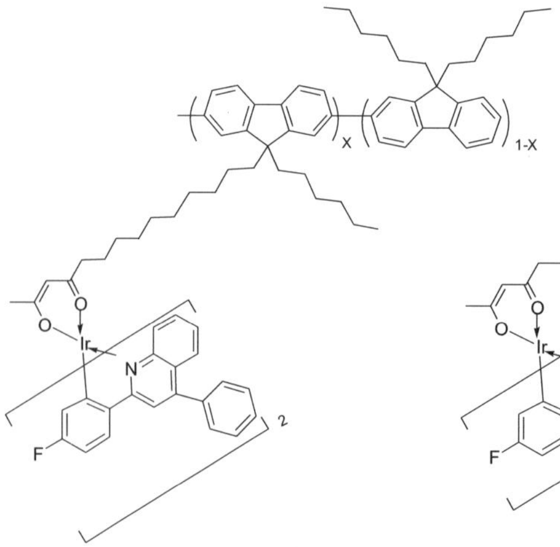

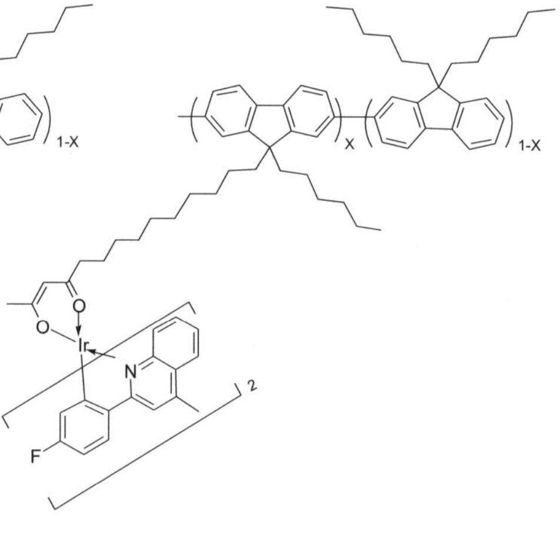

of 4.49 cd/A and a maximum power efficiency of 2.35 lm/W at 6.0 V were realized.

2.1.2
Polyfluorenes with Neutral Iridium(III) Complexes in the Main-Chain

Polyfluorene with Ir(III) complexes in the main-chain can be obtained by chelating iridium with C^N ligand introduced into the polyfluorene mainchain [5, 29, 32]. Oligo- and poly(9,9-dioctylfluorenyl-2,7-diyl) with red or green Ir(III) complexes in the main-chains were first reported by Holmes et al.

(polymer **18** and **19**) [5]. The efficiencies of the green devices are moderate, but the devices still represent improvements over blended composite materials of organometallic phosphors in polyfluorene host. The red light-emitting devices with better triplet energy level matching between iridium complex and fluorene energy levels show significant improvements in device efficiency.

Subsequently, Cao et al. [30, 31] designed and synthesized polymer **20–22** by similar method and the highly efficient saturated red-phosphorescent polymer light-emitting diodes (PLEDs) were achieved on the basis of copolymer **20**. The best device performances are observed with an external quantum efficiency of 6.5% photon/electron (ph/el) at the current density of 38 mA/cm^2, with the emission peak at 630 nm ($x = 0.65$, $y = 0.31$) and the luminance of 926 cd/m^2.

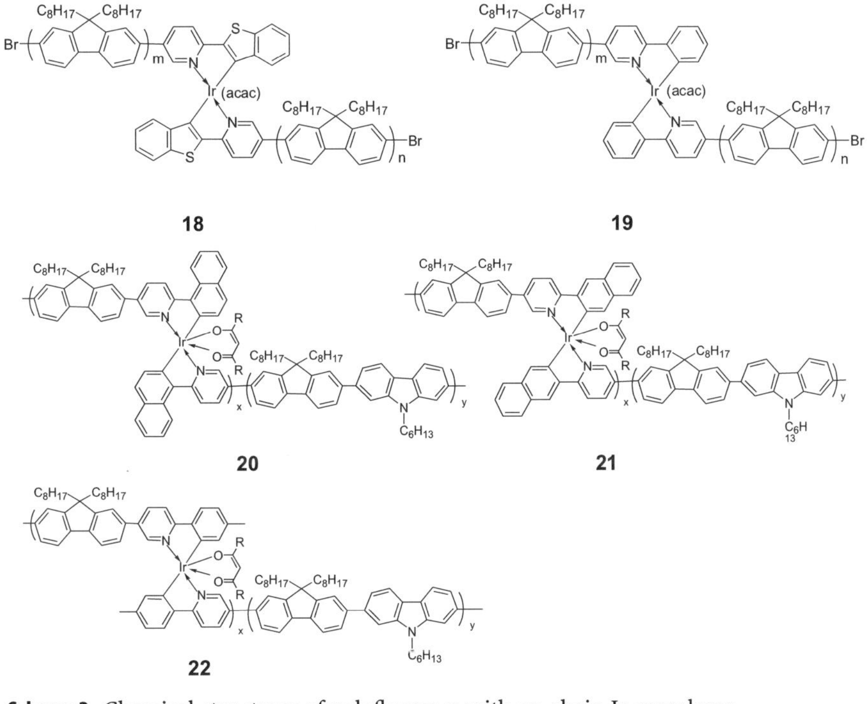

Scheme 3 Chemical structures of polyfluorenes with on-chain Ir complexes

Introducing red-light emitting Ir(III) complex and BT into the main-chain of polyfluorene (polymer **23**) can realize white-light emission. A WPLED with a structure of ITO/PEDOT:PSS/PVK/polymer **23**/CsF/Al showed a maximum external quantum efficiency of 3.7% and the maximum luminous efficiency of 3.9 cd/A at the current density of 1.6 mA/cm^2 with the CIE coordinates of (0.33, 0.34). The maximum luminance of 4180 cd/m^2 is achieved at the current density of 268 mA/cm^2 with the CIE coordinates of (0.31,

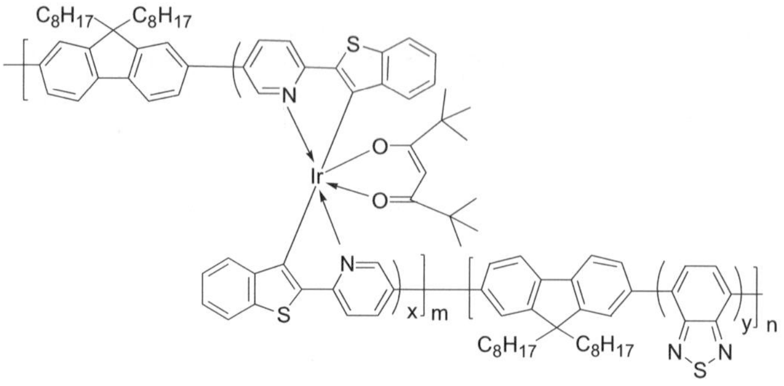

Scheme 4 Chemical structure of white light-emitting polymer based on polyfluorene with Ir complex on the main-chain

0.32) [33]. The white-light emissions from such polymer are stable in the white-light region at all applied voltages, thus indicating that the approach of incorporating singlet and triplet species into the polymer backbone is promising for WPLEDs.

It has been proven that the triplet energy level of polyfluorene can be increased by the incorporation of other units to block the π-conjugation of polyfluorene. For example, introducing pyridine or thiophene into the polyfluorene chains can reduce the effective conjugation, increasing the energy level of the polymer (polymer **24** and **25**) [29]. For **24**, the triplet energy of host poly(9,9-dihexylfluorene-*alt*-2,5-pyridine) is low, resulting in quenching of the excited triplet state of the iridium complex by energy transfer to the triplet of the polymer backbone. This would result in quenching of emission from the phosphor and low quantum efficiencies. However, exchanging the 2,5-linked pyridine group with the 3,4-linked thiophene group resulted in the increase of triplet energy of host (polymer **25**). Thus, the possibility of triplet energy transfer from an excited-state phosphor to the non-emitting triplet state of the backbone is reduced and quantum efficiencies increase significantly.

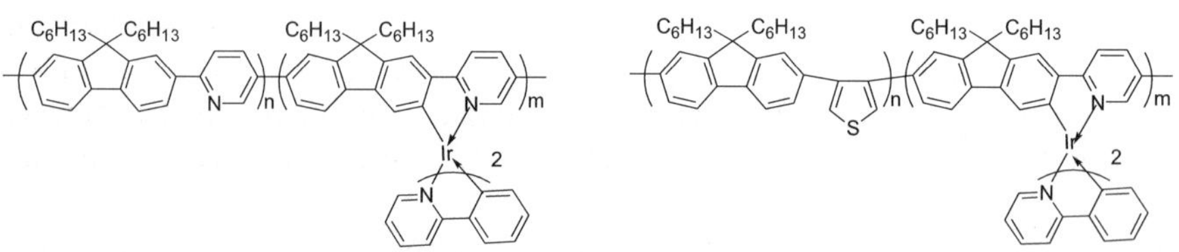

Scheme 5 Chemical structures of polyfluorenes with on-chain Ir complexes containing pyridine or thiophene on the main-chain

Another way to introduce Ir(III) complexes into the main-chain of polyfluorenes was realized by Suzuki polycondensation of fluorene segments and β-diketone ligand chelated with Ir(III) chloride-bridged dimmer (polymer 26 and 27) [34, 35]. A saturated red-emitting polymer light-emitting diode was achieved from the device ITO/PEDOT/polymer 27 + PBD (40%)/Ba/Al with the maximum external quantum efficiency of 0.6% at the current density (J) of 38.5 mA/cm^2 and the maximum luminance of 541 cd/m^2 at 15.8 V.

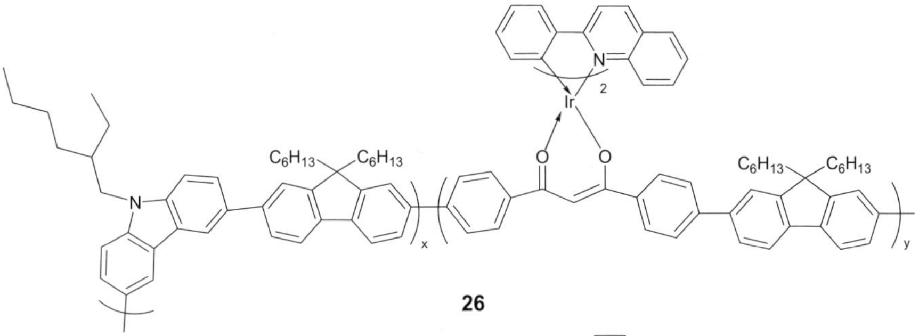

Scheme 6 Chemical structures of polyfluorenes with Ir complexes on the main-chain

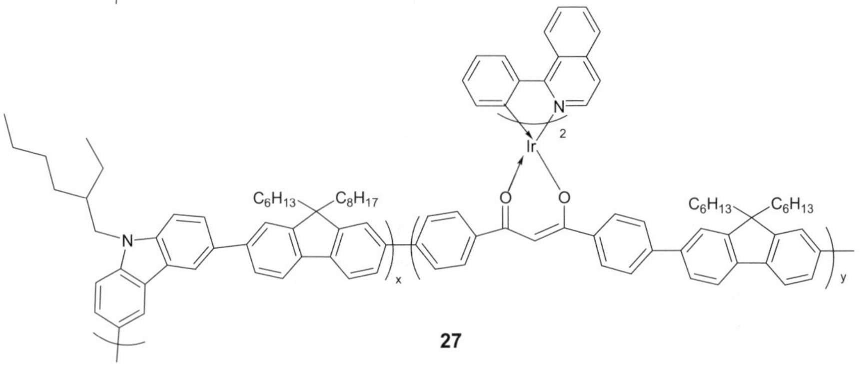

2.2
Polyfluorenes with Charged Iridium(III) Complexes

Recently, charged Ir(III) complexes have been attracting increasing attention because they provide emission centres and ionic conductivity in the same species and have some features which may make them one of the best candidates for lighting, display, and light-emitting electrochemical cells (LEC) applications [36–43].

Our group [44, 45] first synthesized a series of polyfluorene derivatives (polymer 28–30) with charged red-light emitting Ir(III) complexes in the backbones, in which the phosphorescent chromophores were molecularly dispersed within the composite material. Saturated red-light emission can be

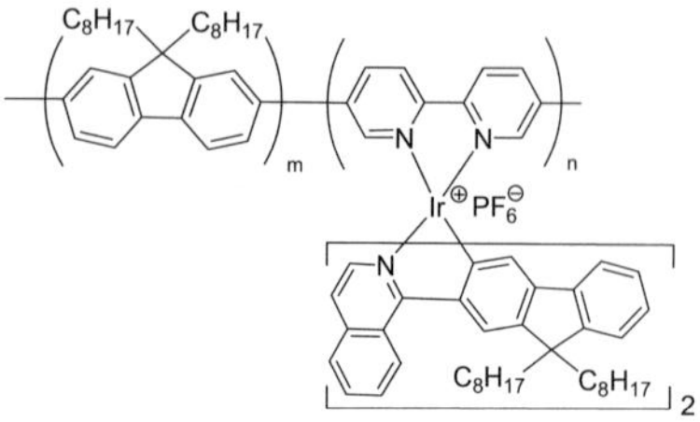

Scheme 7 Chemical structures of polyfluorenes with charged Ir(III) complexes in the main-chain

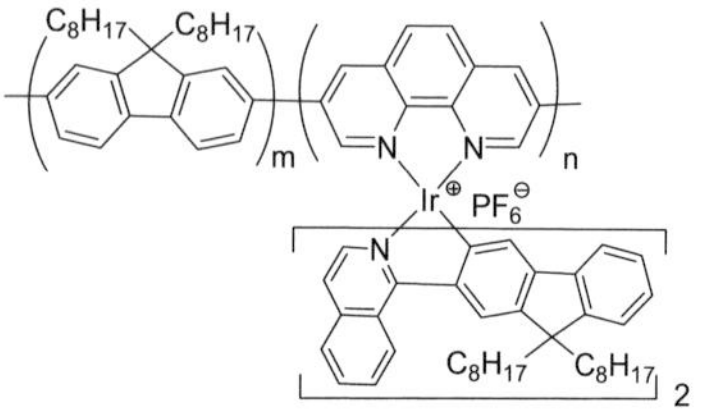

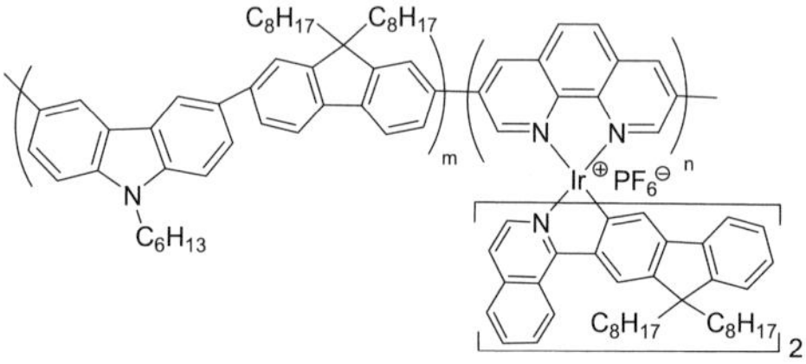

realized by choosing appropriate ligands. Cz units were also introduced into the backbones of polymer. Here, Cz would play multi-roles: as hole transport moiety on the polymer/ITO interface, as charge trapping site, and as a barrier for back energy transfer from the triplet state of Ir(III) complexes to that of host [4, 46–48]. Almost complete energy transfer from the host fluorene segments to the guest Ir(III) complexes was achieved in solid state even at the low feed ratio of complexes. Intra- and inter-chain energy transfer mechanisms coexisted in the energy migration process of this host–guest system, and the intramolecular energy transfer might be a more efficient process. Chelating polymers showed more efficient energy transfer than the corresponding blended system and exhibited good thermal stability, redox reversibility, and film formation as well. Preliminary results concerning polymer light-emitting devices indicate that these materials offer promising opportunities in optoelectronic applications.

Polyfluorene with charged Ir(III) complexes in the side-chain can be obtained by introducing N^N ligand into side-chain of polymer and then chelating with Ir(III) chloride-bridged dimmer. Yang and Peng et al. [49] synthesized a series of copolymers with charged Ir(III) complexes in the side-chain using fluorene and Cz as segments of backbone (polymer **31**). Excellent device performances were obtained using this kind of polymer as emission materials. A maximum external quantum efficiency of 7.3% and a luminous efficiency of 6.9 cd/A were achieved at a current density of 1.9 mA/cm^2.

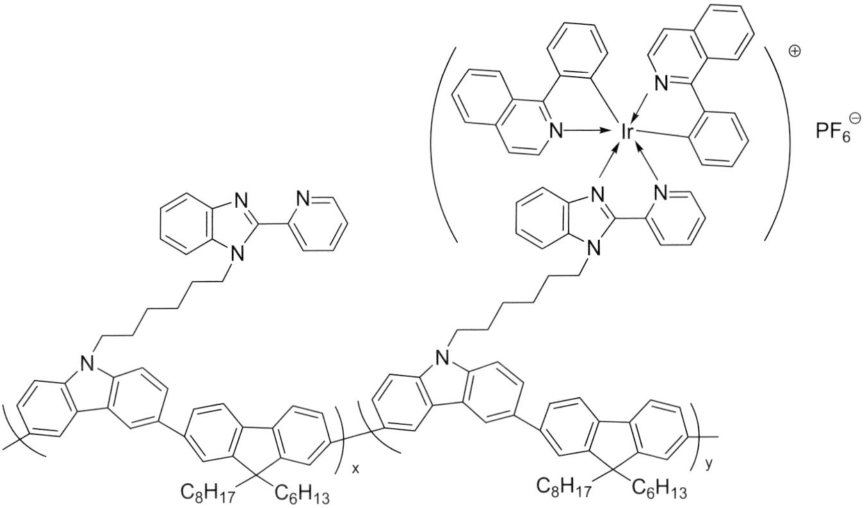

Scheme 8 Chemical structure of polyfluorene derivatives with charged Ir(III) complex in the side-chain

3
Polyfluorenes Containing Platinum(II) Complexes

Pt(II) complexes have been studied intensively, because they show high luminescent quantum efficiency and its planar structure favors Pt–Pt interactions, which usually leads to the self-assembly behavior [50, 51]. Cao et al. [52] first reported the synthesis and photo- and electroluminescent properties of copolymers (polymer **32**) containing tetraphenylporphyrin Pt(II) (PtTPP) complexes in the polymer main-chain derived from direct metalation reaction of the copolymers from metal-free poly(fluorene-co-tetraphenylporphyrin) (route 1 in Scheme 9). Moreover, it provides flexibility to shift from one metal ion to another with a single type of metal-free porphyrin copolymer [53]. The copolymers gave deep red emission. The highest external quantum efficiency of the devices based on the Pt(II) complex (1 mol % PtTPP in copolymer) was 0.43% with emission peaking at 676 nm. This kind of polymer can also be obtained through the copolymerization of monomeric porphyrin–platinum(II) complexes and dialkylsubstituted fluorene monomers by Suzuki coupling reaction (route 2 in Scheme 9). The device based on the porphyrin–platinum(II) copolymer (with 5 mol % PtTPP in the copolymer) showed the highest external quantum efficiency of 1.95% in an ITO/poly(3,4-ethylenedioxythiophene)/PVK/70 : 30 (w/w) polymer:PBD/Ba/Al device configuration. In comparison with the copolymers

Scheme 9 Synthesis routes of polyfluorenes with on-chain Pt(II) complexes by post-polymerization metalation (route 1) and copolymerization from Pt(II) metal complexes (route 2)

synthesized via a postpolymerization metalation route, copolymerization from Pt(II) complexes proved to be a more efficient synthetic route for high-efficiency electrophosphorescent polymers [54].

Recently, Anzenbacher et al. [55, 56] first demonstrated molecular wire behavior occurring in solid state and in functional OLEDs in donor–bridge–acceptor triads consisting of Al(III) triquinolinolate (Alq$_3$), oligofluorene

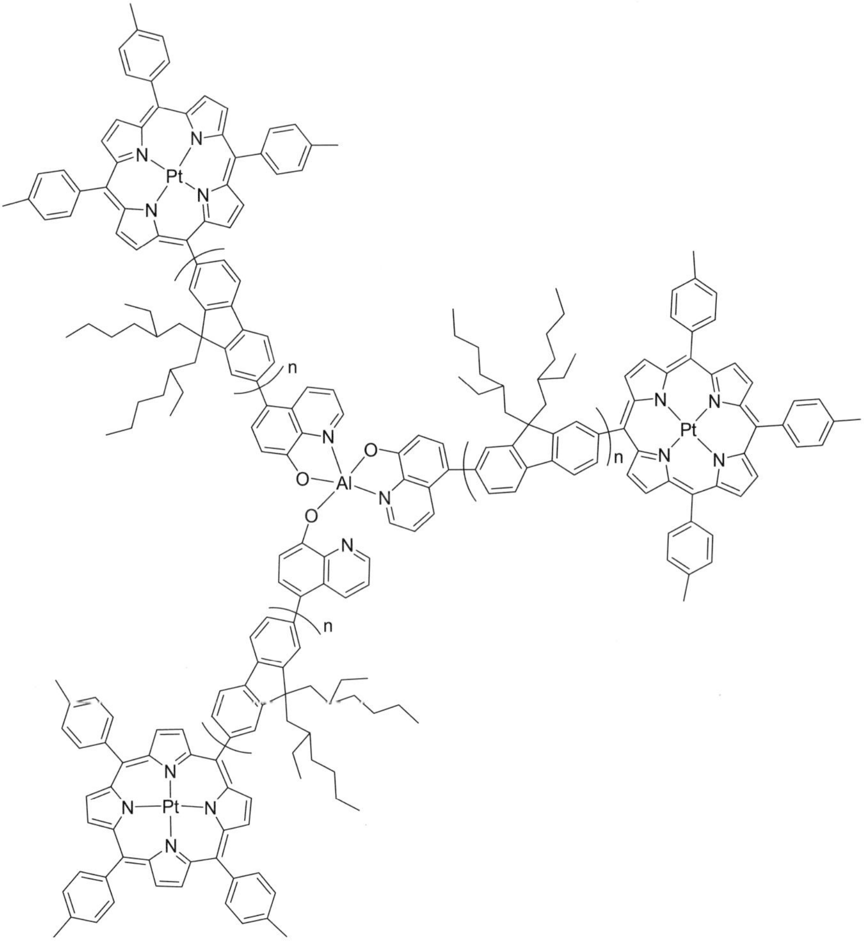

33

Scheme 10 Chemical structure of polyfluorene derivatives with on-chain Pt(II) complexes showing molecular wire behavior

bridge, and Pt(II) tetraphenylporphyrin (PtTPP), in which the energy transfer is facilitated by energy alignment of the components (polymer **33**). Alq$_3$- and fluorene fragments appear to form a single fluorophore owing to strong electronic coupling facilitating the singlet exciton migration to the porphyrin. The materials show effective singlet and triplet energy transfer, and exhibit red emission with high color purity.

The use of fluorescent conjugated polymers is an established method for achieving a large degree of amplification for the chemosensing of analytes in both solution and solid states. Swager et al. [57] have sought to investigate the potential of phosphorescent conjugated polymers as chemosensing materials. The longer lifetimes of triplet excitons in conjugated materials may lead to increased sensitivity to trace analytes. They synthesized copolymers (polymer **34** and **35**) containing fluorene and cyclometalated square-planar platinum complex. The polymeric nature of this conjugated material gives a sensitivity improvement for dissolved oxygen quantification, demonstrating the potential usage of phosphorescent conjugated polymers as chemosensing materials.

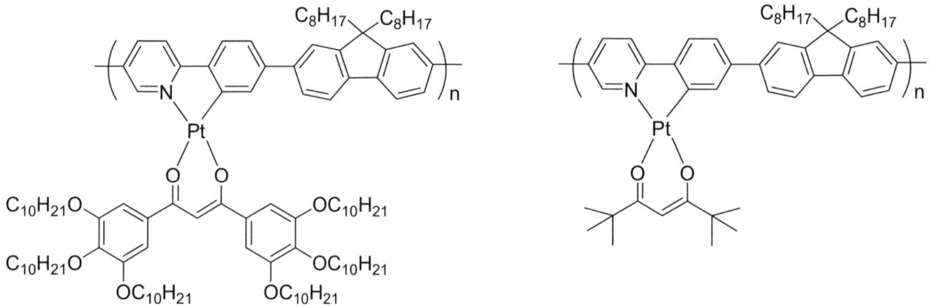

Scheme 11 Chemical structures of polyfluorenes with on-chain Pt(II) complexes used in chemosensors

4
Polyfluorenes Containing Europium(III) Complexes

Rare earth compounds are excellent chromophores that exhibit intense emission with a narrow spectral bandwidth (full width at half-maximum of 5–20 nm) and relatively long decay lifetime (10^{-2}–10^{-6} s) [58]. Since the emission from rare earth ions originates from transitions between the f levels of rare earth atom that are well protected from environmental perturbations by the filled $5s^2$ and $5p^6$ orbitals, the resulting emission spectra are expected to be sharp and narrow. They are the most widely used materials in inorganic light-emitting diodes [59]. A great deal of effort has also been devoted to the application of rare earth complexes in organic memory devices and OLEDs [60].

Ling and his coworkers [61–64] designed and synthesized a series of conjugated copolymers (polymer **36–39**) containing fluorene and Eu(III) complexes chelated to the main chains through a three-step process, involving Suzuki coupling copolymerization, hydrolysis of benzoate units, and postchelation. The photoluminescence spectra of the copolymer consisted of two emission bands, one in the 350–550 nm region and another at around 612 nm, corresponding to the $\pi^*-\pi$ transitions of the fluorene moieties and the $f-f$ transitions of the europium ions, respectively. In the copolymer films casting from solutions, emission from the fluorene moieties could be suppressed, and the absorbed excitation energy was transferred effectively to the Eu(III) complexes in the copolymer. Nearly monochromatic red emission was detected under UV excitation at room temperature.

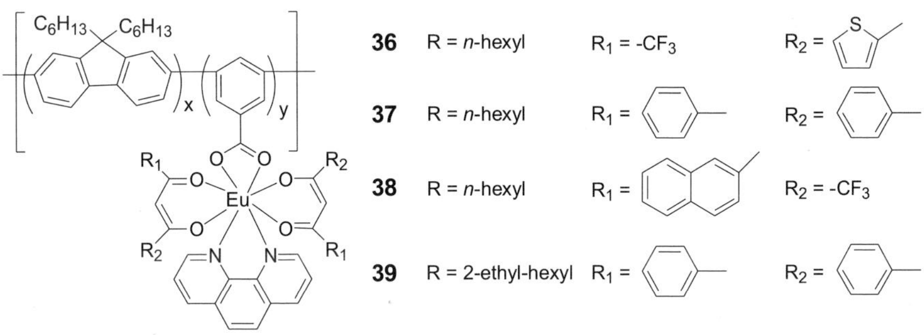

Scheme 12 Chemical structures of polyfluorenes with on-chain Eu(III) complexes

Subsequently, they first demonstrated a conjugated copolymer of 9,9-dialkylfluorene and Eu-complexed benzoate for write-once read-many-times (WORM) memory application in a sandwich structure of Al/polymer/ITO [65]. In these active polymers, the fluorene moiety served as the backbone and electron donor, while the europium complex served as the electron acceptor. An electrical bistability phenomenon was observed on this device: low conductivity state for the as-fabricated device and high conductivity state after device transition by applying a voltage. At the low conductivity state, the device showed a charge injection controlled current and at the high conductivity state, the device showed a space charge limited current. At the same applied voltage, the device exhibited two distinguishable conductivities. Thus, the device can be used as a WORM electronic memory, with a high ON/OFF current ratio up to 10^7, stable ON-states and OFF-states up to 10^8 read cycles at a read voltage of 1 V, and projected stability up to 10 years at a constant voltage stress of 1 V.

5 Polyfluorenes Containing Rhenium(I) or Ruthenium(II) Complexes

Complexes *fac*-(L)Re(I)$(CO)_3$Cl have attracted more and more attention due to the fact that *fac*-(L)Re-(I)$(CO)_3$Cl complexes are one of many typical metal-to-ligand charge-transfer (MLCT) materials with highly desirable properties, including microsecond excited-state lifetimes, intense visible absorption and emission, good redox chemistry properties, and high stability [66–69]. Interesting materials incorporating Re(I) complexes into π-conjugated polymer backbones have been reported [70–72]. Using 2,7-(9,9′-dihexylfluorene) and 2,2′-bipyridine (bpy) copolymers as the backbones, Ma et al. have synthesized a series of π-conjugated copolymers (polymer **40–46**) incorporating (bpy)Re$(CO)_3$Cl in the backbones of the copolymers through a highly active replacement reaction of bpy elements and CO in Re$(CO)_5$Cl complexes [73]. They demonstrate that polymers incorporating metal complexes in the backbones may not be suitable for application in the OLEDs, but may be suitable for application in the photovoltaic fields. The incorporation of Re(I) into the polymer backbone results in the red shift of absorption and the presence of new peak at about 520 nm in emission spectra [74].

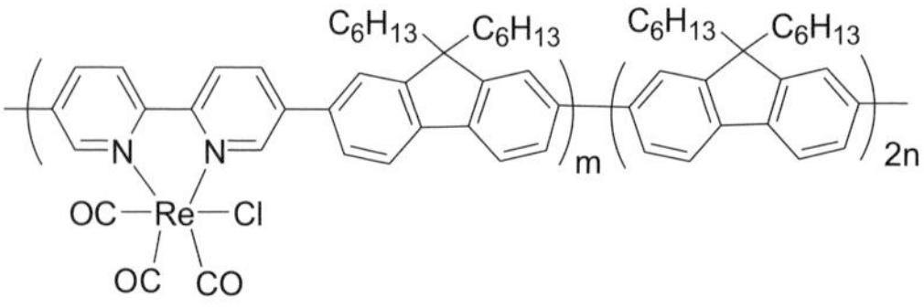

40 m / 2(m+n) = 0.1

41 m / 2(m+n) = 0.2

42 m / 2(m+n) = 0.5

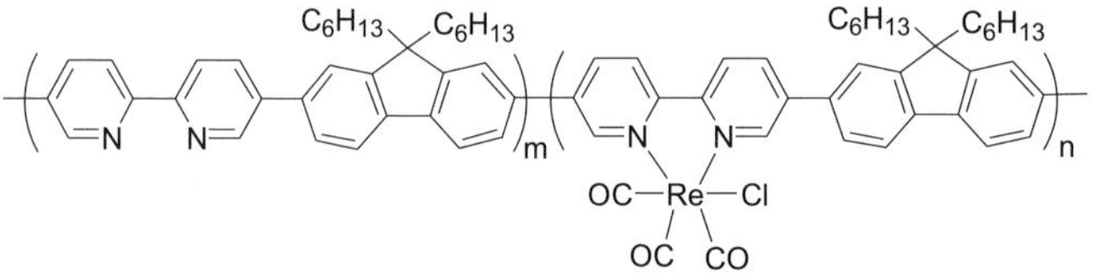

43 m / n = 100 / 5

44 m / n = 100 / 10

45 m / n = 100 / 50

46 m / n = 100 / 100

Scheme 13 Chemical structures of polyfluorenes with on-chain Re(I) complexes

Hsu et al. [75, 76] reported a new method for incorporating metal complexes into polyfluorenes to prepare phosphorescent polymers (polymer **47** and **48**). A pyridine end-capped polyfluorene has been synthesized. The pyridine was used to form a polymer metal complex with 2,2-bipyridyl(tricarbonyl)rhenium(I) chloride. Using the end-capping approach not only can control the molecular weight of polymer, but also avoid the interference of the metal complex and conjugated polymer in energy transfer. They can

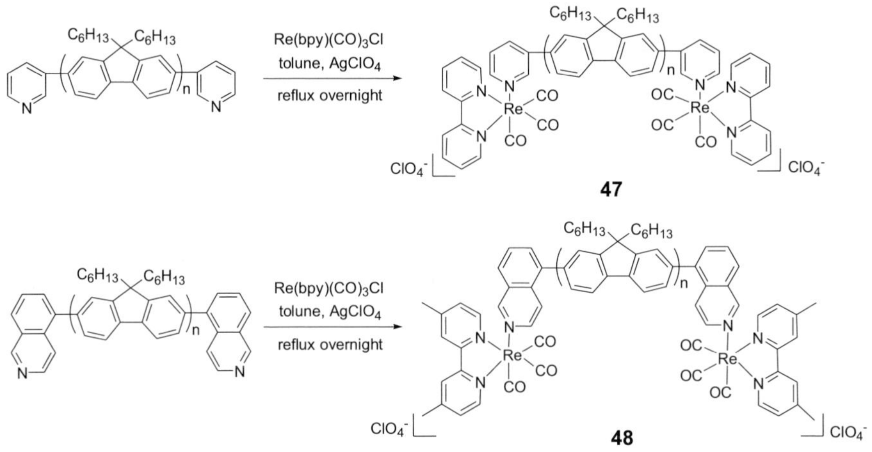

Scheme 14 Synthesis of end-capped polyfluorenes with on-chain Re(I) complexes

change the end-capping monomers and the organic–metal complex to make differently colored phosphorescent polymers and even make a white-light phosphorescent polymer on a single polymer.

Polyfluorene with on-chain ruthenium complex (polymer **49**) has been synthesized by Suzuki polycondensation [77]. The photoluminescence of the copolymer was slightly blue-shifted as the concentration of dipyridylamine increased. The introduction of dipyridylamine and the ruthenium complex into the polymer significantly improved the photoluminescence efficiency.

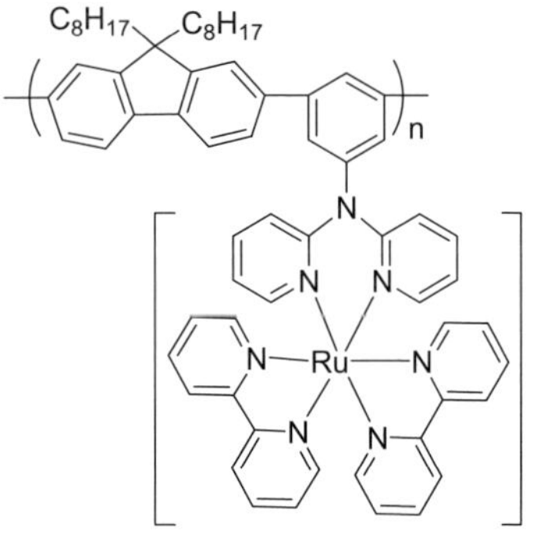

Scheme 15 Chemical structure of polyfluorenes with on-chain Ru(II) complex

6
Conclusions and Outlook

In this review, the synthesis, properties, and applications in optoelectronic fields of polyfluorenes with on-chain metal centers have been briefly summarized. Metal complexes involving iridium(III), platinum(II), europium(III), rhenium(I), and ruthenium(II) complex coupled with polyfluorene are surveyed. Efficient energy transfer from polymer main-chain to metal-centers can occur in these host–guest systems. These kinds of novel polymers are usually applied in the fields of phosphorescent OLEDs, memory devices, and sensors. In particular, the realization of efficient energy transfer and phosphorescence offers a huge potential for future optoelectronic devices based on these kinds of materials.

Acknowledgements This work was supported by the National Natural Science Foundation of China under Grants 60235412, 90406021, and 50428303 as well as Nanjing University of Posts and Telecommunications under Grant NY206070.

References

1. Baldo MA, O'Brien DF, You Y, Shoustikov A, Sibley S, Thompson ME, Forrest SR (1998) Nature 395:151
2. Ma YG, Zhang HY, Shen JC (1998) Synth Met 94:245
3. Lamansky S, Kwong RC, Nugent M, Djurovich PI, Thompson ME (2001) Org Electron 2:53
4. Chen X, Liao JL, Liang Y, Ahmed MO, Tseng HE, Chen SA (2003) J Am Chem Soc 125:636
5. Sandee AJ, Williams CK, Evans NR, Davies JE, Boothby CE, Kohler A, Friend RH, Holmes AB (2004) J Am Chem Soc 126:7041
6. Wang Z, McWilliams AR, Evans CEB, Lu X, Chung S, Winnik MA, Manners I (2002) Adv Funct Mater 12:415
7. Scott JC, Kaufman JH, Brock PJ, Dipietro R, Salem J, Goitia JA (1996) J Appl Phys 79:2745
8. Peng Q, Xie MG, Huang Y, Lu ZY, Cao Y (2005) Macromol Chem Phys 206:2373
9. Zhan X, Liu Y, Yu G, Wu X, Zhu DB, Sun R, Wang D, Epstein AJ (2001) J Mater Chem 11:1606
10. Grimsdale AC, Müllen K (2006) Adv Polym Sci 199:1
11. Yu WL, Pei J, Huang W, Heeger AJ (2000) Adv Mater 12:828
12. Liu B, Yu WL, Lai YH, Huang W (2001) Chem Mater 13:1984
13. Zeng G, Yu WL, Chua SJ, Huang W (2002) Macromolecules 35:6907
14. Huang W, Meng H, Yu WL, Gao J, Heeger AJ (1998) Adv Mater 10:593
15. Xin Y, Wen GA, Zeng WJ, Zhao L, Zhu XR, Fan QL, Feng JC, Wang LH, Wei W, Peng B, Cao Y, Huang W (2005) Macromolecules 38:6755
16. Jiang CY, Yang W, Peng JB, Xiao S, Cao Y (2004) Adv Mater 16:537
17. Gong X, Ostrowski JC, Bazan GC, Moses D, Heeger AJ, Liu MS, Jen AKY (2003) Adv Mater 15:45

18. Niu YH, Chen BQ, Liu S, Yip H, Bardecker J, Jen AKY, Kavitha J, Chi Y, Shu CF, Tseng YH, Chien CH (2004) Appl Phys Lett 85:1619
19. Chen FC, He GF, Yang Y (2003) Appl Phys Lett 82:1006
20. Rothe C, Monkman AP (2002) Phys Rev B 65:073201
21. Adachi C, Baldo MA, Forrest SR, Lamansky S, Thompson ME, Kwong RC (2001) Appl Phys Lett 78:1622
22. Duan JP, Sun PP, Cheng CH (2003) Adv Mater 15:224
23. Lamansky S, Djurovich P, Murphy D, Abdel-Razzaq F, Lee HE, Adachi C, Burrows PE, Forrest SR, Thompson ME (2001) J Am Chem Soc 123:4304
24. Jiang J, Jiang C, Yang W, Zhen H, Huang F, Cao Y (2005) Macromolecules 38:4072
25. Evans NR, Devi LS, Mak CSK, Watkins SE, Pascu SI, Köhler A, Friend RH, Williams CK, Holmes AB (2006) J Am Chem Soc 128:6647
26. Klessinger M, Michl J (1995) Excited States and Photochemistry of Organic Molecules. VCH Publishers, New York
27. Jiang JX, Xu YH, Yang W, Guan R, Liu ZQ, Zhen HY, Cao Y (2006) Adv Mater 18:1769
28. Mei CY, Ding JQ, Yao B, Cheng YX, Xie ZY, Geng YH, Wang LX (2007) J Poly Sci Part A: Poly Chem 45:1746
29. Schulz GL, Chen X, Chen SA, Holdcroft S (2006) Macromolecules 39:9157
30. Zhen HY, Jiang CY, Yang W, Jiang JX, Huang F, Cao Y (2005) Chem Eur J 11:5007
31. Zhen HY, Luo C, Yang W, Song W, Du B, Jiang J, Jiang C, Zhang Y, Cao Y (2006) Macromolecules 39:1693
32. Ito T, Suzuki S, Kido J (2005) Polym Adv Technol 16:480
33. Zhen HY, Xu W, Yang W, Chen QL, Xu YH, Jiang JX, Peng JB, Cao Y (2006) Macromol Rapid Commun 27:2095
34. Zhang K, Chen Z, Zou Y, Yang C, Qin J, Cao Y (2007) Organometallics 26:3699
35. Zhang K, Chen Z, Yang CL, Gong SL, Qin JG, Cao Y (2006) Macromol Rapid Commun 27:1926
36. Tamayo AB, Garon S, Sajoto T, Djurovich PI, Tsyba IM, Bau R, Thompson ME (2005) Inorg Chem 44:8723
37. Marin V, Holder E, Hoogenboom R, Schubert US (2004) J Poly Sci Part A: Poly Chem 42:4153
38. Zhao Q, Liu SJ, Shi M, Wang CM, Yu MX, Li L, Li FY, Yi T, Huang CH (2006) Inorg Chem 45:6152
39. Plummer EA, van Dijken A, Hofstraat HW, Cola LD, Brunner K (2005) Adv Funct Mater 15:281
40. Slinker JD, Koh CY, Malliaras GG, Lowry MS, Bernhard S (2005) Appl Phys Lett 86:173506
41. Lowry MS, Goldsmith JI, Slinker JD, Rohl R, Pascal RA, Malliaras GG, Bernhard S (2005) Chem Mater 17:5712
42. Kim JI, Shin IS, Kim H, Lee JK (2005) J Am Chem Soc 127:1614
43. Neve F, Deda ML, Crispini A, Bellusci A, Puntoriero F, Campagna S (2004) Organometallics 23:5856
44. Liu SJ, Zhao Q, Chen RF, Deng Y, Fan QL, Li FY, Wang LH, Huang CH, Huang W (2006) Chem Eur J 12:4351
45. Liu SJ, Zhao Q, Xia YJ, Deng Y, Lin J, Fan QL, Wang LH, Huang W (2007) J Phys Chem C 111:1166
46. Wang XD, Qgino K, Tanaka K, Usui H (2004) IEEE J Quantum Elect 10:121
47. Vaeth KM, Tang CW (2002) J Appl Phys 92:3447
48. Dijken A, Bastiaansen JJAM, Kiggen NMM, Langeveld BMW, Rothe C, Monkman A, Bach I, Stössel P, Brunner K (2004) J Am Chem Soc 126:7718

49. Du B, Wang L, Wu HB, Yang W, Zhang Y, Liu RS, Sun ML, Peng JB, Cao Y (2007) Chem Eur J 13:7432
50. Brooks J, Babayan Y, Lamansy S, Djurovich PI, Tsyba I, Bau R, Thompson ME (2002) Inorg Chem 41:3055
51. Haskins-Glusac K, Pinto MR, Tan CY, Schanze KS (2004) J Am Chem Soc 126:14964
52. Hou Q, Zhang Y, Li FY, Peng JB, Cao Y (2005) Organometallics 24:4509
53. Hou Q, Zhang Y, Yang RQ, Yang W, Cao Y (2005) Synth Met 153:193
54. Zhuang WL, Zhang Y, Hou Q, Wang L, Cao Y (2006) J Poly Sci Part A: Poly Chem 44:4174
55. Montes VA, Pérez-Bolívar C, Agarwal N, Shinar J, Anzenbacher P (2006) J Am Chem Soc 128:12436
56. Montes VA, Pérez-Bolívar C, Estrada LA, Shinar J, Anzenbacher P (2007) J Am Chem Soc ASAP. Article DOI: 10.1021/ja073491x
57. Thomas SW, Yagi S, Swager TM (2005) J Mater Chem 15:2829
58. Sinha APB (1971) In: Rao CNR, Ferraro JR (eds) Spectroscopy in Inorganic Chemistry. Academic, New York, vol 2
59. Jüstel T, Nikol H, Ronda C (1998) Angew Chem Int Ed 37:3084
60. Kido J, Okamoto Y (2002) Chem Rev 102:2357
61. Ling QD, Kang ET, Neoh KG, Huang W (2003) Macromolecules 36:6995
62. Song Y, Tan YP, Teo EYH, Zhu CX, Chan DSH, Ling QD, Neoh KG, Kang ET (2006) J Appl Phys 100:084508
63. Ling QD, Song Y, Teo EYH, Lim SL, Zhu CX, Chan DSH, Kwong DL, Kang ET, Neoha KG (2006) Electrochem Solid-State Lett 9:G268
64. Song Y, Ling QD, Zhu C, Kang ET, Chan DSH, Wang YH, Kwong DL (2006) IEEE Electron Device Lett 27:154
65. Ling QD, Song Y, Ding SJ, Zhu CX, Chan DSH, Kwong DL, Kang ET, Neoh KG (2005) Adv Mater 17:455
66. Wrighton M, Morse DL (1974) J Am Chem Soc 96:998
67. Lees AJ (1998) Coord Chem Rev 177:3
68. Lees AJ (2001) Coord Chem Rev 211:255
69. Chen JB, Wang RT (2002) Coord Chem Rev 231:109
70. Ley KD, Whittle CE, Bartberger MD, Schanze KS (1997) J Am Chem Soc 119:3423
71. Yu SC, Gong X, Chan WK (1998) Macromolecules 31:5639
72. Chan WK, Ng PK, Gong X, Hou S (1999) J Mater Chem 9:2103
73. Zhang M, Lu P, Wang XM, He L, Xia H, Zhang W, Yang B, Liu LL, Yang L, Yang M, Ma YG, Feng JK, Wang DJ, Tamai N (2004) J Phys Chem B 108:13185
74. Zhang M, Lu P, Ma YG, Li GW, Shen JC (2003) Synth Met 135–136:211
75. Lee PI, Hsu SLC, Chung CT (2006) Synth Met 156:907
76. Lee PI, Hsu SLC (2007) J Poly Sci Part A: Poly Chem 45:1492
77. Dinakaran K, Chou CH, Hsu SL, Wei KH (2004) J Poly Sci Part A: Poly Chem 42:4838

Fluorene-Based Conjugated Oligomers for Organic Photonics and Electronics

J. U. Wallace · S. H. Chen (✉)

Chemical Engineering Department and Laboratory for Laser Energetics, University of Rochester, 240 East River Rd., Rochester, NY 14623-1212, USA
shch@lle.rochester.edu

1	Introduction	147
2	Material Synthesis	148
2.1	Synthetic Approaches to Oligofluorenes	148
2.2	Synthetic Incorporation of Comonomer Units	152
2.3	Synthesis of Fluorene-Based Oligomers with Other Functionalities	154
2.4	Polymers Containing Flourene Oligomers in Repeat Units	156
3	Morphological Properties	157
3.1	Thermal Stability and Solubility	157
3.2	Crystallization Versus Glass Transition	157
3.3	Liquid Crystallinity	159
4	Photophysical Properties	162
4.1	Efficient Blue Emission	162
4.2	Full Color Light Emission	162
4.3	Studies of Excited Electronic States	163
4.4	Polarized Photoluminescence	163
5	Electronic Properties	165
5.1	Electrochemistry: Energy Levels and Properties of Ionic States	165
5.2	Bipolar Charge-Carrier Transport	166
6	Photonic and Electronic Applications	168
6.1	Organic Light-Emitting Diodes	168
6.2	Solid-State Organic Lasers	171
6.3	Organic Field Effect Transistors	173
6.4	Organic Solar Cells	174
7	Fluorene-Based Oligomers to Probe Polyfluorenes	176
7.1	Fluorene-Fluorenone Co-oligomers	176
7.2	Insight into Degradation Processes	177
8	Summary	179
	References	180

Abstract Recent advances in fluorene-based conjugated oligomers are surveyed, including molecular design, material synthesis and characterization, and potential application to organic photonics and electronics, such as light-emitting diodes, solid-state lasers, field effect transistors, and solar cells.

Keywords Fluorene-based conjugated oligomers ·
Material synthesis and characterization · Molecular design ·
Organic photonics and electronics

Abbreviations

AcOH	Acetic acid
ASE	Amplified spontaneous emission
cd	Candela
CIE	Commission Internationale de l'Eclairage
CuI	Copper Iodide
CV	Cyclic voltammetry
DMSO	Dimethylsulfoxide
E_a	Electron affinity
EL	Electroluminescence
Et_2O	Diethylether
EQE	External quantum efficiency
ϕ	Photoluminescent quantum efficiency
G	Glassy state
g_e	Dissymmetry factor
H_2SO_4	Sulfuric acid
HCl	Hydrochloric acid
HOMO	Highest occupied molecular orbital
HWE	Horner–Wadsworth–Emmons reaction
ICl	Iodine monochloride
ITO	Indium tin oxide
I	Isotropic state
I_p	Ionization potential
IR	Infrared
KOH	Potassium hydroxide
KO-*t*-Bu	Potassium *tert*-butoxide
LCD	Liquid crystal display
LiF	Lithium fluoride
LUMO	Lowest unoccupied molecular orbital
MALDI-TOF	Matrix-assisted laser desorption/ionization time of flight
Mg:Ag	Magnesium:silver alloy (20 : 1)
n-BuLi	*n*-Butyllithium
Nem	Nematic state
NaOH	Sodium hydroxide
Ni(COD)	Bis(1,5-cyclooctadiene)nickel(0)
NMR	Nuclear magnetic resonance
NTSC	National Television System Committee
OFET	Organic field effect transistor
OLED	Organic light emitting diode
$PdCl_2$(dppf)	[1,1′-Bis(diphenylphospino)ferrocene] dichloropalladium(II)
$PdCl_2(PPh_3)_2$	Bis(triphenylphosphino) dichloropalladium(II)
$Pd(OAc)_2$	Palladium(II) acetate
$Pd(PPh_3)_4$	Tetrakis(triphenylphosphino)palladium(0)
PEDOT/PSS	Poly(3,4-ethylenedioxythiophene)/poly(styrenesulfonate)
S	Orientational order parameter

SiMe₃	Trimethylsilyl
T_c	Clearing (or isotropization) temperature
T_g	Glass transition temperature
TGA	Thermal gravimetric analysis
THF	Tetrahydrofuran
TPBI	1,3,5-Tri(phenyl-2-benzimidazole)-benzene
UPS	Ultraviolet photoemission spectroscopy
UV	Ultraviolet
UV-Vis	Ultraviolet-Visible
wt. %	Weight percent

1
Introduction

Fluorene is a common polycyclic aromatic hydrocarbon found even in engine exhaust (see Fig. 1 for its structure). Fluorene has long been known for the intense violet fluorescence, after which it was named [1]. Fluorene-based conjugated oligomers can emit deep blue light at an efficiency of up to 99% in solution and 90% in neat solid films [2], which is rare in this spectral region.

The decoupling provided by bonding groups to fluorene's 9-position is also noteworthy, as these groups are placed at 90° angles to the π-conjugated system comprising fluorene molecules (also see Fig. 1). This allows a fluorene compound's solubility and aggregation behavior to be modulated independently of its electronic properties [3].

These properties have been exploited extensively in polyfluorenes that have grown to be a major player in organic electronics and photonics [4–8], exhibiting efficient polarized light emission [9–13], long-lasting blue electroluminescence [14], impressive lasing gain [15–18], and promising performance in organic solar cells [19, 20].

The field of organic electronics has also seen a growing focus on oligomeric materials, with sizes and properties intermediate between small

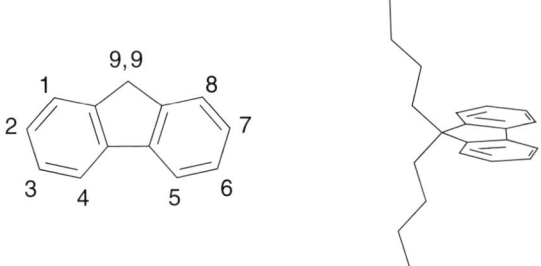

Fig. 1 Fluorene structure showing numbering of sites available for substitution, as well as an illustration of the decoupling provided by bonding groups to its 9-position

molecules and polymers [21]. Using correlations from oligomers to gain insight into their analogous polymers has been a long tradition in polymer chemistry, as numerous properties, such as glass transition and electronic band gap, vary monotonically with molecular weight before reaching their asymptotic values [22–29]. Such correlations, however, are known to be imperfect [30–33]. In addition to offering structure–property relationships, monodisperse oligomers are attractive as active materials in their own right, with properties such as higher field-effect mobilities [34], stronger chiroptical activities [35], greater optical amplification of quenching [36], and better lasing efficiency [37] than their polymeric analogues.

These unique properties of oligomers make them an ideal complement to conjugated polymers in organic electronics and photonics. While the synthesis of long oligomers becomes rather laborious in comparison to that of polymers, the more rigorous purification techniques available to oligomers are advantageous for practical applications [38]. Oligomers are generally devoid of chain folds and kinks [39] and chain defects [37] that tend to compromise polymers' performance. Moreover, polydispersity can render polymer properties difficult to reproduce [40, 41]. Because of their lower melt viscosities, oligomers are easier to process than polymer analogues, and many oligomers can also be readily vacuum-deposited.

This review focuses on the synthesis, characterization, and performance of monodisperse fluorene-based conjugated oligomers with potential applications to organic electronics and photonics. Included herein are oligomers and co-oligomers that contain at least two fluorene units in their backbone structures, while excluding those with a lone spirofluorene unit without other fluorene units in the backbone, as these have been covered elsewhere [42, 43].

2
Material Synthesis

2.1
Synthetic Approaches to Oligofluorenes

The synthesis of fluorene oligomers is intensive in terms of number of steps, but the unique properties and the new insight into structure–property relationships are well worth the effort. A number of approaches have been developed from the initial brute force fractionation of polydisperse systems to the more elegant deterministic synthesis of monodisperse systems.

Synthesis obviously begins with the monomers themselves. Figure 2 shows four of the most common types of substituents on fluorene, illustrated with typical pendants, accompanied by synthetic pathways. Alkyl pendants are readily attached with sodium hydroxide in DMSO [44], e.g., converting compound 1 into 2, while the aryl pendants, such as in compound 5, require

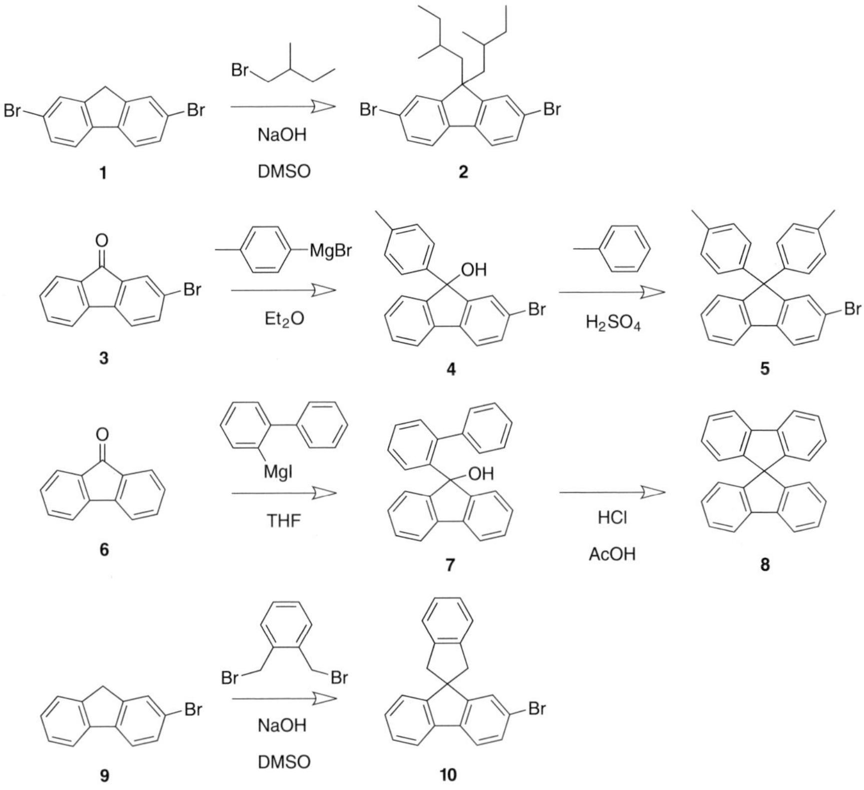

Fig. 2 Various substitution schemes of fluorene monomers shown for representative structures

two steps, a Grignard reaction to form compound **4** and an acid-catalyzed condensation reaction [2]. Spiro-linked fluorene derivatives (compound **8**) have been known for some time [45] and are readily synthesized starting with a fluorenone, such as compound **6** [46]. Other types of spiro-configured fluorene monomers have also been synthesized, as illustrated with compound **10** [47]. Starting with bromine substituted fluorene or bromination that results in mono- or di-functionalized fluorene (e.g., compounds **9** and **1**, respectively), oligofluorenes are constructed in subsequent reactions. A wide variety of functionalizations of spirofluorene have been summarized in a recent review article [42].

Various carbon–carbon cross-coupling reactions are employed to arrive at oligomers from fluorene monomers, including nickel-cataylzed Yamamoto coupling between aryl halides [48], palladium(0)-catalyzed Suzuki coupling between boronic acids/esters and halides [49], and ferric-chloride-catalyzed Scholl reaction, also called Friedel–Crafts arylation, of bare fluorenes [50, 51].

◀ **Fig. 3** Common reactions and cross-couplings in oligofluorene synthesis shown for example structures, where $PdCl_2(dppf)$ is [1,1′-bis(diphenylphospino)ferrocene] dichloropalladium(II), $Pd(PPh_3)_4$ is tetrakis(triphenylphosphino)palladium(0), and Ni(COD) is bis(1,5-cyclooctadiene)nickel(0)

Yamamoto and Suzuki coupling, and other common reactions in fluorene oligomer synthesis are illustrated in Fig. 3.

Yamamoto coupling was used for one of the first isolations of a series of oligofluorenes, but even with end-capping it required extensive and time-consuming fractionation [52]. This approach has been recently used to synthesize polydisperse oligofluorenes for ease of synthesis [53] to demonstrate interesting properties. More relevant to deterministic synthesis, Yamamoto coupling has been used for symmetric coupling of monofunctional oligomeric intermediates. For example, Yamamoto coupling was used to join two monobromo-octafluorene units (compound **14**) into a symmetric hexadecamer (compound **15**) [35].

However, the most prominent reaction used to synthesize fluorene oligomers is Suzuki, or more fully Suzuki–Miyaura, coupling. This high-yield reaction between two different functionalities, one a boronic acid or ester (such as compound **12**) and the other a halide or similar leaving group (such as compound **2**), has allowed considerable control in building up oligomers in a stepwise manner. An obvious method is simply to use a monomer with a single boronic acid group for reaction with an excess of a dibromofluorene monomer, as illustrated in the synthesis of compound **13** [54]. This product can then be converted into a boronic acid for subsequent coupling and lengthening of the oligomer. Excess boronic acid or ester can also be used, but often leads to a side reaction where the desired remaining boronic acid/ester group is converted to a hydrogen atom [55], resulting in a low yield.

More advanced approaches utilize either a difference in reactivity in Suzuki coupling or a protecting group during the Suzuki coupling steps. The high selectivity of Suzuki coupling for aryl iodides [56] or aryl diazonium salts [57] over aryl bromides has been effectively used in oligofluorene synthesis, with difunctional intermediates. In addition, nitro groups are excellent masking groups for amino groups during Suzuki couplings [58], which can be converted into diazonium salts or used for other chemistry. A more generic protecting group is the trimethylsilyl moiety, which acts as a dormant iodide or bromide [35]. It can be introduced selectively to one side of a dibromofluorene unit (as with **16** into **17**), remains intact under Suzuki conditions, and can be easily converted to an iodide by iododesilylation using iodine monochloride (of **17** into **18**), or to a bromide by similar reaction with molecular bromine.

While adding one unit at a time to reach the final products results in a repetitive divergent approach [59], purification can become an issue as the separation between longer oligomers becomes increasingly challenging.

Thus, the iterative divergent–convergent strategy has been applied to the synthesis of oligofluorenes for ease of purification, by adding larger pieces at a time, at the cost of a greater number of intermediates. This approach has allowed access to oligomers of all lengths from dimer to dodecamer, with hexadecamer also synthesized in concert [35]. This can even be accomplished with mixed, or nonuniform, pendants placed in a controlled manner to realize other desirable properties [60]. The versatility of this approach to test a wide variety of structures relatively rapidly is of great advantage in establishing structure–property relationships, but is not the most efficient way to reach a single large target or simply to synthesize a series of uniform oligomers.

2.2
Synthetic Incorporation of Comonomer Units

The incorporation of other co-monomers into fluorene backbones can require a number of additional chemical reactions. Suzuki coupling is still immensely useful in this regard, but is not always easy or possible to use with various sensitive groups. The Miyaura [61] reaction (see reaction of **11** into **12**) to introduce boronic esters via palladium catalysis instead of harsh butyl lithium treatment is quite useful to access an array of co-monomers for Suzuki coupling [57], but still does not encompass many desirable targets. Palladium(II)-catalyzed Kumada coupling between aryl Grignard reagents and halides provides another option [62], as illustrated for the synthesis of **20** from **19**. Relatively early work to incorporate ethynylene groups utilized the Pd/Cu-catalyzed Sonogashira reaction and 3-methyl-1-butyn-3-ol masked reagents [63]. Shortly after that, vinylene and various phenylene vinylene groups were incorporated [64] with one of two prominent reactions: Heck coupling (to synthesize **22**) of vinyl groups to aryl halides, or Horner–Wadsworth–Emmons reaction (shown in making **23**) of aldehydes with phosphonic acid esters. Stille coupling [65] is valuable with thiophene units due to the ease of introducing the requisite trialkyltin groups to couple with halofluorenes [64]. These coupling reactions are illustrated in Fig. 4.

Two approaches using co-monomers have allowed for additional flexibility in synthesis of fluorene oligomers. The first takes advantage of the fact that Kumada, Heck, and Stille couplings are done under anhydrous conditions to allow coupling reactions to proceed while leaving boronic esters intact for subsequent steps [66]. In the second, an alternating series of reactions, drastically changing the polarity of the intermediates back and forth, substantially facilitates the purification, opening the door to a direct divergent method which is

Fig. 4 Additional reactions and cross-couplings that are commonly used to synthesize fluorene co-oligomers, where $Pd(OAc)_2$ is palladium(II) acetate and $PdCl_2(PPh_3)$ is bis(triphenylphosphino) dichloropalladium(II)

Fluorene-Based Conjugated Oligomers for Organic Photonics and Electronics 153

Fig. 5 Example structures of fluorene co-oligomers and functionalized oligofluorenes ▶

much more efficient, and requires fewer extraneous intermediates [67]. This method is used to synthesize oligomers up to undecafluorene.

In addition to the aforementioned ethynylene, vinylene, and phenylene vinylene groups, a wide variety of groups have been introduced as co-monomer units. Thiophenes, bithiophenes [68], dibenzothiophene [69], and longer oligothiophene segments have all been incorporated [70]. Some additional examples are shown in Fig. 5. An exotic unit incorporating a thiophene as part of a spirofluorene has been synthesized (compound **27**) [71]. Anthracene (e.g., compound **28**) [72, 73] and perylene [74] are useful for sky-blue emission from fluorene oligomers. Fluorenone has been incorporated with fluorene units to study the resultant co-oligomers as possible defect sites in polyfluorenes [75–79]. Benzothiadiazole is a useful green emitter and electron-deficient moiety [64], and can also give rise to red emission when connected to fluorene via phenylene-vinylene segments, as in compound **29**. Thiophene-S,S-dioxide and dibenzothiophene-S,S-dioxide are other electron-deficient co-monomers [69, 80, 81] used with fluorene.

2.3
Synthesis of Fluorene-Based Oligomers with Other Functionalities

A number of functionalities have been included with oligofluorenes that are not strictly co-monomers. One creative approach uses boronic acid end groups to form immobilized networks of oligofluorenes [82], in contrast to the conventional functionalizations of oligofluorenes with convenient polymerizable groups to form insoluble networks [55]. Hydrophilic, or even ionic, side chains to the 9-position of fluorene have been used to realize self-condensed nanoparticles of fluorene (with compound **30**) [83] and to realize sensing of oligonucleotides [84]. Hydrogen-bonding oligofluorenes have also been synthesized through the attachment of ureidopyrimidinone end groups (compounds **31–33** in Fig. 5) [58].

There are also a number of compounds with central cores to which oligofluorene arms are directly attached. These have shown differences in thermal [85] and optical [86] properties compared to linear oligomers, including much stronger nonlinear absorption [87] when put on a single benzene core or into a dendritic structure. On a truxene core, electrochromism and improved fluorescence have also been observed [88]. Compounds with oligofluorene pendants to functional cores have also been synthesized. With oligofluorenes bonded to an emissive core, energy transfer from the pendants to the core is permitted while preventing molecular aggregation [89, 90]. Attachment of oligofluorenes to a charge-transporting core enables the energy levels of the hybrid molecule to be modulated [91].

27, **28**, **29**, **30**

31-33, n = 3, 5, 7

The most prevalent trend in modifying fluorene oligomers is to introduce hole and sometimes electron-transporting moieties to enhance their performance in organic light-emitting diodes. The most common method is end-capping with triarylamine groups through either Suzuki coupling or palladium-catalyzed amination [92] of halogen-decorated oligomers, as seen in a number of papers [93–100]. One paper puts one of each type of charge transport moiety on the oligofluorene [101]. Some also introduce these charge-transporting moieties as a central core with arms attached in direct conjugation [91, 102] or as a central unit in the oligomer [103]. Direct conjugation with these moieties does influence their electronic and optical properties, which can be avoided by employing spiro-linked transport moieties [104], 9-position attachment of these arylamines [105], or alkyl spacers between the moieties [106, 107].

For oligomers, as opposed to polymers [9], the inclusion of functional moieties tends to disrupt the mesomorphic and self-organizing properties of the resulting multifunctional materials with one notable exception. Attachment of oligofluorenes through a flexible alkyl spacer on the 2-position of the oligomer to a charge transport core allows the pendant oligomers to align into nematic films through deformation of the spacer [106]. Synthetically, this is accomplished by Suzuki coupling of the boronic acid of the oligofluorene to allyl bromide, followed by activation of the allyl group with hydroboration for Suzuki coupling to a halogen-bearing functional core. The advantages of this approach will be discussed more fully in the appropriate sections below.

2.4
Polymers Containing Flourene Oligomers in Repeat Units

While polyfluorenes and their copolymers are usually easier to synthesize than oligomers, a number of papers have focused on incorporating oligofluorenes as structural units in polymers and copolymers. Such systems take advantage of the desirable properties of these oligomers while retaining the advantages of full-fledged polymers. Most of these studies target terfluorene [108–115] as an optimum element in terms of the ease of synthesis and excellent blue light emission characteristics. Two of these reports also discuss testing the performance of a pentafluorene [114, 115], and one paper looks at a series of oligomeric segments containing two to seven fluorene units in the polymer backbone [113]. Two groups use fluorene co-oligomers [116, 117], while yet another uses only a hexafluorene [118]. Most studies utilize oligofluorene as a unit in the main chain [110–114] or in the side chain [115–117], and also as a macroinitiator for atom transfer radical polymerization to form rod-coil block copolymers [108, 109, 118]. In addition, the variety of polymer types is impressive: polyesters [110, 111], polyether [112], polycarbonate [113], poly(aryl ether)s [114, 115], polynorbornene [116], and polystyrene (as the backbone [117] or the other block [108, 109, 118]).

3
Morphological Properties

With the ease and variety of substitutions at their C-9 positions, fluorene units offer considerable control over a conjugated oligomer's solubility and molecular packing behavior. As a result, a wide range of morphological properties can be realized through molecular design.

3.1
Thermal Stability and Solubility

Fluorene-based conjugated oligomers are capable of a high degree of thermal stability. Thermal gravimetric analysis (TGA) has shown that fluorene oligomers decompose at temperatures higher than 400 °C, specifically reaching 5% mass loss in the range of 400–450 °C at a heating rate of 20 °C min^{-1} under nitrogen [2]. Another group of researchers measured decomposition temperatures of end-dendronized oligofluorenes within a few degrees of 600 °C, without specifying the mass loss in their TGA experiments [97]. In addition, many fluorene oligomers can be purified by sublimation and can be vacuum deposited as thin films. In fact, a tetrafluorene derivative can be vacuum deposited without experiencing decomposition [119]. However, oligomers comprising more than four fluorene units cannot be sublimed and must be solution processed.

Fluorene oligomers, especially those functionalized at the C-9 position, are readily soluble in common organic solvents. The most striking example is in a study of a series of oligofluorenes' static and dynamic solution properties, where they remain soluble at concentrations greater than 500 g L^{-1} in toluene, or > 50 wt. %, and even up to 800 g L^{-1} in the case of a terfluorene [120]. Routinely, fluorene oligomers are purified in solvent mixtures consisting mostly of nonpolar hexanes, and then solution-cast into thin films.

3.2
Crystallization Versus Glass Transition

While many polymers form glasses, oligomers are more likely to crystallize [21, 23]. Polyfluorene itself is often semi-crystalline, but, depending on the details of its substitution, it can range from mostly glassy to highly crystalline [4]. Fluorene oligomers fall on one side or the other of this semi-crystalline middle ground, being either intrinsically glass-forming or highly crystalline.

Fluorene oligomers can form morphologically stable organic glasses. Annealing samples of certain nematic oligofluorenes above their glass transition temperatures, T_g do not form any crystallites detectable by electron diffraction even after 96 h [60]. In addition, a fully spiro-configured terfluorene has

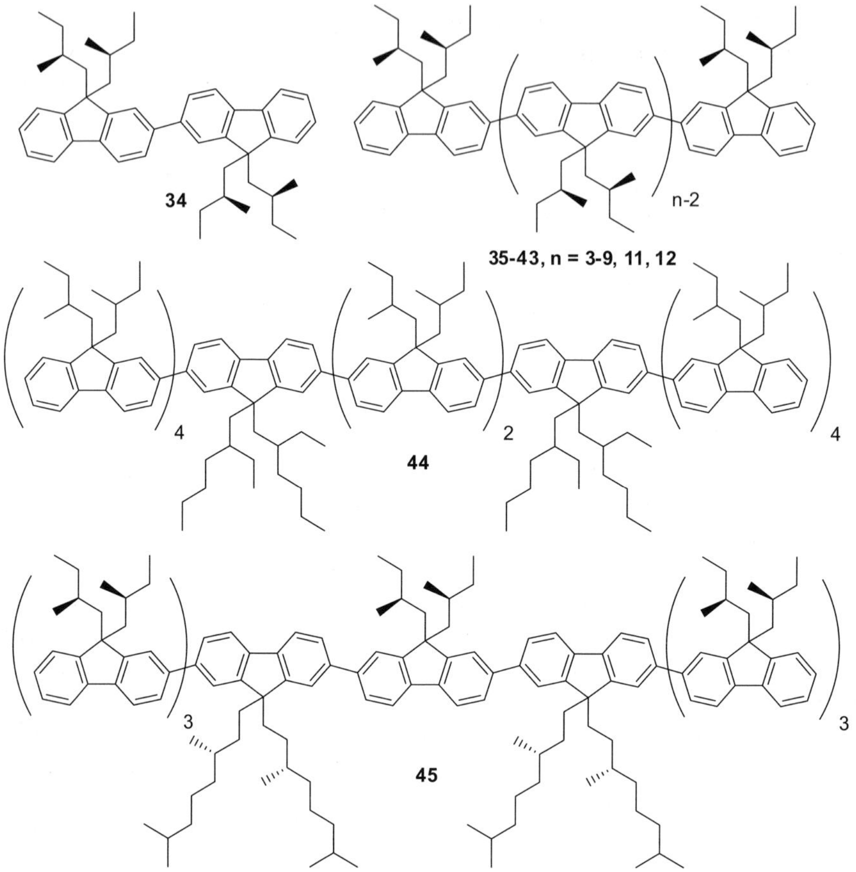

Fig. 6 Structures of uniform chiral oligofluorenes and representative nematic and chiral oligofluorenes with mixed pendants

shown a T_g of 296 °C, the highest reported for organic molecular systems at the time [47]. Later that year, a more extended spiro-configured pentafluorene set the new record with a T_g of 330 °C, showing no crystallization or melting transition up to 550 °C [121].

An illustrative example is a series of chiral oligofluorenes, uniformly substituted with 2S-methylbutyl pendants [35], shown in Fig. 6 as compounds **34–43**. Glass transitions without any crystallization are seen from two units up to five units; unstable glasses that crystallize on heating but not cooling are seen from six to eight units; and no glass transitions but strong crystallization is seen from nine units up to the polymer (equivalent to 70 units on average). As expected, the relevant phase transition temperatures increase with oligomer length. An important strategy to prevent crystallization is the use of mixed pendants, and especially the inclusion of branched alkyl pendants, providing stable glassy morphologies in oligofluorenes up to 12 units in length, as with compound 44 [60, 122].

A few other approaches have been taken to ensure that a given fluorene oligomer will form a stable glass. Kelly et al. have done considerable work with photopolymerizable groups on fluorene-containing chromophores, some of which classify as oligomers of fluorene, to form crosslinked networks [123, 124]. Strohriegl et al. have also developed such systems from their broad experience in crosslinkable liquid crystals [55]. Another approach to a crosslinked fluorene oligomer network is to dehydrate terminal boronic acids, with the added advantage of being reversibly recovered [82]. In addition to crosslinking, attaching fluorene oligomers in a rigid manner to a tetrahedral core has also been shown to result in formation of stable glassy phases [85].

The details of morphology in the glassy state of fluorene oligomers have also been examined. Wegner et al. modeled the conformations of oligofluorenes, showing that inversion of their helical motifs can occur and that excimer formation is discouraged by steric hindrance [125]. The same group has also investigated motions in glassy and melt states with dielectric spectroscopy [30] as well as orientational dynamics during their glass formation with dynamic light scatting and dynamic NMR spectroscopy [126]. Wu et al. used spectroscopic ellipsometry to show that the anisotropy present in their vacuum-deposited oligofluorene films disappears upon thermal annealing [127, 128]. Aggregation has been investigated using solvent interactions in water-soluble, self-condensed oligofluorene nanoparticles, where the choice of solvent can influence their packing [83]. Such aggregation is seen in red-shifted emission spectra, which is evident to a large degree in a six-armed star structure [129].

In contrast to these glass-forming oligomers, highly crystalline fluorene oligomers are also desirable for certain applications, in particular as active layers in organic field effect transistors. Some fluorene oligomers will form highly structured polycrystalline films upon vacuum deposition onto substrates heated to 140 °C [130]. These crystalline domains can grow as large as ten microns [130], and the details of the crystal structures of these fluorene oligomers are very sensitive to their substitution [131]. In one carefully examined case, the crystal structure was found to depend on the thickness of the deposited layer [132].

3.3
Liquid Crystallinity

One very attractive feature of extended fluorene compounds is their mesomorphism, or liquid crystalline self-organization [8]. Their chemical structure classifies them as hairy-rod liquid crystals with a rigid conjugated backbone decorated with flexible alkyl side chains. In order of descend-

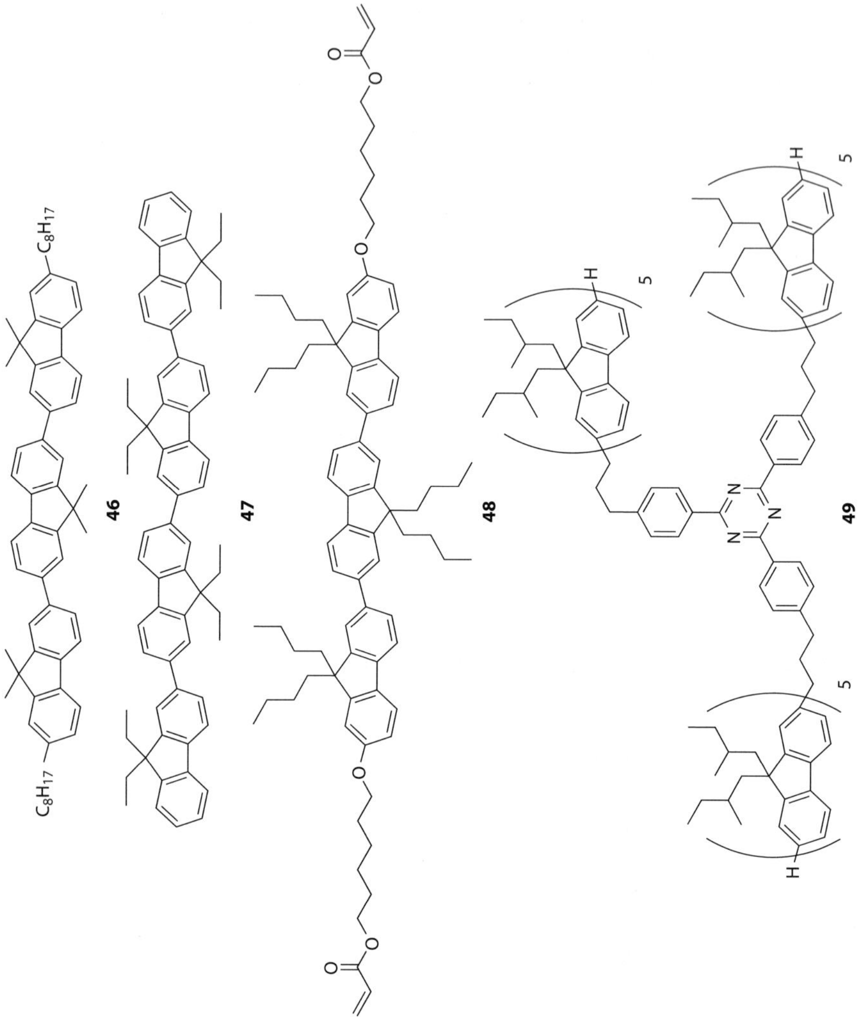

Fig. 7 Additional example structures of liquid crystalline oligofluorenes

ing molecular ordering, fluorene oligomers primarily exhibit three different mesophases: smectic, cholesteric (also known as chiral-nematic), and nematic [133]. The glass transitions of many oligofluorenes allow these unique and useful liquid crystalline morphologies to be frozen into the solid state without crystallization in order to be exploited for a range of optical and electronic applications. To take advantage of these ordered morphologies, alignment layers are used, including the most commonly employed rubbed polymer films [134], but also photoalignment layers which are effective with some oligofluorenes [135, 136].

Oligofluorenes exhibiting smectic mesomorphism organize into layers, resulting in one-dimensional positional order as well as orientational order. The shortest oligofluorene to exhibit any mesomorphism is a smectic trimer (compound **46**, Fig. 7) with only methyl groups as side chains and *n*-octyl as end chains, which is unfortunately highly crystalline [137]. Wegner et al. synthesized a series of oligofluorenes, from four to seven units in length, substituted with ethylhexyl side chains, and they observed excellent quality X-ray and electron diffraction patterns of their layered structures [138]. The broadest mesophase range of these oligomers is the heptamer with a T_g of 17 °C and a clearing temperature (T_c, or transition to isotropic liquid) of 221 °C.

Cholesteric oligofluorenes have a low enough viscosity to form highly oriented films, exhibiting very high degrees of chiroptical activity. Uniformly substituted with 2S-methylbutyl side chains, cholesteric mesomorphism begins with five units and persists up to the polymer [35], but crystallization is often encountered. With mixed chiral pendants (as in compound **45**, Fig. 6), glassy-cholesteric morphology is exclusively realized, and selective reflection bands are readily preserved into the solid state. The strong chiroptical activity observed was found to be dominated by their helical stacking in thin films, as opposed to molecular level helical conformations [122]. The cholesteric mesophase range of the mixed-pendant nonamers (such as compound **45**) was broad with T_g around 86 °C and T_c around 283 °C.

Glassy nematic oligofluorenes form large monodomain anisotropic thin films. Depending on their molecular aspect ratio, an order parameter, *S*, of 0.84 can be achieved. A very broad nematic range is exhibited by the dodecamer (compound **44**) from its T_g of 123 °C to beyond 375 °C [60]. Fluorene co-oligomers with large groups in their centers, can still exhibit broad nematic mesophase ranges of over 250 °C [64]. The shortest nematic pure oligofluorene consists of four units, as in compound **47** shown in Fig. 7, but its short ethyl pendants render it crystalline at room temperature [119]. A number of crosslinkable nematic oligofluorenes exhibit a moderate to high degree of order [55, 123, 124], such as compound **48** shown in Fig. 7.

Lastly, an interesting case is that of hybrid oligofluorene compounds [106, 139], one of which is shown as the last example in Fig. 7 (compound **49**). Here, pentafluorenes are attached to an electron-transporting core via flexible alkyl spacers. Unlike all the other fluorene oligomers modified with charge transport moieties, this type of compound manages to retain mesomorphism in the presence of the alkyl spacers which allow the oligomeric pendants to orient themselves. In fact, their order parameters in thin films assume the same values as that of the stand-alone oligofluorene. These hybrid compounds demonstrate the versatility of the core-pendant concept in affording multifunctional glassy liquid crystals [140], where oligofluorene pendants and a large non-mesogenic functional core are incorporated in a single molecular entity.

4
Photophysical Properties

4.1
Efficient Blue Emission

One of the reasons fluorene is appealing is its efficient emission of ultraviolet or deep blue light. Terfluorenes have several emission maxima at around 393, 412, and 441 nm [2]. The photoluminescent quantum yields, ϕ, of two ter(9,9-diaryl)fluorenes (compounds **50** and **51**, Fig. 8) are both 99% in solution and 90% in solid films [2]. However, most oligofluorenes have a photoluminescence quantum yield in the range from 40 to 70% [35, 54, 60, 141].

In addition, among oligofluorenes there is virtually no "long wavelength" or green emission, even after extended thermal annealing [47, 54, 60]. In contrast, polyfluorenes are known to suffer from undesirable green emission [142–144]. This may well be due to the higher purity accessible with oligomers, as very carefully purified monomers can also result in polyfluorenes showing stable blue emission [145].

4.2
Full Color Light Emission

The fluorene backbone also enables efficient emission of light across the visible spectrum with an incorporation of other chromophores. The photoluminescent quantum yield in solution can be as high as 75 to 82% from the sky blue to the orange with a slightly lower efficiency in the red at 47% [64]. A number of fluorene co-oligomers have emission at wavelengths across the visible spectrum, from sky-blue [72–74, 95, 146–149] to green [67, 98, 124, 150, 151], yellow [72] and red [69, 90, 152]. Attached as ligands to iridium, oligofluorenylpyridines are capable of red phosphorescence with emission efficiencies up to 24% [153].

Energy transfer, either within the oligomer or to neighboring groups, governs the efficiency of the full-color emission. Fluorene-ethynylene-pyrene teroligomers were synthesized and characterized to demonstrate that pyrene's

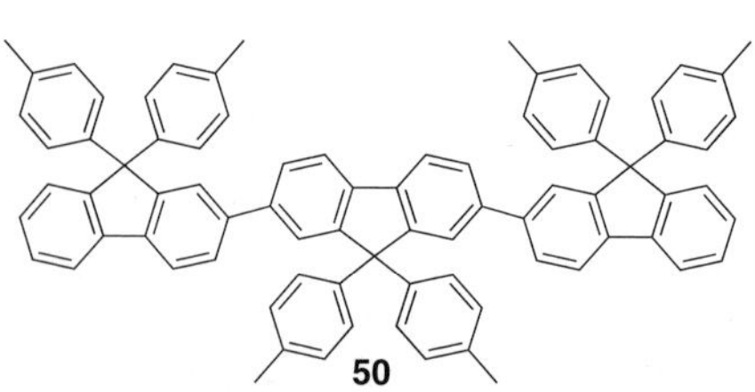

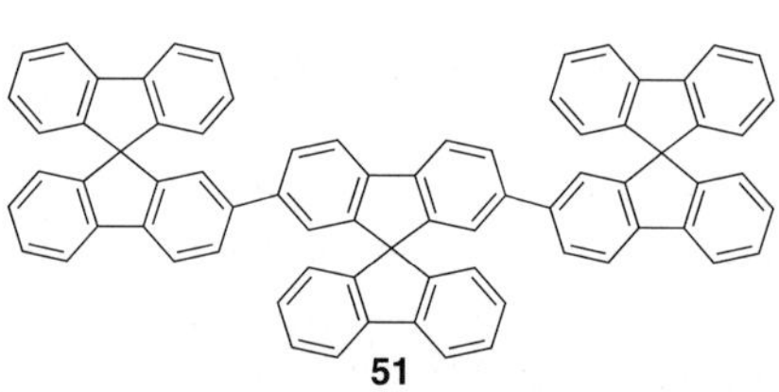

Fig. 8 Two terfluorenes with very high photoluminescent quantum yields in the solid state

position is critical to the efficiency with which it accepts intramolecular energy transfer, the center position being the most favorable [146]. A variety of energy acceptors were also tested through close proximity, either physically or, more preferably, through hydrogen bonding [58]. The solvent effects on such transfer processes have also been examined [80]. In addition, Förster resonance energy transfer from oligofluorenes to co-oligomers has been quantified and subsequently exploited to generate white light emission [74]. Finally, fluorene oligomers have been used as conjugated molecular wires to mediate energy transfer between different moieties [89].

4.3
Studies of Excited Electronic States

The chain length dependence of the photophysical properties of oligofluorenes has been studied in detail. As the length increases, the energies of their fluorescence, phosphorescence, triplet–triplet absorption, and their fluorescent lifetime all decrease, while the magnitude of the transition dipole moment increases [154, 155]. Furthermore, the triplet state was found to be only slightly more confined than the singlet state [155].

A number of studies have examined various other aspects of the photophysical processes involved in fluorene oligomers' light emission. A theoretical model has reproduced the observed vibrational structures in their absorption and emission spectra [156]. The contributions of two conformations to light emission [157] and their conformational dynamics [158] have also been reported. Moreover, model systems have been used to look at the effects of their packing in the solid state on their emission spectra [159]. In addition, the efficiency of intersystem crossing from singlet to triplet excitons has been measured to be approximately 6% [160], but it is sensitive to heavy atoms such as sulfur [161]. Measurements of electric-field-assisted dissociation have indicated the singlet fraction for charge recombination may be as high as 35% [162, 163], instead of the statistical 25%. On a very different note, the nonlinear optical properties, and in particular two-photon absorption, of oligofluorenes have been characterized [56, 87]. A number of other theoretical studies have been conducted on fluorene oligomers [100, 164–168].

4.4
Polarized Photoluminescence

Liquid crystalline fluorene oligomers can emit polarized light with proper alignment of their molecular axes and, hence, of their transition dipole moments. For example, linearly polarized emission results from uniaxial alignment of fluorene oligomers. One common way to characterize linearly polarized emission is to measure its emission dichroism, viz. the emission

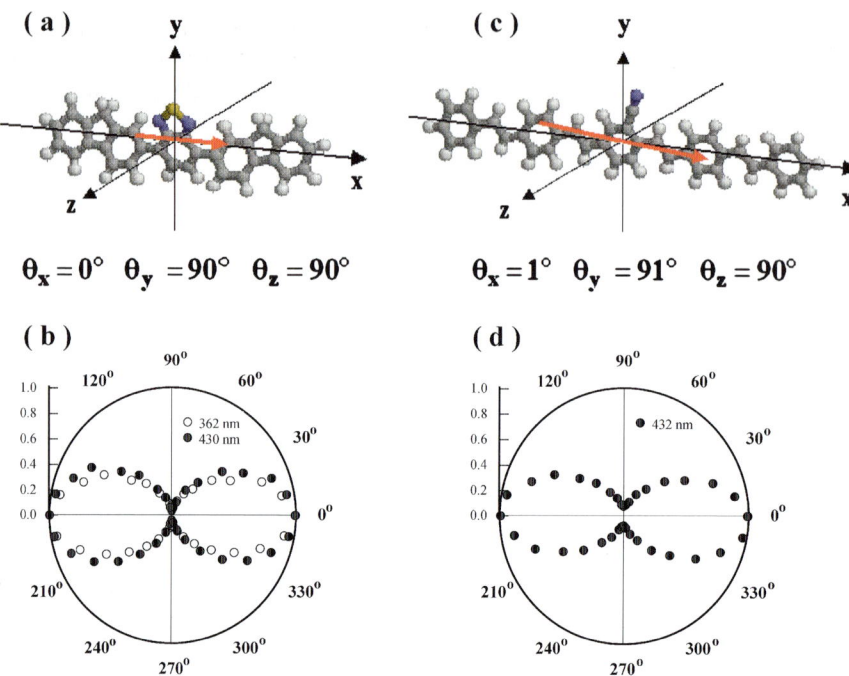

Fig. 9 Quantum mechanical calculations performed using the Gaussian 98 package to locate electron transition dipoles of the central segments: **a** 4,7-bis(2,2′-fluorene)-2,1,3-benzothiadiazole and **c** 1,4-bis[2-(4-[2-(4-phenyl)ethenyl]phenyl)ethenyl]-2-cyanobenzene, of two co-oligomers; and the UV-Vis absorption dichroism at absorption maxima of glassy-nematic films of co-oligomers containing these two segments, (**b**) and (**d**), respectively, indicating that the central segments responsible for absorption are aligned with the nematic director placed along 0 and 180° in the polar coordinates. Used with permission [64]

intensity parallel over that perpendicular to the director. A dodecafluorene (compound **44**) has exhibited a very high emission dichroism of 17 : 1, but even a heptamer with short enough pendants can reach a ratio of 14 : 1 [60]. Crosslinkable liquid crystalline fluorene oligomers have also yielded a high emission dichroism [55, 124]. For colors other than blue, it is important that the transition dipole moment of the chromophore inserted into an oligofluorene is along the molecular axis of the entire oligomer. Molecular modeling has shown its feasibility for a variety of inserted chromophores (for two examples, see Fig. 9), and the resultant emission dichroic ratios are from 9 : 1 to 14 : 1 across the visible spectrum [64]. Other nematic liquid crystalline oligomers have been synthesized, but generally have a lower degree of order and, thus, smaller emission dichroism, of 4 : 1 to 7 : 1 [169].

Circularly polarized light can be emitted by chiral oligofluorenes. The chiral side chains cause the oligofluorene backbones to preferentially twist into one helical conformation over the other. This chiral preference has been revealed by molecular simulations as well as exciton coupling signatures in circular dichroism [35, 122], resulting in a low degree of circularly polarized light absorption and emission. However, when annealed into a well-aligned helical stack characteristic of cholesteric mesomorphism, the degree of circular polarization, quantified by the dissymmetry factor, g_e, increases by nearly an order of magnitude to 0.7 [122]. Note that a g_e value of ± 2 represents perfect circularly polarized light emission. For circularly polarized light emitted by a thin film g_e value of 0.7 is a very high value which, coupled with a circular dichroism of up to 12 degrees, testifies to the supramolecular chirality achievable with cholesteric mesomorphism manifested by chiral oligofluorenes.

5
Electronic Properties

5.1
Electrochemistry: Energy Levels and Properties of Ionic States

The electronic properties of a material are highly relevant to a variety of organic electronic applications. These include the energy levels of the material, the properties of their ionic states, and their charge transport properties. The energy levels of a compound refer to its highest-occupied molecular orbital (HOMO) and lowest-unoccupied molecular orbital (LUMO). These roughly correspond to its ionization potential (I_p) and its electron affinity (E_a), respectively.

Oligofluorenes' LUMO and HOMO energy levels are slightly farther apart than polyfluorenes' because of the shorter conjugation lengths and larger band gaps. A study by Wegner et al. illustrates the oligomer length dependence as its properties approach that of the analogous polymer [170]. In appropriate solvents, oligofluorenes exhibit electrochemical stability as demonstrated by the reversibility of the cyclic voltammetry (CV) scans for both oxidation and reduction (see Fig. 10 for a representative example). In addition, the electrochemically inactive substituents at fluorene's C-9 position have an insignificant effect on the energy levels associated with the backbone [171], as shown by ultraviolet photoemission spectroscopy (UPS), which is expected of electronic decoupling. The exception to this rule is a spirofluorene linkage, which does influence the energy levels to some degree through spiroconjugation [171].

For fluorene co-oligomers, the band gap often changes with concomitant changes to the energy levels of the compound [69]. Inclusion of electron-rich co-monomer units, predominantly raises the HOMO level [147, 172], while electron-deficient moieties mostly lower the LUMO level [69, 80, 173]. Attachment through direct conjugation with electronically neutral cores has more

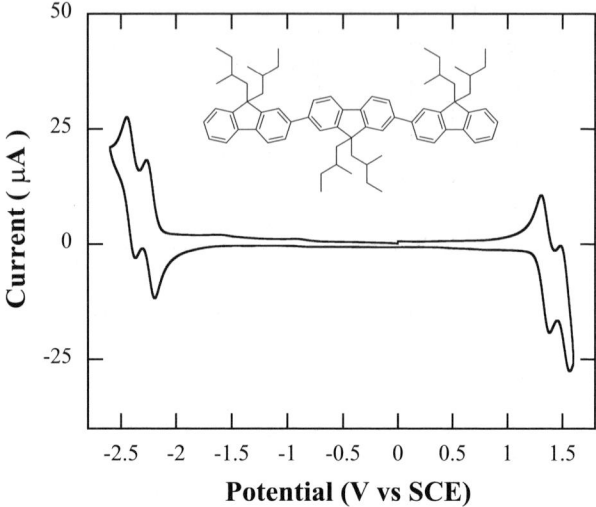

Fig. 10 Cyclic voltammetric (CV) reduction and oxidation scans of a terfluorene at 2.5×10^{-4} M in acetonitrile and toluene (1:1 v/v) with 0.1 M tetrabutylammonium tetrafluoroborate as the supporting electrolyte using a glassy carbon electrode

subtle effects through complex conjugation behavior between the various segments [174, 175].

In addition, through end-capping [93–100], substitution at the C-9 position [105], direct connection to such a core [91, 102], or non-conjugated attachment through a flexible spacer [106, 176], functionalization with electron-rich moieties contributes additional oxidation peaks in the CV scans. Similar non-conjugated attachment of electron-deficient moieties results in additional peaks in the reduction scans [106, 176]. Functionalization has also been performed with both moieties on a single oligomer, one electron-rich and the other electron-deficient, showing additional peaks in both oxidation and reduction scans [101].

The properties of ionized, or charged, oligofluorenes have been characterized and modeled [177–179]. One of these studies has found that charge carriers on oligo- or poly-fluorene backbones delocalize to a greater extent than a neutral exciton [177]. In addition, the fluorene backbone is more planar in the charged state than in the neutral state [178, 179].

5.2
Bipolar Charge-Carrier Transport

Oligofluorenes are exceptional in their charge transport properties, exhibiting relatively high bipolar charge-carrier mobilities in comparison to other non-crystalline organic materials. An interesting study of the dependence

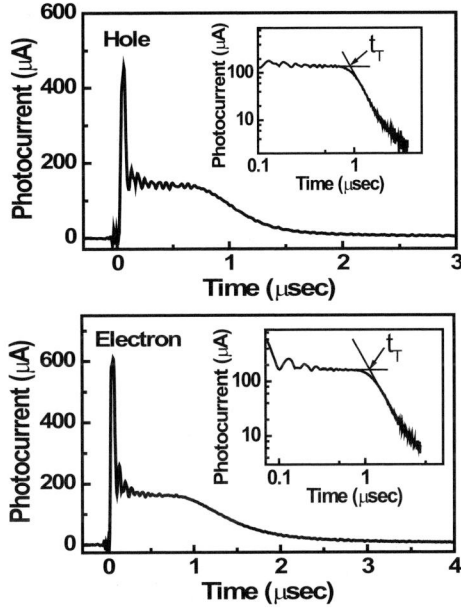

Fig. 11 Representative time-of-flight transients for 2,2″-bi-9,9′-spirobifluorene (2.1 μm thick): **a** hole, $E = 4.83 \times 10^4$ V/cm, **b** electron, $E = 9.53 \times 10^4$ V/cm. *Insets* of **a** and **b** are double logarithmic plots of **a** and **b**, respectively. Used with permission [180]

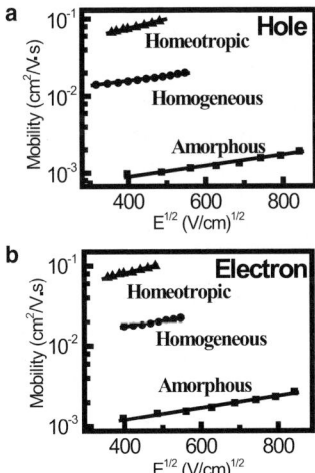

Fig. 12 **a** Hole mobilities and **b** electron mobilities vs. $E^{1/2}$ for a tetrafluorene (compound **47**) in the amorphous state, the homogeneously aligned glassy-nematic state (charge moving perpendicular to the average long molecular axis), and in the homeotropically aligned glassy-nematic state (charge moving along the average long molecular axis). Used with permission [182]

of the carrier mobilities on oligomer length was undertaken by Wu et al., showing the bipolar nature of these compounds [180]. Figure 11 shows the non-dispersive electron and hole transients for a spirobifluorene dimer. Similar bipolar properties are also present in alkyl-substituted oligofluorenes, as seen in a terfluorene measured by photocurrent time-of-flight, and used as a charge generation layer to aid in the measurement of a number of other compounds for both hole and electron mobility [181].

The liquid crystalline order of oligofluorenes also sets them apart from many other materials. The advantage of order in charge transport is apparent in a liquid crystalline tetrafluorene (compound 47), which exhibits over an order of magnitude increase in both its electron and hole mobilities upon homogeneous alignment into a glassy liquid-crystal film [119]. By aligning it in a homeotropic fashion, further work with this compound enhances the bipolar transport by another order of magnitude, yielding the hole and electron mobilities both nearly $0.1\,\mathrm{cm}^2\,\mathrm{V}^{-1}\,\mathrm{s}^{-1}$ [182]. The results also indicate the charge transport to be highly anisotropic, being preferentially along the oligofluorene backbone by roughly a factor of five, see Fig. 12.

6
Photonic and Electronic Applications

6.1
Organic Light-Emitting Diodes

The efficient photoluminescence, relatively high bipolar charge transport, morphological stability, and ease of processing all render fluorene oligomers potentially useful to organic light-emitting diodes, or OLEDs. Oligofluorenes emit light of a deep blue color with CIE (Commision Internationale de l'Eclairage) coordinates roughly in the range of (0.16, 0.04) to (0.17, 0.12), very close to the NTSC (National Television System Committee) standard blue color characterized at (0.14, 0.08) used in televisions and monitors [183]. This deep blue color is useful for achieving a wide color gamut in displays and for improving the quality of white OLEDs [184]. Many of these oligomers can be readily vacuum deposited into complex device structures, but others employ low-cost solution processing, most often by spin-coating. In addition, the mesomorphism of some of these oligomers provides opportunities for intrinsically polarized light emission from an OLED.

The highest efficiency OLED utilizing a spiro-terfluorene (compound 51) in a "double confinement" device structure [185], reaches an external quantum efficiency (EQE) of 5.3%, or $1.53\,\mathrm{cd}\,\mathrm{A}^{-1}$, with CIE coordinates of (0.158, 0.041). Others using similar "double confinement" structures have achieved

EQEs of 1.8 to 4.1% [186–189]. Incorporating electron-deficient moieties allows for a simpler device structure with an EQE of 1.6% [104], which can be an advantage or a disadvantage over non-functionalized oligofluorenes [103, 104], depending on the details of the device structure [190]. To attain higher efficiencies with simpler, solution-processed devices, hole-transport moieties are often incorporated into these oligomers with EQEs ranging from 1.4 to 3.7% [91, 102, 107, 173, 191, 192]. Note that EQE is much more fair for the comparison of efficiency, as cd A^{-1} varies sharply with CIE color coordinate in the blue region of the spectrum [193].

Device lifetime, especially for blue-emitting devices, is a key concern. One oligofluorene derivative, a ditolylamine end-capped terfluorene, has been demonstrated in a device with a time to half luminescence of 500 h at 1100 cd m^{-2}, which is extrapolated to over 10 000 h at a typical display brightness of 200 cd m^{-2} [189]. A study on a series of anthracene-containing fluorene oligomers suggested that the difference in hole mobility through the emitter layer was responsible for the observed difference in OLED device lifetime [148].

In addition to blue-emitting devices, fluorene oligomers have been used to emit colors across the visible spectrum. Shifting the color slightly into the sky-blue has been accomplished with phenylene-vinylene [149], perylene [74], and anthracene [73, 148] units, reaching efficiencies up to 3.2 cd A^{-1} with a CIE of (0.17, 0.19) [73]. Green-emitting fluorene oligomers have reached an impressive efficiency of 11 cd A^{-1} [124] as a neat fluorescent layer, by incorporating fluorene-vinylene [67], oligothiophenes [124, 150], and thiophene-S,S-dioxide [151], among others. Red-emitting derivatives (such as compound **29**) have employed thiophenes and benzathiadiazole units [64, 69]. However, the most impressive red-emitting oligofluorene is a terfluorene-pyridine ligand to iridium, which results in red electrophosphorescence with an EQE of 2.8% [153]. Three papers span most of the visible spectra with their wide variety of incorporated units [64, 69, 74]. In addition, an efficient white OLED has been fabricated using a terfluorene and rubrene to reach an efficiency of 6.15 cd A^{-1} and a CIE of (0.32, 0.37) [187]. Lastly, a bifluorene has been used to realize electroluminescence in the ultraviolet region with a notable efficiency of 3.6% EQE [194].

Liquid crystalline fluorene oligomers have shown excellent performance in polarized OLEDs. Linearly polarized OLEDs could be used as backlights for LCD displays, saving energy by allowing more of the light to be utilized without losing it to the initial polarizer [195]. In addition, stereoscopic displays and projection displays are other opportunities for linearly polarized OLEDs. For a dodecafluorene (compound **44**), efficiencies of up to 1.07 cd A^{-1}, with CIE coordinates of (0.159, 0.062), have been realized with linear polarization ratios (integrated over its emission spectrum) up to 24.6 (or 31.2 at 448 nm) [196], as can be seen in its electroluminescence spectra in Fig. 13, along with the device structure as the inset. A simplified structure resulted

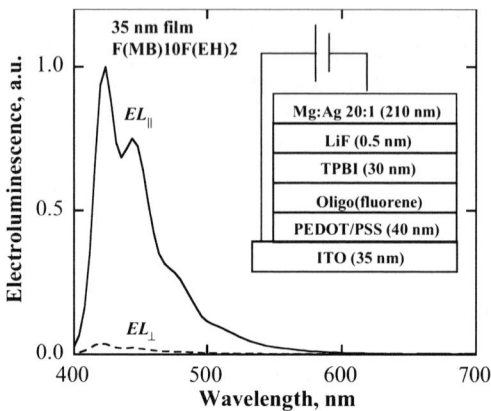

Fig. 13 Polarized electroluminescence (EL) spectra and device structure of an OLED containing a 35 nm thick dodecafluorene (compound **44**) film at a current density of 20 mA cm^{-2}. Used with permission [196]

in a loss in efficiency, down to 0.18 cd A^{-1}, but a sizeable gain in polarization ratio of 50.1 at 448 nm [197]. For green, red, and white OLEDs, polarization ratios of 17.7 (at 504 nm), 17.9 (at 620 nm), and 16.6 (integrated), and efficiencies of 6.4, 1.5, and 4.5 cd A^{-1}, respectively, have been achieved with fluorene co-oligomers [64, 74].

The performance achieved by these devices was record-setting, with each of their characteristics independently superior to reported values from any other polarized OLEDs, including those using polyfluorenes [9–13]. Recently, an integrated polarization ratio of 31 has been reported but with an efficiency of 0.3 cd A^{-1} for a deep blue OLED comprising poly(9,9-dioctylfluorene) [198]. Two other papers have reported polarization ratios at their emission maxima of 25.7 [199] and 29 [200], which are still slightly lower than 31.2 reported for the blue dodecafluorene device [196].

The fluorene oligomers in these record-setting devices have also been used to investigate polarized OLEDs in more detail. The first parameter of interest was the chain length. As expected, the longer the oligomer length, the higher the order and resultant polarization ratio, from 11.8 for a pentamer to the 31.2 of the dodecamer [196]. The dependence of the emitter layer thickness was also investigated for the dodecamer (compound **44**), showing an increasing polarization ratio at a decreasing film thickness [196]. The effect of thermal annealing was also explored, and it was found that pristine, isotropic devices had the same performance as the annealed, polarized devices [197], in sharp contrast to many polarized OLEDs containing polyfluorenes [11–13].

In addition, chiral oligofluorenes have exhibited circularly polarized electroluminescence with the highest polarization to date, as characterized by the dissymmetry factor, without addition of optical elements external to the OLED device. This device had an efficiency of 0.94 cd A^{-1} in the deep blue, with a dissymmetry factor of 0.35, which remained over 0.3 across the majority of its emission spectrum [122]. Circularly polarized OLEDs allow the use of circular polarizers on OLED displays to block glare from ambient light, thus increasing the display's contrast without losing output intensity [201]. For colors other than blue, doping could readily result in circularly polarized emission across the visible spectrum, including white light as was done with linearly polarized OLEDs [64, 74].

6.2
Solid-State Organic Lasers

Organic materials have been used as lasing media, both as laser dyes in solution or, more recently, in the solid-state. The use of oligofluorenes as dye

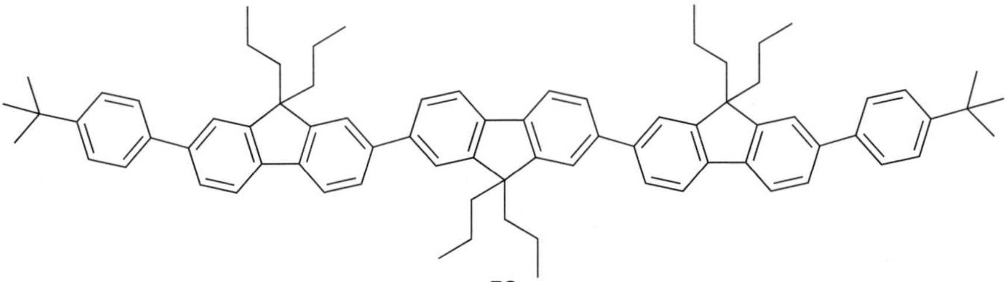

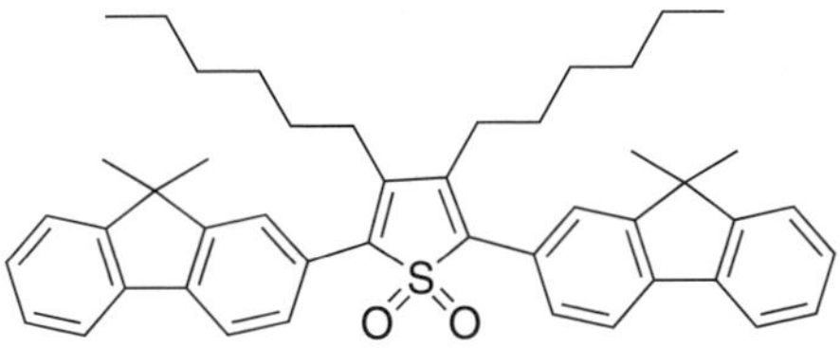

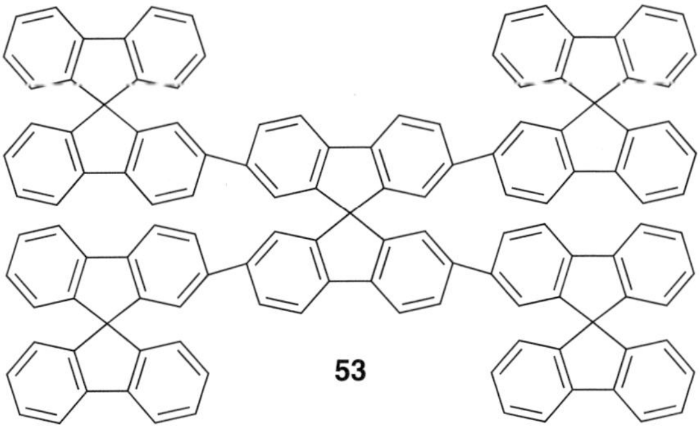

Fig. 14 Representative fluorene oligomers used in lasing and amplified spontaneous emission (ASE)

lasers was among the first indications of their excellent performance in lasers. When excited by excimer lasers or appropriate flashlamps, bifluorenes [202] up to tetrafluorenes [203] showed excellent lasing performance in the UV to deep blue. Outstanding among them is a laser dye now known as Exalite 428 (compound **52**), marketed by Exciton Inc., shown in Fig. 14 [203]. It exhibits relatively high conversion efficiency and a long lifetime (in terms of energy input to a single batch of laser dye solution).

As for organic solid-state lasers, polyfluorenes took an early lead, demonstrating their favorable characteristics [15] and culminating in relatively efficient photo-pumped lasing across the visible spectrum [16–18]. Work with fluorene oligomers has focused on their suitability as lasing media, as characterized through amplified spontaneous emission, ASE, which indicates their potential as excellent solid-state lasers when implemented in optical resonators. Salbeck et al. were the first to report the use of oligofluorenes as neat materials for solid-state ASE [121]. These structures are highly spiro-functionalized and have very high glass transition temperatures. Compound **53**, shown in Fig. 14, exhibits an ASE threshold of 3.2 µJ cm^{-2} [121], which is lower than that of the corresponding spiro-oligophenylene [204].

Using a fluorene-S,S-dioxide-thiophene co-oligomer (compound **54**, Fig. 14), Lattante et al. have realized a gain of 8 cm^{-1} at 550 nm [81]. The threshold was higher, but comparable to other small molecules used as organic active media. An aryl-substituted terfluorene (compound **50**), which exhibits anisotropic optical properties upon vacuum deposition [127], has shown excellent stimulated emission characteristics. When in the anisotropic state, the gain reaches an impressive 78 cm^{-1} [205], an improvement of nearly an order of magnitude over that in its isotropic state. In addition, controlled annealing of compound **50** allows for one-time tuning of the ASE wavelength subsequent to device fabrication, from 426 to 413 nm, shown in Fig. 15 [128].

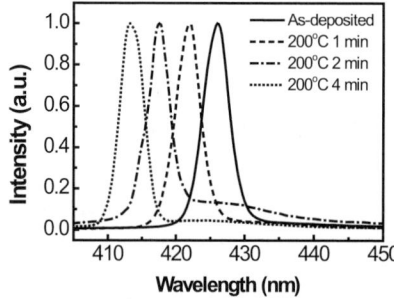

Fig. 15 Tunable amplified spontaneous emission (ASE) spectra of a 47-nm-thick terfluorene (compound **50**) film measured as deposited and after annealing at 200 °C for 1, 2, and 4 min. Used with permission [128]

6.3
Organic Field Effect Transistors

Many high-performance organic field effect transistors, abbreviated as OFETs, are composed of small molecular weight compounds, oligomers being prominent among them [34]. In particular, oligothiophenes are attractive, but their stability against oxidation and, therefore, ambient exposure is poor. Instead of oligothiophenes, fluorene-thiophene co-oligomers have emerged as more oxidatively stable materials in p-type OFETs. This is evidenced by oligomer FTTF (compound **55**), depicted in Fig. 16, which shows no decrease in hole mobility or on/off ratio after two months of ambient exposure [206]. As demonstrated by Noh et al., if present, fluorenone defects are detrimental to device performance despite the improved oxidative stability [207, 208]. However, without intense UV irradiation, these fluorenone defects are not observed [206].

In addition to their relative stability under ambient conditions, these fluorene-thiophene oligomers are attractive because of their good mobility and the potential for further improvement through understanding their structure–property relationships. Various derivatives have all shown mobilities slightly above 0.1 cm^2 V^{-1} s^{-1}, which is on the order of amorphous silicon, such as DHFTTF [209], DDFTTF [131], CHFTTF [130], C8FTTF [210], and C12FTTF [210], all shown in Fig. 16 as compounds **56–60**. Other fluorene co-oligomers incorporating a dithienothiophene unit (such as compound **61**)

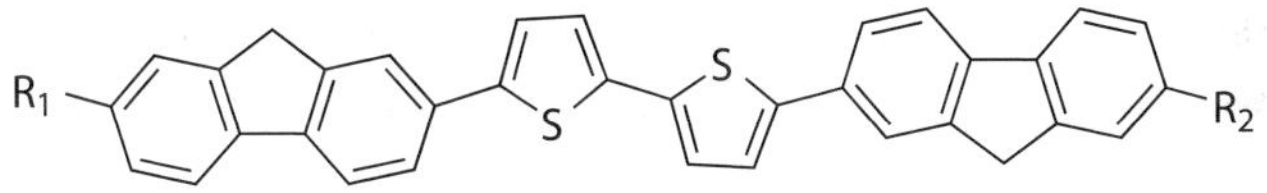

55, FTTF: $R_1 = R_2 = H$
56, DHFTTF: $R_1 = R_2 = C_6H_{13}$
57, DDFTTF: $R_1 = R_2 = C_{12}H_{25}$

58, CHFTTF: $R_1 = R_2 =$

59, C8FTTF: $R_1 = H; R_2 = C_8H_{17}$
60, C12FTTF: $R_1 = H; R_2 = C_{12}H_{25}$

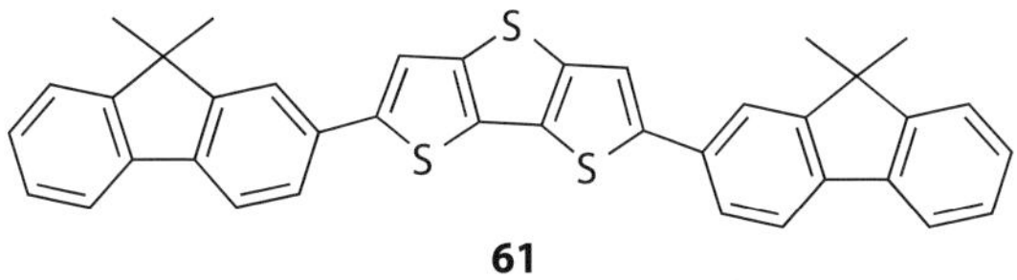

61

Fig. 16 Structures of crystalline fluorene co-oligomers used in organic field effect transistors (OFETs)

exhibit slightly lower mobilities [211, 212], while oligofluorenes can reach a field effect mobility of 1.2×10^{-2} cm^2 V^{-1} s^{-1} [213].

From these studies a number of structure-property relationships have emerged. First, bithiophene is the optimal length for the central thiophene group [209]. The presence of alkyl end groups is very important for good crystal packing which is essential to reaching high mobilities [131]. Cyclohexyl groups in particular can form large crystal domains up to 10 µm in size [130]. It is interesting to note that an alkyl end group is only needed on one end to reach relatively high mobilities, as in compounds **59** and **60** [210]. Returning to the crystal structures, the film thickness was found to affect molecular packing, and thus the mobility, of one of these compounds [132]. Noh et al. indicated that biphenyl groups are more stable than fluorene groups against oxidation by air in the presence of UV light, as the C-9 position of fluorene appears to be the most vulnerable site [214].

Thanks to their mesomorphism, oligofluorenes are of interest for OFETs which exhibit anisotropic field effect mobility. Indeed, interest in liquid crystalline materials in OFETs is on the rise [215]. Such anisotropy can help to diminish cross-talk as carriers move more slowly in the direction perpendicular to the channel than parallel to it. Glassy liquid crystalline films of oligofluorenes have shown their promise in this regard, reaching a degree of anisotropy in their field effect mobilities of 6.3 to 1. With a long enough oligomer, their performance surpassed that of the analogous polymer system [213], reaching a mobility of 1.2×10^{-2} cm^2 V^{-1} s^{-1} for a dodecamer (compound **44**). The highest mobility for these nematic oligofluorenes is seen along the long axis of these molecules in a monodomain film where the shortest number of rate-limiting hops would occur, followed in decreasing order by a polydomain film, a monodomain film measured perpendicular to the nematic director, and a disordered amorphous film. This anisotropic field effect mobility would become potentially useful in practice should the mobility of similar oligomers be raised by one to two orders of magnitude.

6.4
Organic Solar Cells

Fluorene oligomers have seen only limited use in organic solar cells to date. If they incorporate any fluorene units at all, most efforts have focused on polyfluorene derivatives [19, 20]. For example, Bryce et al. synthesized a series of oligofluorene-fullerene derivatives to study their fundamental photophysical and charge transfer properties [216] (compounds **62–67** in Fig. 17), but have not pursued such structures for organic solar cells yet.

Kelly et al. remain the primary group involved in using fluorene oligomers for solar cell applications. Their initial effort utilized a fluorene oligomer, which incorporated a perylene diimide central unit (compound **68**) [217],

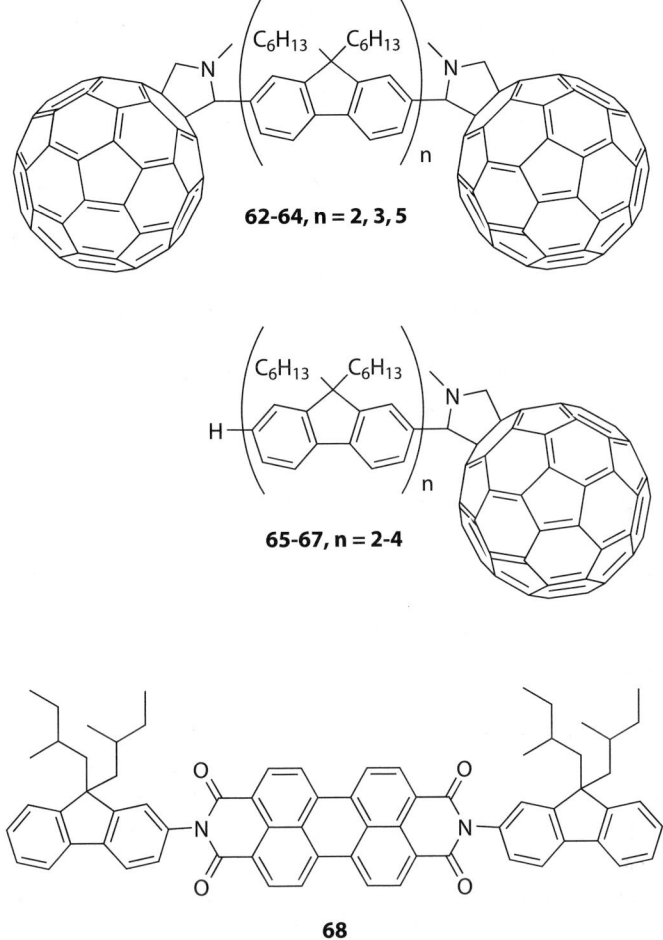

Fig. 17 Fluorene co-oligomers potentially useful for fabrication of organic solar cells

to achieve a vertically segregated interface with low molecular weight crosslinkable compounds. Subsequent work using a co-oligomer of fluorene and thiophene as the p-type component, along with a fluorene-perylene diimide oligomer as its n-type counterpart, results in a slightly higher conversion efficiency of 0.8% in a distributed bilayer structure [218]. Both of these studies involved the utilization of liquid crystalline mesomorphism to control the morphology, showing promise for future improvement.

7
Fluorene-Based Oligomers to Probe Polyfluorenes

In addition to many of the unique properties and advantages of fluorene oligomers surveyed in this article, they also serve as model systems for understanding the behavior of polyfluorenes, which have been widely explored for organic electronics. Many research groups have synthesized a series of oligomers to learn about the analogous polymeric systems. One of the early synthetic efforts with oligofluorenes was undertaken to determine the effective conjugation length of polyfluorene based on the position of the absorption maximum [52]. More recent studies have also looked at the energy levels of oligofluorenes in comparison to polyfluorene [170], delocalization lengths of charge carriers [177], and detailed photophysical studies [155], among other less direct but still useful surveys of oligomer series. Departures from such correlations, however, are well known [21, 30–33].

7.1
Fluorene-Fluorenone Co-oligomers

One particular area that has garnered considerable attention is the green emission that often plagues polyfluorene light-emitting materials [142, 143]. While there is still debate over the prevalence of two different mechanisms, it is settled that fluorenone defects do play a role in this green emission band in the absence of aggregation, as clearly illustrated in a paper by Becker et al. [144]. This has spawned a number of studies on fluorene oligomers containing fluorenone units. Their synthesis is straightforward with simple Suzuki couplings to the halo- or dihalo-fluorenone units [75], although treatment of these ketones with butyllithium should obviously be avoided in subsequent steps.

The first use of fractionated oligofluorenes to explore the origin of green emission from polyfluorenes reported green emission from both material classes in solution and neat film [219], in contrast to the more recent experimentation with monodisperse oligofluorenes [47, 54, 60]. Time-resolved spectroscopy was used in an attempt to exclude the influence of excimers, which became a repeated theme in future articles. Another group synthesized the first fluorene-fluorenone co-oligomer (compound **69** in Fig. 18) to examine the origin of green emission and the electroluminescence of these compounds, finding the fluorenone unit to be a decent green emitter in an OLED [77]. Wegner et al. added to this understanding by doping co-oligomers consisting of one, two, and three fluorene units at both the 2- and 7-positions of a central fluorenone into a highly defect-free polyfluorene [78], with another group confirming that an equivalent pentamer is an accurate reproduction of the photophysics of the emitting species in degraded green-emitting polyfluorene samples [79]. Synthetically, a number of placements of

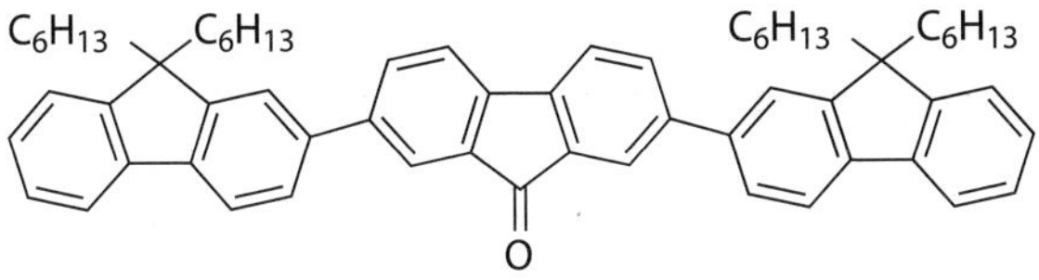

Fig. 18 Two fluorene-fluorenone co-oligomers and an additional model compound at an intermediate stage of oxidation

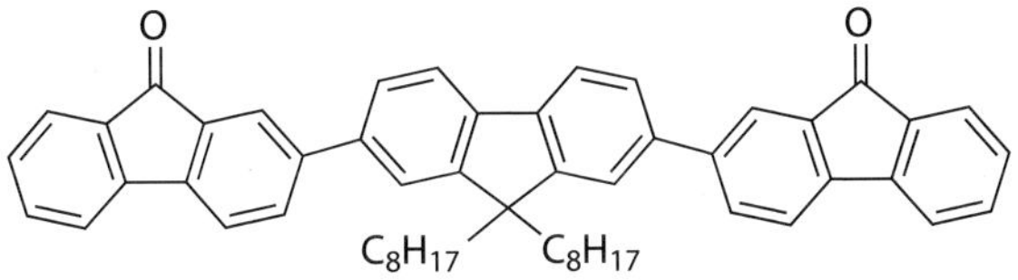

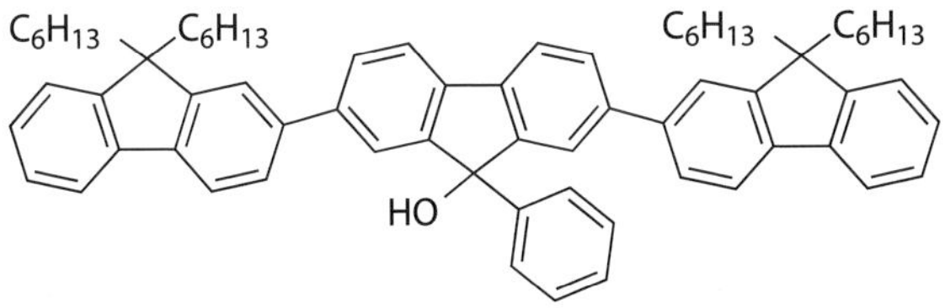

fluorenone in such co-oligomers (e.g., compound **70**) are predictable in terms of the intensity of green emission based on the number of ketone groups present [75]. Another interesting oligomer was synthesized with a hydroxyl group at the C-9 position of fluorene on the central unit (compound **71**), also shown in Fig. 18. This compound showed only a slight red shift of the blue emission without any green emission [76]. While aggregation may still play a role in the undesired green emission from polyfluorenes, fluorene-based oligomers have helped to elucidate that fluorenone is a primary cause.

7.2
Insight into Degradation Processes

Not departing too far from fluorenones, Liu et al. have done an elegant study on the photodegradation of fluorene oligomers [220]. The structures they use vary from that of compound **51** to compound **72**, shown in Fig. 19, by varying the amount of alkyl versus spiro-substitution. These compounds are carefully characterized and found to contain no detectable fluorenone as synthesized. Also shown are a number of defect sites they propose and subsequently detect after photoirradiating the samples. Figure 20 shows the matrix-assisted laser

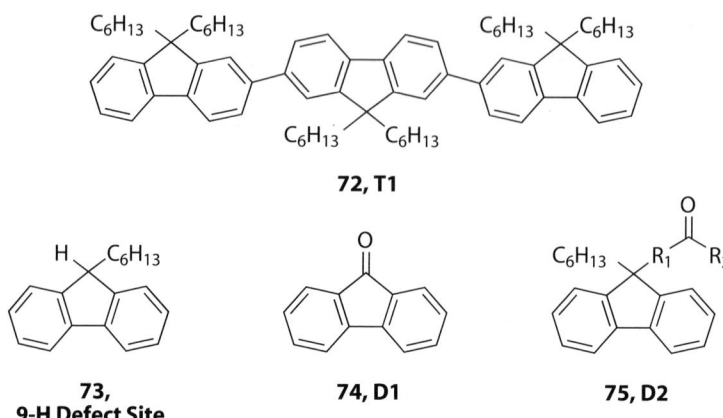

Fig. 19 A representative terfluorene used to study photodegradation, with a number of monomers showing the types of defect sites found experimentally after intense photo-irradiation

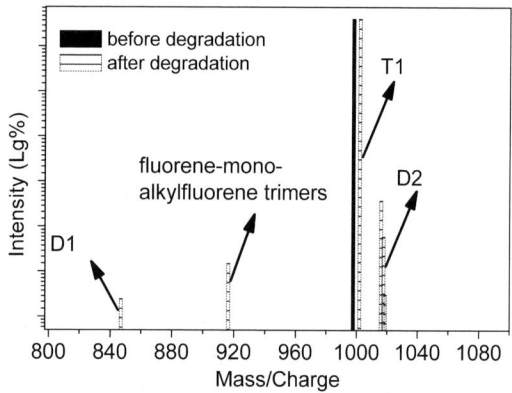

Fig. 20 MALDI-TOF mass spectra of **T1** (compound **72**) and the soluble part after 30 min of UV irradiation by a 125 W high-pressure mercury-arc light source. The *arrows* point towards the possible products with different masses (with **73–75** as defect sites). Used with permission [220]

desorption/ionization time of flight (MALDI-TOF) spectra of compound **72** and the detected degradation products, containing the defect sites shown in Fig. 19 by their mass (and confirmed by IR spectroscopy). They propose that photodegradation occurs via a radical chain reaction process that is propagated by the alkyl side chains, which also explains the stability of aryl-substituted fluorenes as they exhibit almost no degradation (only seen in a slight increase in the tail of its photoluminescence).

Kondakov et al. have reported the analysis by high performance liquid chromatography of degradation products of a phosphorescent OLED based on evaporable materials [221]. This work has inspired confidence in using fluorene-based oligomers, evaporable or solution-processable, as model compounds for continuing effort to address the OLED device instability issue through the isolation and identification of degradation products. Similarly, the ability to isolate and identify the products of electrochemical degradation involving the more tractable oligofluorenes than polyfluorenes [222] would be a fruitful line of investigation.

8
Summary

Fluorene-based conjugated oligomers represent an emerging class of materials that complement the properties of their polymer analogues, bringing a number of unique strengths to the arsenal of organic optoelectronics. A wide diversity of structures is readily accessible with versatile synthetic strategies supported by various cross-coupling reactions. Through molecular design, fluorene-based conjugated oligomers can be tailored from highly crystalline to morphologically stable amorphous materials. Oligofluorenes are quite attractive for the fabrication of blue-emitting devices because of their high photoluminescence quantum yields of up to 90% in the solid state. Emission of green, red, and white light is also feasible via insertion of proper chromophores in oligofluorenes aided by Förster energy transfer. Furthermore, their fast bipolar charge transport is a unique asset to organic semiconductors. With sufficiently high molecular aspect ratios, fluorene oligomers have been shown to exhibit nematic, cholesteric, and smectic liquid crystalline mesomorphism with elevated glass transition and clearing temperatures. Thanks to their low viscosities in the absence of chain entanglement and their ability to form self-organized solid films across a large area, liquid crystalline conjugated oligomers have found potential applications where a high degree of optical or electronic anisotropy is advantageous or essential. Through supramolecular chirality, oligofluorenes with enantiomeric aliphatic pendants are capable of extremely high chiroptical activities in thin film. In a nutshell, fluorene-based conjugated oligomers are expected to play a prominent role in organic electronics and photonics, from polarized and unpolarized light-emitting diodes, through tunable organic lasers, environmentally stable charge transport in organic transistors, and emerging application in organic solar cells. Additionally, fluorene-based conjugated oligomers have been instrumental to furnishing fundamental insight into the chemistry and instability issues confronting polyfluorenes as the prime candidate for blue light emission.

References

1. Birks JB (1970) Photophysics of Aromatic Molecules. Wiley-Interscience, New York
2. Wong K-T, Chien Y-Y, Chen R-T, Wang C-F, Lin Y-T, Chiang H-H, Hsieh P-Y, Wu C-C, Chou CH, Su YO, Lee G-H, Peng S-M (2002) J Am Chem Soc 124:11576
3. Setayesh S, Grimsdale AC, Weil T, Enkelmann V, Müllen K, Meghdadi F, List EJW, Leising G (2001) J Am Chem Soc 123:946
4. Bernius MT, Inbasekaran M, O'Brien J, Wu W (2000) Adv Mater 12:1737
5. Leclerc M (2001) J Polym Sci A 39:2867
6. Neher D (2001) Macromol Rapid Commun 22:1365
7. Scherf U, List EJW (2002) Adv Mater 14:477
8. Knaapila M, Stepanyan R, Lyons BP, Torkkeli M, Monkman AP (2006) Adv Funct Mater 16:599
9. Miteva T, Miesel A, Knoll W, Nothofer H-G, Scherf U, Muller DC, Meerholz K, Yasuda A, Neher D (2001) Adv Mater 13:565
10. Sainova D, Zen A, Nothofer H-G, Asawapirom U, Scherf U, Hagen R, Bieringer T, Kostromine S, Neher D (2002) Adv Funct Mater 12:49
11. Whitehead KS, Grell M, Bradley DDC, Jandke M, Strohriegl P (2000) Appl Phys Lett 76:2946
12. Grell M, Knoll W, Lupo D, Miesel A, Miteva T, Neher D, Nothofer H-G, Scherf U, Yasuda A (1999) Adv Mater 11:671
13. Nothofer H-G, Miesel A, Miteva T, Neher D, Forster M, Oda M, Lieser G, Sainova D, Yasuda A, Lupo D, Knoll W, Scherf U (2000) Macromol Symp 154:139
14. Leadbeater M, Patel N, Tierney B, O'Connor S, Grizzi I, Towns C (2004) SiD Int Symp Dig Tech Pap p 162
15. Xia R, Heliotis G, Bradley DDC (2003) Appl Phys Lett 82:3599
16. Heliotis G, Xia R, Bradley DDC, Turnball GA, Samuel IDW, Andrew P, Barnes WL (2003) Appl Phys Lett 83:2118
17. Heliotis G, Xia R, Turnball GA, Andrew P, Barnes WL, Samuel IDW, Bradley DDC (2004) Adv Funct Mater 14:91
18. Heliotis G, Xia R, Bradley DDC, Turnball GA, Samuel IDW, Andrew P, Barnes WL (2004) J Appl Phys 96:6959
19. Arias AC, MacKenzie JD, Stevenson R, Halls JJM, Inbasekaran M, Woo EP, Richards D, Friend RH (2001) Macromolecules 34:6005
20. Arias AC, Corcoran N, Banach M, Friend RH, MacKenzie JD, Huck WTS (2002) Appl Phys Lett 80:1695
21. Müllen K, Wegner G (1998) Electronic Materials: The Oligomer Approach. Wiley-VCH, New York
22. Fried JR (1995) Polymer Science and Technology. Prentice Hall PTR, Englewood Cliffs
23. Martin RE, Diederich F (1999) Angew Chem Int Ed 38:1350
24. Percec V, Pugh C (1987) Oligomers. In: Mark HF, Bikales NM, Overberger CG, Menges G (eds) Encyclopedia of Polymer Science and Engineering, vol 10. Wiley, New York, p 432
25. Rothe M, Rothe J (1989) Physical Data of Oligomers. In: Brandup J, Immergut EH (eds) Polymer Handbook, 3rd edn, vol IV. Wiley, New York, p 1
26. Tour JM (1996) Chem Rev 96:537
27. Staudiger H, Luthy M (1925) Helv Chim Acta 8:41
28. Papadopoulos P, Floudas G, Chi C, Wegner G (2004) J Chem Phys 120:2368
29. Jenekhe SA (1990) Macromolecules 23:2848

30. Hutchison GR, Zhao YJ, Delley B, Freeman AJ, Ratner MA, Marks TJ (2003) Phys Rev B 68:035204
31. Guay J, Kasai P, Diaz A, Wu R, Tour J, Dao LH (1992) Chem Mater 4:1097
32. Elandaloussi EH, Frére P, Richomme P, Orduna J, Garin J, Roncali J (1997) J Am Chem Soc 119:10774
33. Mulazzi E, Ripamonti A, Athouël L, Wery J, Lefrant S (2002) Phys Rev B 65:085204
34. Murphy AR, Fréchet JM (2007) Chem Rev 107:1066
35. Geng Y, Trajkovska A, Katsis D, Ou JJ, Culligan SW, Chen SH (2002) J Am Chem Soc 124:8337
36. Gaylord BS, Wang S, Heeger AJ, Bazan GC (2001) J Am Chem Soc 123:6417
37. Fakis M, Polyzos I, Tsigaridas G, Giannetas V, Persephonis P, Spiliopoulos I, Mikroyannidis J (2004) Opt Mater 27:503
38. Katz HE, Bao Z, Gilat SL (2001) Acc Chem Res 34:359
39. Meskers SCJ, Janssen RAJ, Haverkort JEM, Wolter JH (2000) Chem Phys 260:415
40. Kline RJ, McGehee MD, Kadnikova EN, Liu J, Fréchet JM (2003) Adv Mater 15:1519
41. Menon A, Dong H, Niazimbetova ZI, Rothberg LJ, Galvin ME (2002) Chem Mater 14:3668
42. Pudzich R, Fuhrmann-Lieker T, Salbeck J (2006) Adv Polym Sci 199:83
43. Saragi TPI, Spehr T, Siebert A, Fuhrmannn-Lieker T, Salbeck J (2007) Chem Rev 107:1011
44. Woo EP, Inbasekaran M, Shiang WR, Roof GR (1997) Int Patent WO 9 705 184
45. Clarkson RG, Gomberg M (1930) J Am Chem Soc 52:2881
46. Wu R, Schumm JS, Pearson DL, Tour JM (1996) J Org Chem 61:6906
47. Geng Y, Katsis D, Culligan SW, Ou JJ, Chen SH, Rothberg LJ (2002) Chem Mater 14:463
48. Yamamoto T, Morita A, Miyazaki Y, Maruyama T, Wakayama H, Zhou Z-H, Nakamura Y, Kanbara T, Sasaki S, Kubota K (1992) Macromolecules 25:1214
49. Miyaura N, Suzuki A (1995) Chem Rev 95:2457
50. Scholl R, Seer C, Weitzenböck R (1910) Ber Deutsch Chem Ges 43:2202
51. Fukuda M, Sawada K, Yoshino K (1993) J Polym Sci A 31:2465
52. Klaerner G, Miller RD (1998) Macromolecules 31:2007
53. Scheler E, Strohriegl P (2007) Liq Cryst 34:667
54. Katsis D, Geng YH, Ou JJ, Culligan SW, Trajkovska A, Chen SH, Rothberg LJ (2002) Chem Mater 14:1332
55. Thiem H, Jandke M, Hanft D, Strohriegl P (2006) Macromol Chem Phys 207:370
56. Anémian R, Mulatier J-C, Andraud C, Stéphan O, Vial J-C (2002) Chem Commun 1608
57. Jo J, Chi C, Höger S, Wegner G, Yoon DY (2004) Chem Eur J 10:2681
58. Dudek SP, Pouderoijen M, Abbel R, Schenning APH, Meijer EW (2005) J Am Chem Soc 127:11763
59. Lee SH, Tsutsui T (2000) Thin Solid Films 363:76
60. Geng Y, Culligan SW, Trajkovska A, Wallace JU, Chen SH (2003) Chem Mater 15:542
61. Ishiyama T, Murata M, Miyaura N (1995) J Org Chem 60:7508
62. Hayashi T, Konishi M, Kobori Y, Kumada M, Higuchi T, Hirotsu K (1984) J Am Chem Soc 106:158
63. Lee SH, Nakamura T, Tsutsui T (2001) Org Lett 3:2005
64. Geng Y, Chen ACA, Ou JJ, Chen SH, Klubek K, Vaeth KM, Tang CW (2003) Chem Mater 15:4352
65. Stille JK (1986) Angew Chem Int Ed Engl 25:508
66. Zhang X, Tian H, Liu Q, Wang L, Geng Y, Wang F (2006) J Org Chem 71:4332
67. Liu Q, Liu W, Yao B, Tian H, Xie Z, Geng Y, Wang F (2007) Macromolecules 40:1851

68. Zhang X, Qu Y, Bu L, Tian H, Zhang J, Wang L, Geng Y, Wang F (2007) Chem Eur J 13:6238
69. Li ZH, Wong MS, Fukutani H, Tao Y (2005) Chem Mater 17:5032
70. Wong KT, Wang C-F, Chou CH, Su YO, Lee G-H, Peng S-M (2002) Org Lett 4:4439
71. Xie L-H, Fu T, Hou X-Y, Tang C, Hua Y-R, Wang R-J, Fan Q-L, Peng B, Wei W, Huang W (2006) Tetrahedron Lett 47:6421
72. Zhao Z, Yu S, Xu L, Wang H, Lu P (2007) Tetrahedron 63:7809
73. Kim Y-H, Shin D-C, Kim S-H, Ko C-H, Yu H-S, Chae Y-S, Kwon S-K (2001) Adv Mater 13:1690
74. Chen ACA, Culligan SW, Geng Y, Chen SH, Klubek KP, Vaeth KM, Tang CW (2004) Adv Mater 16:783
75. Jaramillo-Isaza F, Turner ML (2006) J Mater Chem 16:83
76. Pogantsh A, Zaami N, Slugovc C (2006) Chem Phys 322:399
77. Kulkarni AP, Kong X, Jenekhe SA (2004) J Phys Chem B 108:8689
78. Chi C, Im C, Enkelmann V, Ziegler A, Lieser G, Wegner G (2005) Chem Eur J 11:6833
79. Zhou X-H, Zhang Y, Xie Y-Q, Cao Y, Pei J (2006) Macromolecules 39:3830
80. Dias FB, Pollock S, Hedley G, Pålsson L-O, Monkman A, Perepichka II, Perepichka IF, Tavasli M, Bryce MR (2006) J Phys Chem B 110:19229
81. Lattante S, Anni M, Salerno M, Lagonigro L, Cingolani R, Gigli G, Pasini M, Destri S, Porzio W (2006) Opt Mater 28:1072
82. Li Y, Ding J, Day M, Tao Y, Lu J, D'iorio M (2003) Chem Mater 15:4936
83. Koizumi Y, Seki S, Tsukuda S, Sakamoto S, Tagawa S (2006) J Am Chem Soc 128:9036
84. Wang S, Liu B, Gaylord BS, Bazan GC (2003) Adv Funct Mater 13:463
85. Jeeva S, Moratti SC (2007) Synthesis 2007:3323
86. Zhou X-H, Yan J-C, Pei J (2003) Org Lett 5:3543
87. Barsu C, Andraud C (2005) J Nonlinear Opt Phys 14:311
88. Kanibolotsky AL, Berridge R, Skabara PJ, Perepichka IF, Bradley DDC, Koeberg M (2004) J Am Chem Soc 126:13695
89. Montes VA, Pérez-Bolívar C, Agarwal N, Shinar J, Anzenbacher P Jr (2006) J Am Chem Soc 128:12436
90. Li B, Li J, Fu Y, Bo Z (2004) J Am Chem Soc 126:3430
91. Lai W-Y, Zhu R, Fan Q-L, Hou L-T, Cao Y, Huang W (2006) Macromolecules 39:3707
92. Yang BH, Buchwald SL (1999) J Organomet Chem 576:125
93. Promarak V, Saengsuwan S, Jungsuttiwong S, Sudyoadsuk T, Keawin T (2007) Tetrahedron Lett 48:89
94. Promarak V, Punkvuang A, Sudyoadsuk T, Jungsuttiwong S, Saengsuwan S, Keawin T, Sirithip K (2007) Tetrahedron 63:8881
95. Li ZH, Wong MS, Tao Y, Lu J (2005) Chem Eur J 11:3285
96. Xiong MJ, Li ZH, Wong MS (2007) Aust J Chem 60:608
97. Li ZH, Wong MS (2006) Org Lett 8:1499
98. Grisorio R, Dell'Aquila A, Romanazzi G, Suranna GP, Mastrorilli P, Cosma P, Acierno D, Amendola E, Ciccarella G, Nobile CF (2006) Tetrahedron Lett 62:627
99. Zhang Q, Chen JS, Cheng YX, Geng YH, Wang LX, Ma DG, Jing XB, Wang FS (2005) Synthetic Meth 152:229
100. Liu Y-L, Feng J-K, Ren A-M (2007) J Phys Org Chem 20:600
101. Li ZH, Wong MS, Fukutani H, Tao Y (2006) Org Lett 8:4271
102. Liu Q-D, Lu J, Ding J, Day M, Tao Y, Barrios P, Stupak J, Chan K, Li J, Chi Y (2007) Adv Funct Mater 17:1028
103. Wang H-Y, Feng J-C, Wen G-A, Jiang H-J, Wan J-H, Zhu R, Wang C-M, Wei W, Huang W (2006) New J Chem 30:667

104. Wong K-T, Chen R-T, Fang F-C, Wu C-C, Lin Y-T (2005) Org Lett 7:1979
105. Kong Q, Zhu D, Quan Y, Chen Q, Ding J, Lu J, Tao Y (2007) Chem Mater 19:3309
106. Chen AC-A, Wallace JU, Wei SK-H, Zeng L, Chen SH (2006) Chem Mater 18:204
107. Tang S, Liu M, Lu P, Xia H, Li M, Xie Z, Shen F, Gu C, Wang H, Yang B, Ma Y (2007) Adv Funct Mater 17:2869
108. Chochos CL, Kallitsis JK, Gregoriou VG (2005) J Phys Chem B 109:8755
109. Tsolakis P, Kallitsis JK (2003) Chem Eur J 9:936
110. Belletête M, Ranger M, Beaupré S, Leclerc M, Durocher G (2000) Chem Phys Lett 316:101
111. Beaupré S, Ranger M, Leclerc M (2000) Macromol Rapid Commun 21:1013
112. Chochos CL, Papakonstandopoulou D, Economopoulos SP, Gregoriou VG, Kallitsis JK (2006) J Macromol Sci A 43:419
113. Burnell T, Cella JA, Donahue P, Duggal A, Early T, Heller CM, Liu J, Shiang J, Simon D, Slowinska K, Sze M, Williams E (2005) Macromolecules 38:10667
114. Jiang G, Wu J, Yao B, Geng Y, Cheng Y, Xie Z, Wang L, Jing X, Wang F (2006) Macromolecules 39:7950
115. Jiang G, Yao B, Geng Y, Cheng Y, Xie Z, Wang L, Jin X, Wang F (2006) Macromolecules 39:1403
116. Zaami N, Slugovc C, Pogantsch A, Stelzer F (2004) Macromol Chem Phys 205:523
117. Hu X, Chuai Y, Wang F, Lao C, Luan WQ, Baik W, Huang L, Zou D (2006) Japan J Appl Phys 45:579
118. Yu X-F, Lu S, Ye C, Li T, Liu T, Liu S, Fan Q, Chen E-Q, Huang W (2006) Macromolecules 39:1364
119. Chen L-Y, Hung W-Y, Lin Y-T, Wu C-C, Chao T-C, Hung T-H, Wong K-T (2005) Appl Phys Lett 87:112103
120. Somma E, Loppinet B, Chi C, Fytas G, Wegner G (2006) Phys Chem Chem Phys 8:2773
121. Salbeck J, Schörner M, Fuhrman T (2002) Thin Solid Films 417:20
122. Geng Y, Trajkovska A, Culligan SW, Ou JJ, Chen HMP, Katsis D, Chen SH (2003) J Am Chem Soc 125:14032
123. Aldred MP, Eastwood AJ, Kitney SP, Richards GJ, Vlachos P, Kelly SM, O'Neill M (2005) Liq Cryst 32:1251
124. Woon KL, Contoret AEA, Farrar SR, Liedtke A, O'Neill M, Vlachos P, Aldred MP, Kelly SM (2006) J Soc Inf Disp 14:557
125. Marcon V, Van der Vegt N, Wegner G, Raos G (2006) J Phys Chem B 110:5253
126. Somma E, Chi C, Loppinet B, Grinshtein J, Graf R, Fytas G, Spiess HW, Wegner G (2006) J Chem Phys 124:204910
127. Lin H-W, Lin C-L, Chang H-H, Lin Y-T, Wu C-C, Chen Y-M, Chen R-T, Chien Y-Y, Wong K-T (2004) J Appl Phys 95:881
128. Lin H-W, Lin C-L, Wu C-C, Chao T-C, Wong K-T (2005) Appl Phys Lett 87:071910
129. Han Y, Fei Z, Sun M, Bo Z, Liang W-Z (2007) Macromol Rapid Commun 28:1017
130. Locklin J, Li D, Mannsfeld SCB, Borkent E-J, Meng H, Advincula R, Bao Z (2005) Chem Mater 17:3366
131. Shin TJ, Yang H, Ling M-M, Locklin J, Yang L, Lee B, Roberts ME, Mallik AB, Bao Z (2007) Chem Mater 19:5882
132. DeLongchamp DM, Ling MM, Jung Y, Fischer DA, Roberts ME, Lin EK, Bao Z (2006) J Am Chem Soc 128:16579
133. Demus D, Richter L (1980) Textures of Liquid Crystals, Revised edn. VEB Deutscher Verlag für Grundstoffindustrie, Leipzig
134. Ishihara S (2005) IEEE J Disp Technol 1:30

135. Trajkovska A, Kim C, Marshall KL, Mourey TH, Chen SH (2006) Macromolecules 39:6983
136. Kim C, Wallace JU, Trajkovska A, Ou JJ, Chen SH (2007) Macromolecules 40:8924
137. Güntner R, Farrell T, Scherf U, Miteva T, Yasuda A, Nelles G (2004) J Mater Chem 14:2622
138. Chi C, Lieser G, Enkelmann V, Wegner G (2005) Macromol Chem Phys 206:1597
139. Chen AC-A, Wallace JU, Zeng L, Wei SKH, Chen SH (2005) Proc SPIE 5936:59360I-1
140. Fan FY, Culligan SW, Mastrangelo JC, Katsis D, Chen SH, Blanton TN (2001) Chem Mater 13:4584
141. Jiang H-J, Feng J-C, Wang H-Y, Wei W, Huang W (2006) J Fluorine Chem 127:973
142. Bliznyuk VN, Carter SA, Scott JC, Klärner G, Miller RD, Miller DC (1999) Macromolecules 32:361
143. List EJW, Guentner R, de Freitas PS, Scherf U (2002) Adv Mater 14:374
144. Becker K, Lupton JM, Feldmann J, Nehls BS, Galbrecht F, Gao D, Scherf U (2006) Adv Funct Mater 16:364
145. Craig MR, de Kok MM, Hofstraat JW, Schenning APHJ, Meijer EW (2003) J Mater Chem 13:2861
146. Zhao Z, Xu X, Jiang Z, Lu P, Yu G, Liu Y (2007) J Org Chem 72:8345
147. Perepichka II, Perepichka IF, Bryce MR, Pålsson L-O (2005) Chem Commun 3397
148. Culligan SW, Chen ACA, Wallace JU, Klubek KP, Tang CW, Chen SH (2006) Adv Funct Mater 16:1481
149. He F, Xia H, Tang S, Duan Y, Zeng M, Liu L, Li M, Zhang H, Yang B, Ma Y, Liu S, Shen J (2004) J Mater Chem 14:2735
150. Li ZH, Wong MS, Tao Y, Fukutani H (2007) Org Lett 9:3659
151. Destri S, Pasini M, Porzio W, Giovanella U, Gigli G, Pisignano D (2003) Synthetic Meth 135/136:409
152. Huang J, Li C, Xia Y-J, Zhu X-H, Peng J, Cao Y (2007) J Org Chem 72:8580
153. Tavasli M, Bettington S, Bryce MR, Al Attar HA, Dias FB, King S, Monkman AP (2005) J Mater Chem 15:4963
154. Chi C, Im C, Wegner G (2006) J Chem Phys 124:024907
155. Wasserberg D, Dudek SP, Meskers SCJ, Janssen RAJ (2005) Chem Phys Lett 411:273
156. Song J, Liang WZ, Zhao Y, Yang J (2006) Appl Phys Lett 89:071917
157. Belfield KD, Bondar MV, Morales AR, Yavuz O, Przhonska OV (2003) J Phys Org Chem 16:194
158. Hintschich SI, Dias FB, Monkman AP (2006) Phys Rev B 74:045210
159. Tirapattur S, Belletête M, Drolet N, Bouchard J, Ranger M, Leclerc M, Durocher G (2002) J Phys Chem B 106:8959
160. Chen H-L, Huang Y-F, Hsu C-P, Lim T-S, Kuo L-C, Leung M-K, Chao T-C, Wong K-T, Chen S-A, Fann W (2007) J Phys Chem A 111:9424
161. Tsoi WC, O'Neill M, Aldred MP, Vlachos P, Kelly SM (2007) J Chem Phys 127:114901
162. Cabanillas-Gonzalez J, Antognazza MR, Virgili T, Lanzani G (2005) Phys Rev B 71:155207
163. Cabanillas-Gonzalez J, Antognazza MR, Virgili T, Lanzani G, Sonntag M, Strohriegl P (2005) Synthetic Meth 152:113
164. Zhou X, Feng J-K, Ren A-M (2004) Chem Phys Lett 397:500
165. Liao Y, Feng J-K, Yang L, Ren A-M, Zhang H-X (2005) Organometallics 24:385
166. Yang L, Feng J-K, Ren A-M (2005) J Org Chem 70:5987
167. Liang WZ, Zhao Y, Sun J, Song J, Hu S, Yang J (2006) J Phys Chem B 110:9908
168. Jansson E, Jha PC, Ågren H (2007) Chem Phys 336:91

169. Millaruelo M, Oriol L, Piñol M, Sáez PL, Serrano JL (2003) J Photochem Photobiol A 155:29
170. Chi C, Wegner G (2005) Macromol Rapid Commun 26:1532
171. Wu C-I, Lee G-R, Lin C-T, Chen Y-H, Hong Y-H, Liu W-G, Wu C-C, Wong K-T, Chao T-C (2005) Appl Phys Lett 87:242107
172. Promarak V, Punkvuang A, Jungsuttiwong S, Saengsuwan S, Sudyoadsuk T, Keawin T (2007) Tetrahedron Lett 48:3661
173. Li T, Yamamoto T, Lan H-L, Kido J (2004) Polym Adv Technol 15:266
174. Omer KM, Kanibolotsky AL, Skabara PJ, Perepichka IF, Bard AJ (2007) J Phys Chem B 111:6612
175. Oliva MM, Casado J, Navarrete JTL, Berridge R, Skabara PJ, Kanibolotsky AL, Perepichka IF (2007) J Phys Chem B 111:4026
176. Chen AC-A, Madaras MB, Klubek KP, Wallace JU, Wei SK-H, Zeng LC, Blanton TN, Chen SH (2006) Chem Mater 18:6083
177. Koizumi Y, Seki S, Acharya A, Saeki A, Tagawa S (2004) Chem Lett 33:1290
178. Koizumi Y, Seki S, Saeki A, Tagawa S (2007) Radiat Phys Chem 76:1337
179. Fratiloiu S, Grozema FC, Koizuma Y, Seki S, Saeki A, Tagawa S, Dudek SP, Siebbeles LDA (2006) J Phys Chem B 110:5984
180. Wu C-C, Liu T-L, Lin Y-T, Hung W-Y, Ke T-H, Wong K-T, Chao T-C (2004) Appl Phys Lett 85:1172
181. Hung W-Y, Ke T-H, Lin Y-T, Wu C-C, Hung T-H, Chao T-C, Wong K-T, Wu C-I (2006) Appl Phys Lett 88:064102
182. Chen L-Y, Ke T-H, Wu C-C, Chao T-C, Wong K-T, Chang C-C (2007) Appl Phys Lett 91:163509
183. Travis D (1991) Effective Color Displays: Theory and Practice. Academic Press, London
184. D'Andrade BW, Forrest SR (2004) Adv Mater 16:1585
185. Wu C-C, Lin Y-T, Wong K-T, Chen R-T, Chien Y-Y (2004) Adv Mater 16:61
186. Cheng G, Zhang Y, Zhao Y, Liu S, Tang S, Ma Y (2005) Appl Phys Lett 87:151905
187. Zhang Y, Cheng G, Zhao Y, Hou J, Liu S, Tang S, Ma Y (2005) Appl Phys Lett 87:241112
188. Gao ZQ, Li ZH, Xia PF, Wong MS, Cheah KW, Chen CH (2007) Adv Funct Mater 17:3194
189. Saitoh A, Yamada N, Yashima M, Okinaka K, Senoo A, Ueno K (2004) SID Int Symp Dig Tech Pap 35 p 150
190. Chen ACA, Wallace JU, Klubek KP, Madaras MB, Tang CW, Chen SH (2007) Chem Mater 19:4043
191. Levermore PA, Xia R, Lai W, Wang XH, Huang W, Bradley DDC (2007) J Phys D Appl Phys 40:1896
192. Zhu R, Lai W-Y, Wang H-Y, Yu N, Wei W, Peng B, Huang W, Hou L-T, Peng J-B, Cao Y (2007) Appl Phys Lett 90:141909
193. Wyszecki G, Stiles WS (1982) Color Science: Concepts and Methods, Quantitative Data and Formulae. Wiley, New York
194. Chao T-C, Lin Y-T, Yang C-Y, Hung TS, Chou H-C, Wu C-C, Wong K-T (2005) Adv Mater 17:992
195. Cornelissen HJ, Huck HPM, Broer DJ, Picken SJ, Bastiaansen CWM, Erdhuisen E, Maaskant N (2004) SID Int Symp Dig Tech Pap 35 p 1178
196. Culligan SW, Geng Y, Chen SH, Klubek K, Vaeth KM, Tang CW (2003) Adv Mater 15:1176
197. Culligan SW (2006) PhD thesis, University of Rochester

198. Misaki M, Ueda Y, Nagamatsu S, Chikamatsu M, Yoshida Y, Tanigaki N, Yase K (2005) Appl Phys Lett 87:243503
199. Chung S-F, Wen T-C, Chou W-Y, Guo T-F (2006) Japan J Appl Phys 45:L60
200. Sakamoto K, Miki K, Misaki M, Sakaguchi K, Chikamatsu M, Azumi R (2007) Appl Phys Lett 91:183509
201. Vaenkatesan V, Wegh RT, Teunissen J-P, Lub J, Bastiaansen CWM, Broer DJ (2005) Adv Funct Mater 15:138
202. Kauffman JM, Kelley CJ, Ghiorghis A, Neister E, Armstrong L, Prause PR (1987) Laser Chem 7:343
203. Seliskar CJ, Landis DA, Kauffman JM, Aziz MA, Steppel RN, Kelley CJ, Qin Y, Ghiorghis A (1993) Laser Chem 13:19
204. Spehr T, Pudzich R, Fuhrmann T, Salbeck J (2003) Org Electron 4:61
205. Lin H-W, Lin C-L, Wu C-C, Chao T-C, Wong K-T (2007) Org Electron 8:189
206. Meng H, Bao Z, Lovinger AJ, Wang B-C, Mujsce AM (2001) J Am Chem Soc 123:9214
207. Noh Y-Y, Kim D-Y, Yoshida Y, Yase K, Jung B-J, Lim E, Shim H-K, Azumi R (2004) Appl Phys Lett 85:2953
208. Noh Y-Y, Kim D-Y, Yoshida Y, Yase K, Jung B-J, Lim E, Shim H-K, Azumi R (2005) J Appl Phys 97:104504
209. Meng H, Zheng J, Lovinger AJ, Wang B-C, Van Patten PG, Bao Z (2003) Chem Mater 15:1778
210. Tang ML, Roberts ME, Locklin JJ, Ling MM, Meng H, Bao Z (2006) Chem Mater 18:6250
211. Iosip MD, Destri S, Pasini M, Porzio M, Pernstich KP, Batlogg B (2004) Synthetic Meth 146:251
212. Porzio W, Destri S, Giovanella U, Pasini M, Marin L, Iosip MD, Campione M (2007) Thin Solid Films 515:7318
213. Yasuda T, Fujita K, Tsutsui T, Geng Y, Culligan SW, Chen SH (2005) Chem Mater 17:264
214. Noh Y-Y, Azumi R, Goto M, Jung B-J, Lim E, Shim H-K, Yoshida Y, Yase K, Kim D-Y (2005) Chem Mater 17:3861
215. Shimizu Y, Oikawa K, Nakayama K-i, Guillon D (2007) J Mater Chem 17:4223
216. Van der Pol C, Bryce MR, Wielopolski M, Atienze-Castellanos C, Guldi DM, Filippone S, Martin N (2007) J Org Chem 72:6662
217. Carrasco-Orozco M, Tsoi WC, O'Neill M, Aldred MP, Vlachos P, Kelly SM (2006) Adv Mater 18:1754
218. Tsoi WC, O'Neill M, Aldred MP, Kitney SP, Vlachos P, Kelly SM (2007) Chem Mater 19:5475
219. Lupton JM, Craig MR, Meijer EW (2002) Appl Phys Lett 80:4489
220. Liu L, Tang S, Liu M, Xie Z, Zhang W, Lu P, Hanif M, Ma Y (2006) J Phys Chem B 110:13734
221. Kondakov DY, Lenhart WC, Nichols WF (2007) J Appl Phys 101:024512
222. Montilla F, Mallavia R (2007) Adv Funct Mater 17:71

Polyfluorene Photophysics

Andy Monkman (✉) · Carsten Rothe · Simon King · Fernando Dias

Organic Electroactive Polymer Research Group, Durham University, South road, Durham DH1 3LE, UK
a.p.monkman@durham.ac.uk

1	Basic Optical Properties	188
1.1	Absorption	188
1.2	Emission	189
1.3	Fluorescence Lifetime	192
1.4	Phosphorescence	194
1.5	Transient Absorption	197
1.5.1	Singlet Excited State Absorption S_1–S_n	198
1.5.2	Triplet Excited State Absorption T_1–T_n	198
1.5.3	Absorption of the Polyfluorene Charged State (Polaron)	199
1.6	Defect Emission	200
2	Exciton Dynamics	204
2.1	Singlet Migration	204
2.2	Triplet Diffusion	205
2.3	Defect Trapping	207
2.4	Amplified Spontaneous Emission	209
3	Exciton–Exciton Interactions	210
3.1	Singlet–Singlet Annihilation	210
3.2	Triplet–Triplet Annihilation and Delayed Fluorescence	213
3.3	Singlet–Triplet Annihilation	216
4	The "Beta" Phase of Polyfluorene	216
4.1	Absorption and Emission	216
4.2	Effect of Alkyl Chain Length	218
4.3	Amplified Spontaneous Emission	219
	References	220

Abstract The fundamental optical properties and processes that occur in the polyfluorenes are discussed. Details are given on the production of singlet and triplet excitons, their emissive decay channels via fluorescence and phosphorescence and non emissive decay via quenching, together with exciton lifetime and quantum yields of production. The interaction of excitons in these polymers is then discussed at length describing exciton migration dynamics, exciton–exciton annihilation processes and delayed fluorescence and the trapping of excitons at defect sites. Photogenerated charge state production is described along with intra-chain charge transfer states in fluorene co-polymers. Finally, sections describing the optical properties of the beta-phase of polyfluorene are given along with amplified spontaneous emission which occurs at high beta phase exciton populations.

Keywords ASE · Charge pairs · Exciton dynamics · Fluorescence lifetime · Keto defects · Phosphorescence · Polyfluorene beta-phase

1
Basic Optical Properties

As the archetypical "blue" emitting polymer, polyfluorene has been studied extensively using a wealth of optical techniques to elucidate its photophysical properties. Moreover, as it is a simple homopolymer that can be synthesized to very high purity, it offers a unique platform to help us understand many aspects of the excited state behavior of conjugated polymers in general. Here we address the main properties associated with excited state, i.e. exciton, behavior and the many possible interactions between excitons that can occur, although this is by no means an exhaustive review.

1.1
Absorption

Measuring the steady state optical absorption and emission properties of any luminescent polymer is the most basic but fundamental photophysical measurement we can make. Figure 1 depicts both the absorption and emission spectra for a series of polyfluorene oligomers and poly[9,9-di-n-(2-ethylhexyl)fluorene] (PF2/6), along with the fully rigid ladder type MeLPPP.

Upon photo-excitation, any excess vibrational energy of the fluorophores is transferred to their environment on a subpicosecond time scale [1]. Through this process of vibrational cooling, emission originates from the $S_0^n \leftarrow S_1^0$ transition in accordance with Kasha's rule, causing a Stokes shift, i.e. an energy separation between the peak in absorption and peak of emission. After this rapid vibrational relaxation the polymer can undergo further structural relaxation and exciton migration through the energetic disorder [2–4] of the bulk polymer causing an additional ("apparent") Stokes shift. These latter processes also break the mirror symmetry between $S_0^0 \rightarrow S_1^n$ and $S_0^n \leftarrow S_1^0$ spectra (absorption and emission) as absorption is an instantaneous process allowing all possible states to absorb, whereas on the whole only the lowest energy states emit. Thus, absorption spectra of conjugated polymers are commonly structureless and broad, whereas emission spectra can be well-resolved even at room temperature.

From Fig. 1 it can be seen that as the number of repeat units increases both absorption and emission energies red shift but saturate at around 5–7 repeat units, also the Stokes shift decreases for longer oligomers. This is a result of the strong influence of ring rotations on the excited states and the degree to

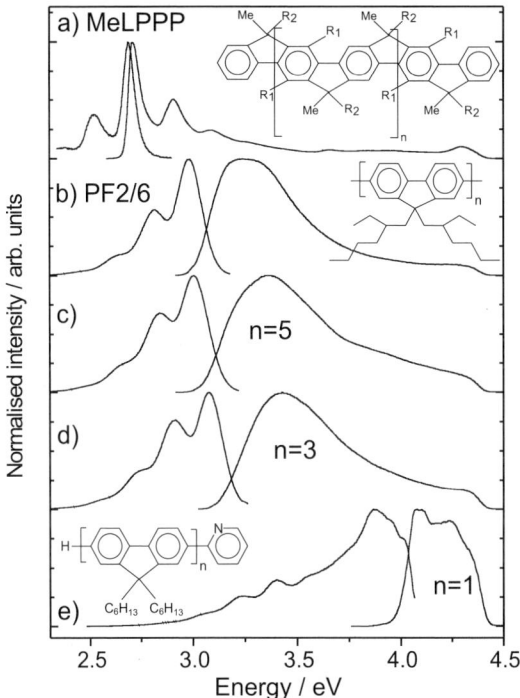

Fig. 1 The steady state absorption and emission spectra of (**a**) methyl ladder type polymer, MeLPPP (**b**) poly[9,9-di-*n*-(2 ethylhexyl)fluorene] (PF2/6, $n \sim 60$) (**c**), pentafluorene (**d**) trifluorene and (**e**) monofluorene measured in dilute toluene solution are compared. The insets depict the chemical structures of each. All spectra were measured in dilute MCH solution

which the rings are free to rotate has a large impact on transition energy, vibronic structure and Stokes shift [4, 5]. This is most clearly exemplified by the fully rigid analogue of PF2/6, ladder type MeLPPP, where no ring rotations can occur. Here the spectra are red shifted with respect to PF2/6 and also both absorption and emission bands are highly structured, but there is virtually zero Stokes shift.

1.2
Emission

Comparing thin film to solution state, we find that there is very little change to the absorption spectrum apart from a small red shift of the whole absorption band, although in solution the band position can be both solvent and time dependent due to conformational changes of the PF backbone in solution [6]. Typical emission spectra of PF2/6 100 nm film at both 295 K and 20 K, excitation at 3.5 eV (10 uJ, 200 ps pulse excitation) are shown in Fig. 2. At low temperature, the clear vibronic replicas of the electronic transition centered at 2.925 eV are seen. At room temperature this progression is still clearly visible but not as well resolved. A constant energy separation of 175 meV is observed which is associated with a C=C bond stretching mode [7], although high-resolution site-specific fluorescence measurements reveal each replica to be made up of at least three distinct modes [2]. One also sees the characteristic increase of quantum yield at low temperature, rising from ca. 0.3 at 295 K, as measured in an integrating sphere at room temperature [8] to 0.7 at 20 K obtained by comparing areas under the emission spectrum for an estimate of the yield at 20 K. It should be noted also that these values vary greatly between different samples of polyfluorenes and the history of the polymer before measurement. Spectra measured in solution are very similar to that seen in films at low temperature [4].

The lack of mirror symmetry between absorption and emission is indicative of nuclear displacements following excitation [9]. Differences in ground and excited state geometry lead to conformational rearrangement and, for solution samples, to local solvent relaxation following excitation. Both Sluch et al. [10] and Karabunarliev et al. [5, 11] report that the excited state potential of conjugated systems favors ordered planar structures much more strongly than in the ground state. A high degree of torsion [12, 13] between the repeat units and a wide distribution of torsional angles translates directly into a high excited state energy [14] and large energetic spread of excited states (density of states distribution, DOS). Rapid energy relaxation down the DOS gives rise to a red-shift of the emission energy. Consistent with theoretical prediction, the experimentally observed Stokes shifts are pronounced for the highly twisted polyfluorenes and polyphenylenevinylenes (PPV) [13]

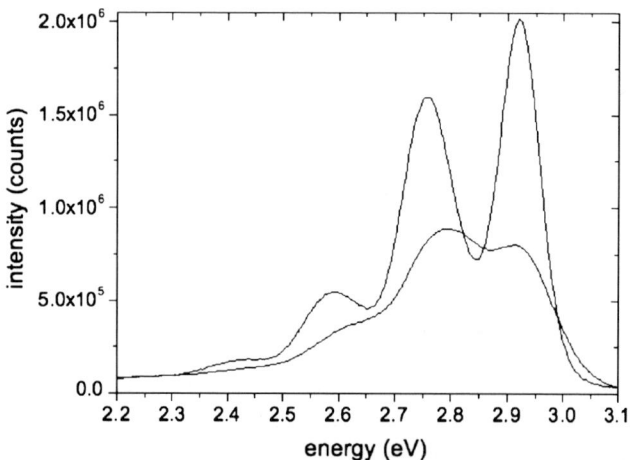

Fig. 2 Room temperature and well resolved 20 K thin film fluorescence spectra of PF2/6

in contrast to the extremely low Stokes shift observed for molecules chemically forced into a planar conformation, i.e. rigid-rod polyparaphenylenes (LPPPs) [15] as seen in Fig. 1. Exciton migration through the DOS will also lead to energetic relaxation and spectral red shifts and will be discussed in Sect. 2.

The planarity of PF2/6 can be controlled by the local environment via its bulky side groups. In Fig. 3 the absorption and emission spectra of dilute PF2/6 in MCH is shown at room temperature and at 77 K (MCH being an excellent glass forming solvent). It is clearly seen that in the low temperature glass, steric distribution and reorganization is hindered and so the absorption band becomes structured and the emission band features sharpen [4].

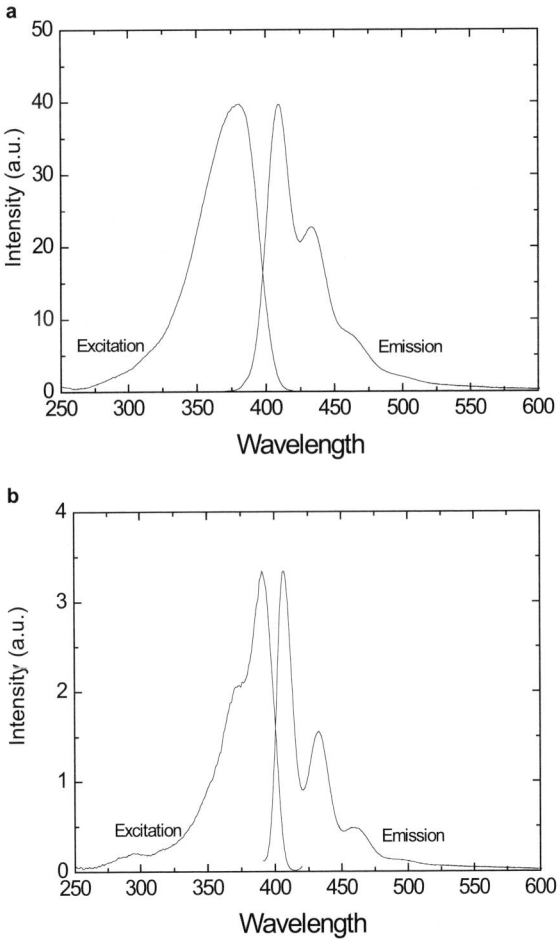

Fig. 3 Excitation and emission spectra of PF2/6 in dilute MCH solution at 295 K (**a**) and 77 K (**b**). Excitation wavelength of the fluorescence measurements was 380 nm

More importantly, the ratio of the intensity of the 0–0 to 0–1 vibrational modes increases. This indicates a decreased Huang–Rhys factor and is ascribed to a low configurational relaxation in the excited state [16]. Thus, torsional motion is hindered by the viscosity of the solvent (glass) and so the excited state has a geometry very similar to that of the ground state. This is also seen in the rigid rod ladder polymer, Fig. 1, where torsional motion is prevented by the inter chain bridging bonds. In this case the line width is narrower but still inhomogeneously broadened, which arises as a result of polydispersity of conjugation lengths and dynamic variation of conjugation lengths [13].

1.3
Fluorescence Lifetime

For most luminescent polymers fluorescence lifetimes lie in the range of 100 ps to 2 ns with multiple decay components, and so the most accurate way to measure this is by the time correlated single photon counting (TCSPC) technique. This is one of the most popular methods to determine excited state lifetimes. The technique makes use of modern pulsed lasers with a very high repetition rate (~80 MHz), typically mode locked Ti:Sapphire lasers. A portion of the excitation pulse is used to generate a START impulse for a time to amplitude converter (TAC) and the remaining light used to excite the sample. The first detected photons emitted by the sample are then sampled by a constant fraction discriminator to generate a pulse that STOPs the TAC. A signal proportional to the elapsed time between START and STOP signals is then produced and the event collected in a memory location with an address proportional to the detection time. After averaging many 1000s of pulses, a histogram of the emission decay is built up which represent the decay curve of the emission. An excellent review on the technique has be written by Becker [17].

Polyfluorenes, like other luminescent polymers show complex fluorescence decays both in thin film and solution, with minor amplitude fast rise and decay components together with a predominant decay time. Although there is still some controversy regarding the origin of the fast components, there is already strong evidence that fast conformational relaxation and energy migration are the cause of this complex behavior. For example, poly(9,9-di-*n*-octylfluorene) (PFO), usually shows a complex decay where sums of two or even three exponential functions are required to achieve excellent fits. This also depends on emission collection wavelength as well.

As a typical example, Fig. 4 shows the decay of PFO in dilute (10^{-5} M) toluene solution with the emission collected at two different wavelengths. A fast component of around 16 ps is found which is more important (greater amplitude) on the onset of the emission band (blue edge), and loses importance when going to longer wavelengths (red edge). Depending on viscosity

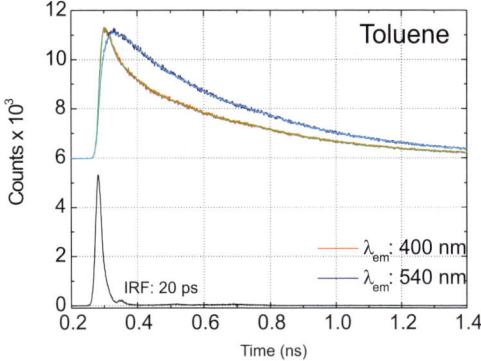

Fig. 4 Time-resolved fluorescence emission decays of PFO in toluene solution obtained at 295 K, with emission collected at 400 nm and 540 nm. Also shown is the instrument response function (IRF) which is deconvoluted with the decay fits to yield a time resolution of 3 ps for the fitted decays

and temperature, this component also appears as a rise time (negative amplitude) when the decay is collected in the emission tail and is the signature of fast conformational relaxations of the polymer backbone. In the rigid ladder polymer, MeLPPP, such fast rise and decay components are not observed, whereas in small fluorene trimers (in dilute solution where processes such as energy migration are not possible) the fast decay and rise components can be readily observed [4, 18, 19]. A minor intermediate component is also observed around 90 ps, which loses importance when going from short to longer emission wavelengths, and is ascribed to weak quenching at impurity sites. A predominant decay time around 360 ps is observed independent of the emission wavelength and attributed to the PFO fluorescence lifetime. Given that the typical PLQY in toluene is 0.8, gives a radiative lifetime of ca. 450 ps. A fast rise of the emission, before decay takes over is clearly observed when the emission is collected at 540 nm. Table 1 summarizes the fitting results, decay times and amplitudes, obtained from time resolved fluorescence decays of PFO in toluene solution and in other organic solvents.

In thin films a similar pattern is observed at short wavelengths, see Fig. 5. The decay is composed of a sum of three exponentials, see Table 2. However, at very long wavelengths, the decay is almost completely dominated by a long component of around 3 ns, attributed to the presence of photooxidized, keto defects or other emissive defects, which are very easily populated by efficient energy migration. This is described in Sect. 1.6. The fast component appears as a rise time of the emission intensity and a component with 349 ps indicates the PFO lifetime.

Table 1 Decay times and amplitudes obtained from time-resolved fluorescence decays of PF2/6 and PFO

Compound	Solvent	λ_{em} (nm)	τ_3 (ps): (A_3)	τ_2 (ps): (A_2)	τ_1 (ps): (A_1)
PF2/6	MCH	400	**40**: (0.11)	–	**371**: (0.89)
		446	**41**: (– 0.29)	–	**373**: (1.00)
PFO	MCH	400	**15**: (0.49)	**71**: (0.07)	**340**: (0.44)
	toluene	400	**16**: (0.54)	**89**: (0.07)	**363**: (0.39)
		540	**16**: (– 0.10)	–	**360**: (1.00)
	chloroform	417			**360**: (1.00)

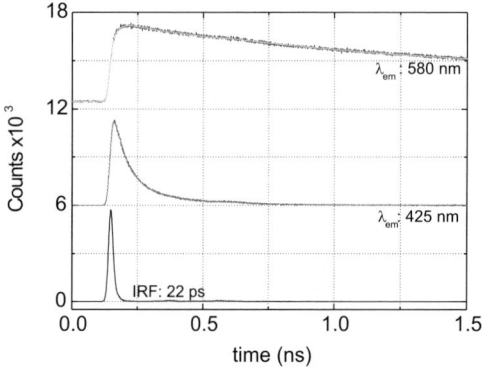

Fig. 5 Time-resolved fluorescence emission decays of PFO films obtained at 295 K, with emission collected at 425 nm and 580 nm. Again IRF also shown

Table 2 Decay times and amplitudes obtained from time-resolved fluorescence decays of PFO in film

		λ_{em} (nm)	τ_3 (ps): (A_3)	τ_2 (ps): (A_2)	τ_1 (ps): (A_1)
PFO	film	425	**36**: (0.57)	**109**: (0.41)	**403**: (0.02)
		580	**28**: (– 0.25)	**349**: (0.28)	**3050**: (0.72)

1.4
Phosphorescence

For all device applications based on organic materials, the energy of the first excited triplet state is of great importance as it is this state that is mainly populated during electrical excitation i.e. via charge recombination [20–22]. An obvious way to measure the triplet energy is by phosphorescence (Ph) detection, i.e. detection of the radiative decay of the first excited triplet state to the singlet ground state. This however, is not easy for most of the polyfluorenes, as the transition is formally spin forbidden. In consequence, the radiative lifetime of the triplet state is, under most experimental conditions much longer than the non-radiative lifetime thus diffusion activated triplet quenching dominates the decay of the state. To get around this, triplet diffusion has to be restricted, usually by performing the measurements at low temperatures, typically below 20 K. Under continuous excitation, the considerably stronger fluorescence long wavelength tail would completely mask the weak phosphorescence signal. Therefore, pulsed excitation in combination with highly sensitive gated spectroscopy has to be applied, where the collection of the spectrum starts some micro- or milliseconds after the excitation pulse [23–25]. Alternatively, a non-optical excitation method can be used such as pulse radiolysis-energy transfer to measure triplet energies, especially in cases where no phosphorescence can be detected [26, 27].

Figure 6 shows a typical delayed luminescence spectrum of PF2/6 thin film after optical excitation. Delayed fluorescence, which appears at the identical spectral position (2.925 eV) as the prompt fluorescence, see Fig. 2, arises as a result of triplet–triplet annihilation, discussed in Sect. 3.2. The phosphorescence spectrum peaks around 2.2 eV, which sets a lower limit for the triplet energy of polyfluorene. Very similar spectra have been recorded in frozen solutions [24, 28], or using electrical instead of optical excitation pulses [25], and for nearly all other polyfluorene derivatives [23, 29]. Triplet formation in polyfluorenes has been confirmed to be solely via the process of inter system crossing [30] and as the spin orbit coupling of the homopolymer is very low explains why triplet yields are very low in polyfluorenes and most other luminescent polymers [31]. Recently, King et al. succeeded in measuring the triplet yield and intersystem crossing rate for a polyspirofluorene directly in both solution and thin film at low temperature, by developing

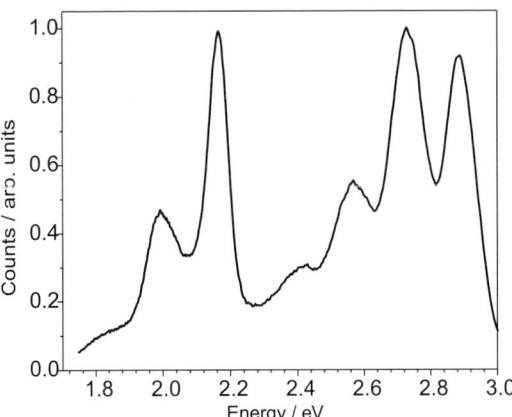

Fig. 6 Delayed luminescence spectrum of a PF2/6 film taken at 20 K with 30 ms delay after excitation and 30 ms detection window. Excitation was provided at 355 nm with 170 ps pulse width using a Nd:YAG laser

a technique based on the femtosecond to nanosecond time resolved measurement of ground state recovery. In solution the triplet yield was found to be 0.05 ± 0.01 with a rate, $k_{ISC} = 5.4 \times 10^7$ s^{-1}, in films this rises to 0.12 ± 0.02 and is temperature independent. The difference is ascribed to triplet formation from photo generated charge state recombination, see Sect. 3.1. This new methodology gives a powerful way to study triplet creation in films for the first time [32].

As noted above, normally the ISC rates in all luminescent polymers and especially the polyfluorenes are very low giving triplet yields of less than 0.05. However, there is one special case when the ISC of the polymer can reach ca. 1. This is in the presence of a heavy atom containing dopant, typically an Ir based metal–organic complex. In these guest–host systems, even at low doping levels, the spin–orbit interaction between polymer chain and complex, mediated by π wavefunction overlap, is a resonantly enhanced long range spin–orbit coupling, effective over very large distances. Thus, the spin–orbit coupling on the polymer back bone in the region close to a dopant is very high causing fast efficient ISC. This leads to rapid fluorescence quenching of the excited polymer and concomitant high triplet population, with a yield approaching 1 [33].

Because of its spin-forbidden nature, the decay rate of the first excited triplet state is extremely low, and very difficult to quantify accurately. What can be measured is the phosphorescence lifetime. Typical long time decays for a poly(bi-spirofluorene) (PSBF) are shown in Fig. 6.

In order to arrive at meaningful (exponential) decay rates in such experiments, non-linear triplet quenching by diffusion, has to be avoided. For example, the initial accelerated decay in Fig. 7 is caused by bimolecular triplet–triplet annihilation dominating the decay of the triplets. Similarly, a faster decay is observed at higher temperature, where triplet exciton diffusion to quenching sites is faster than monomolecular decay. Nevertheless, by using low temperatures and low excitation doses exponential decay kinetics are observed yielding radiative decay rates as low as ~ 1 s^{-1}, which sets an upper limit for the triplet excited state lifetime [28, 34].

Of course the triplet lifetime could also be determined by probing the transient triplet absorption signal as a function of time [24, 30], see Sect. 1.5.2. In practice however, the very low signal to noise ratio of such experiments requires the use of rather high excitation doses, which can make triplet annihilation the dominant decay mechanism even in dilute solutions.

The polarization of the phosphorescence of polyfluorene is also of interest because it provides a clue to the quantum mechanical wavefunction of the triplet state and its relationship with the singlet ground state. In common with most planar aromatic systems, the phosphorescence is found to be predominantly polarized out of the plane of the phenyl rings [35]. This orientation of the triplet state perpendicular to the singlet is particularly important when considering the energy transfer between singlet and triplet states or

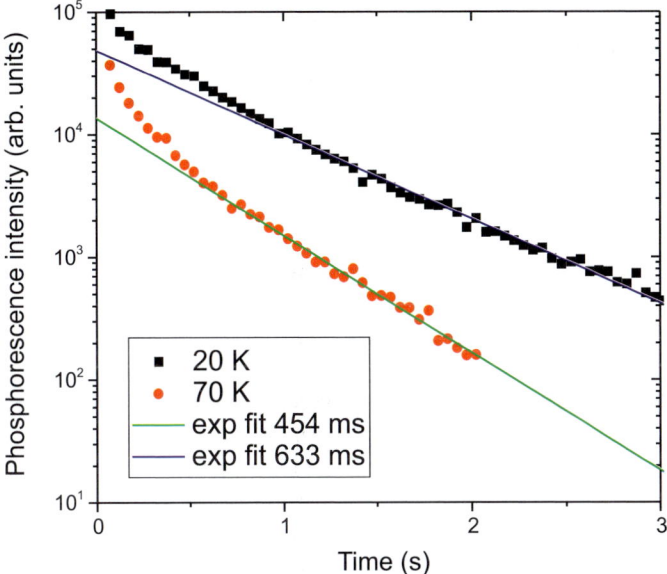

Fig. 7 Phosphorescence intensity as a function of time for a PSBF thin film after pulsed optical excitation and for two temperatures as indicated. The *solid lines* are exponential fits to the late part of the data sets yielding an estimate for the phosphorescence lifetime

the rates of triplet formation by intersystem crossing and the radiative decay rate. The spin forbidden nature of the triplet means that the crossing to the triplet state and its radiative decay are only allowed by mixing of the state with some states of singlet character. The predominantly out of plane nature of the first triplet state compared to the in plane nature of most of the singlet states makes mixing difficult and the rates of triplet formation and radiative decay concomitantly low.

1.5
Transient Absorption

As with the ground state of a molecule, the excited states also have characteristic optical absorption. In a simple system such as the polyfluorene homopolymers, these excited state absorptions can be characterized into three groups, the singlet excited state absorption, the excited state absorption of the first triplet state and the absorption of charged species. The techniques of excited state absorption spectroscopy are particularly useful for investigating non-emissive species such as triplet and charged states and generation from, and interaction with, singlet state. The techniques are also of interest to the device physicist as they are often used to investigate the exciton processes in devices.

1.5.1
Singlet Excited State Absorption S_1–S_n

In common with other conjugated polymers the excited state absorption of the singlet state in polyfluorene homopolymers is a sharp feature peaking just below 1.6 eV shown in Fig. 8. Singlet excited state absorption spectra are typically measured with an ultrafast pump probe system, which can resolve spectra and decays with sub-picosecond resolution [36, 37]. Verification of the feature as the singlet excited state induced absorption has been made by favorable comparison between the decay and excitation dependence of the absorption with the stimulated singlet emission signal and the fluorescence lifetime [38–41]. The excited state absorption of the singlet exciton has been used by many researchers to probe the population of singlet excitons in a number of different types of experiments where it is not possible to use more conventional fluorescence measurements such as at high excitation densities [41, 42], in device structures [39] or most pertinently, where ultrafast time resolution is required [43–45].

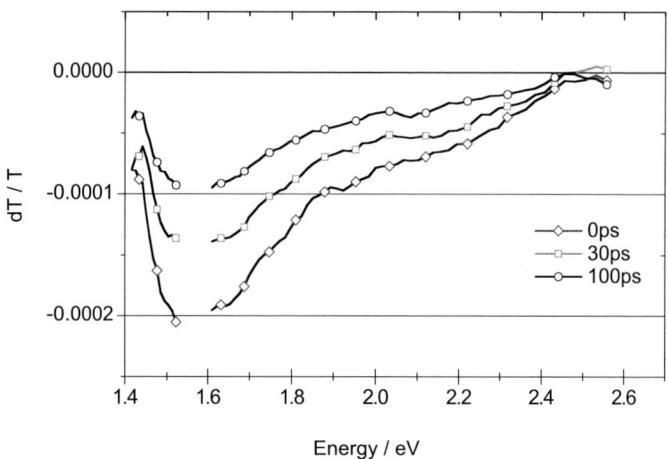

Fig. 8 The excited state absorption of the singlet exciton in PF2/6; the spectra are collected at different delay times after excitation to measure the decay of the state

1.5.2
Triplet Excited State Absorption T_1–T_n

The transient triplet absorption for the common polyfluorene PF2/6 is shown in Fig. 9, the spectra was measured by quasi-cw photoinduced absorption. This lock-in technique allows the absorption of long-lived states to be measured with high sensitivity. It is possible to estimate the lifetime of the state using the quasi-cw experiment [46]; however, more accurate methods in-

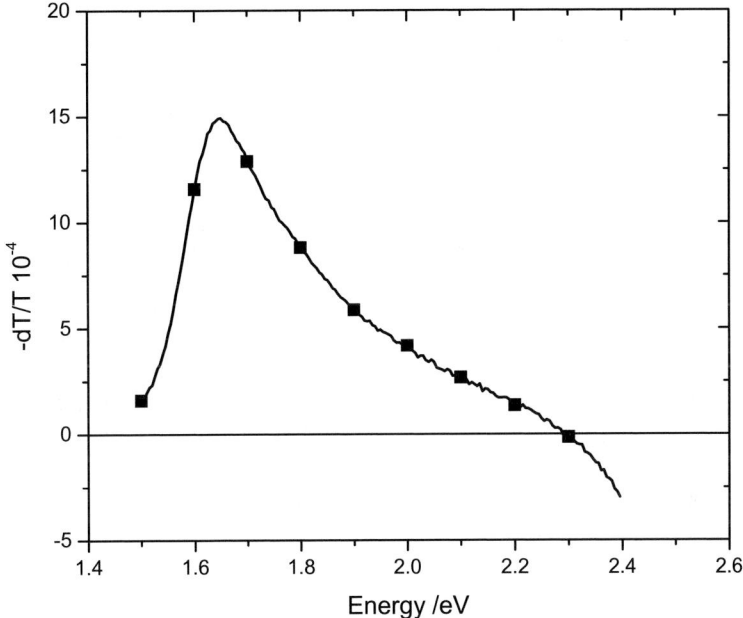

Fig. 9 Excited state absorption spectrum of the triplet exciton of PF2/6

cluding flash photolysis and single wavelength nanosecond photoinduced absorption are preferable given the possibility of multiexponential decays and bimolecular interactions such as triplet–triplet annihilation [24, 28]. Although the peak position at 1.65 eV appears similar to the absorption of the singlet excited state, the feature is confirmed as the absorption of the triplet state by analysis of the lifetime in aerated and degassed solutions. The lifetime is quenched from ∼2 ms to 200 ns when oxygen is present in the solution, confirming that the state is of triplet multiplicity [23, 30, 47, 48]. In frozen solution the dynamics of the triplet induced absorption compare favorably with the lifetime of the phosphorescence, also confirming the feature as the triplet induced absorption.

1.5.3
Absorption of the Polyfluorene Charged State (Polaron)

In addition to singlet and triplet excitons there is the possibility that under certain excitation conditions in thin film samples, notably high excitation powers, formation of other long-lived excited states is possible, these are charged species where enough excess energy after photoexcitation causes the electron and hole of the exciton to separate [49] leading to the formation of a pair of polaron-type charge states on adjacent chains, with the opposite sign of charge to conserve overall charge neutrality [50]. Polarons here being elec-

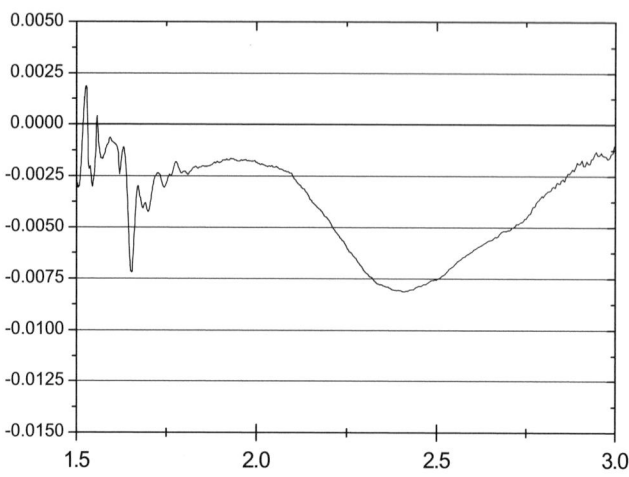

Fig. 10 Transient absorption spectrum of the polyfluorene charged state, in this case free carriers were generated by highly efficient singlet–singlet annihilation; similar spectra are obtained by field induced quenching of polyfluorene singlet states

trons (or holes) coupled to chain deformations [51]. Often these states are "dark" in that they do not directly decay resulting in the emission of light and as such, they are best investigated by absorption techniques. In solution it is also possible to generate charge species, for polyfluorene homopolymers these are a positively charged state of the polymer molecule, i.e. the radical cation, which can be formed by an electron transfer reaction with a strongly electron accepting solvent, such as chloroform. In addition charge states can be formed in thin films of polyfluorene by exciton dissociation in an applied electric field [52] or bimolecular annihilation of singlet excitons [41], the spectrum obtained for the charge states in solid state is also shown in Fig. 10 and this topic is covered in detail in Sect. 3.1.

1.6
Defect Emission

The high fluorescence quantum yield and good thermal stability makes polyfluorenes (PF) one of the more stable blue emitting luminescent polymers. However, it has been observed that light-emitting devices based on polyfluorene tend to degrade showing an undesirable green emission accompanied by an overall decrease of the luminescence intensity [53]. First attributed to interchain interactions such as aggregates and excimers [54], and later explained as resulting from oxidative (keto) defects formed along the polymer backbone and quenching the PF emission [55], the origin of the green emission is still a source of debate within the conjugated polymer community [56, 57]. Figure 11 shows the effect of annealing a PF2/6 film (100 °C

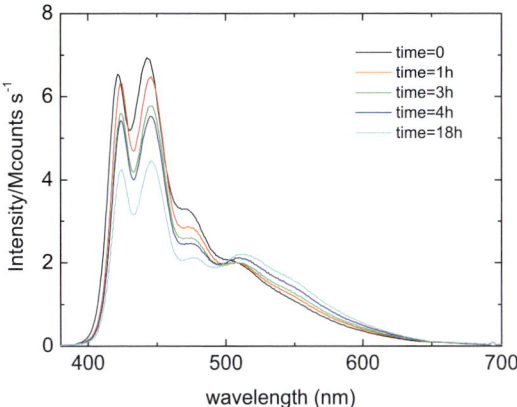

Fig. 11 Fluorescence emission spectra of PFO films obtained at 295 K after being annealed under air

in air), the intensity of the PF emission decreases and a green emission peaking around 530 nm grows over annealing time, showing an isoemissive point around 504 nm.

Synthetic fluorene/fluorenone copolymers [58] show fluorescence emission that matches very well that of degraded films giving strong support to the relation between keto defects and the green emission [57, 59]. Figure 12 shows the emission spectrum of PFO, 9-fluorenone and of a fluorene–fluorenone copolymer with 25% fluorenone groups randomly distributed along the polymer backbone in dilute toluene solution. The green emission does not match that of the simple 9-fluorenone emission, it appears at shorter wavelengths and shows a significantly lower intensity. The PFO emission intensity shows a pronounced decrease when in the presence of fluorenone in the copolymer chain, however this quenching effect does not occur when PFO and 9-fluorenone are simply mixed in solution.

The green emission appears at the expense of PFO emission, indicating that it is quenched by on-chain defects formed between the 9-fluorenone "keto" repeat units and their nearest neighbor fluorene repeat units, forming a delocalized charge transfer state (CTS defect) which decays radiatively to give the green emission [57, 59]. Figure 13, shows the decrease of the PFO emission intensity with increasing the fraction of 9-fluorenone repeat units randomly distributed along the polymer backbone, note the isoemissive point observed around 500 nm, between the PFO and green emissions.

Similar CT states can be formed when PF chains are copolymerized with good charge acceptors. This strategy is now frequently used to improve the charge balance in devices and can have profound impact on the photophysics of these materials via the creation of emissive charge transfer states. Figure 14 shows the absorption and emission spectra, in toluene and chloroform, two solvents with different polarity, of PFO containing dibenzothiophene-

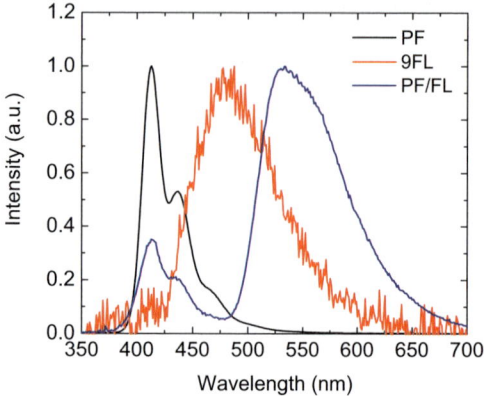

Fig. 12 Normalized fluorescence emission spectra of PFO (*black line*), 9-fluorenone (*red line*) and the fluorene–fluorenone copolymer with 25% 9-fluorenone units (PF/FL$_{0.25}$) (*blue line*)

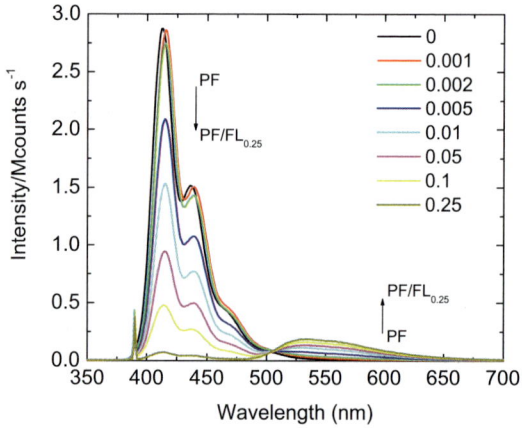

Fig. 13 Fluorescence emission spectra of fluorene–fluorenone copolymers with different 9-fluorenone fractions

S,S-dioxide (S) units randomly distributed along the co-polymer backbone (PFS) [60]. The S units are good electron acceptors and charge transfer between the fluorene moieties (F) and the S units occurs during the PFO excited state lifetime. The charge transfer reaction is controlled by the polarity of the surrounding medium and potentially stabilized by conformational changes of the copolymer backbone. As a result the emission profile of PFS copolymers changes from the typical well-resolved blue emission to a broad and red-shifted emission peaking around 460 nm, when going from non-polar to polar medium.

In the solid state, see Fig. 15, the emission of PFS copolymers also appears dependent on the S content. For high S percentages, the emission appears

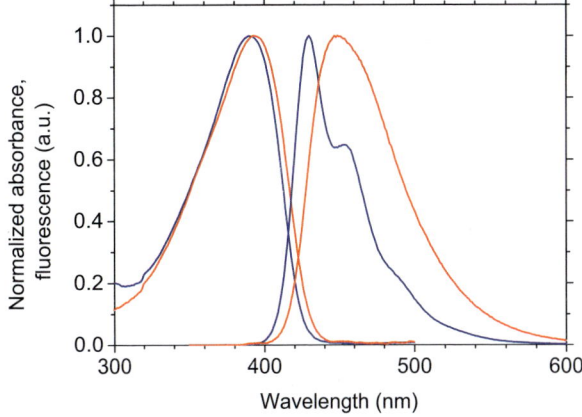

Fig. 14 Normalized absorption and emission spectra of PFS$_{0.3}$ in toluene (*blue line*) and chloroform (*red line*)

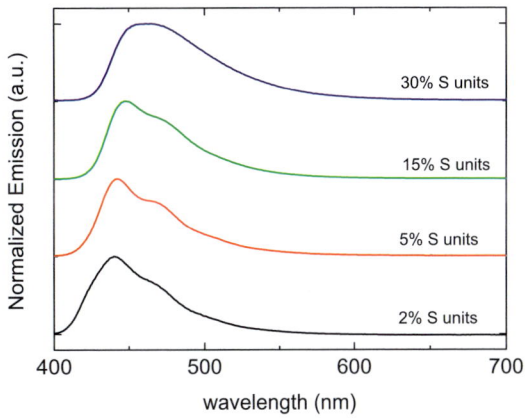

Fig. 15 Normalized fluorescence emission spectra of spin coated thin films on quartz substrates. From *bottom*: PFS$_{0.02}$ (9 mg/mL), PFS$_{0.05}$ (11.2 mg/mL), PFS$_{0.15}$ (9.6 mg/mL), PFS$_{0.3}$ (8.3 mg/mL)

structureless, as it is in a polar medium, but when the S content is low or absent, the emission is structured, as it is in a non-polar environment.

Although the fluorene units are non-polar, the dipole moment in the ground state of the S units is around 5.7 D and the charge transfer reaction increases this value in the excited state. Dipole–dipole interactions between FSF regions in the ground and excited states will thus interact with surrounding S groups of the polymer matrix especially at higher S levels leading to a larger inhomogeneous broadening effect on the emission spectrum of the polymer CT state; at low S levels, the effect is obviously reduced or even non-existent. Thus, the incorporation of co-repeat units that have strong acceptor or donor character can form stable CT excited states which profoundly alter the photophysics of the co-polymer as compared to the parent homopolymer.

2 Exciton Dynamics

2.1 Singlet Migration

The theoretical framework for exciton migration was established by Bässler and co-workers [61] and in the time-dependent experiments in solid state by Richert et al. [62] and Meskers et al. [2]. Here, energy transfer is associated with an exciton "hopping" motion between localized energetic sites along with initial early time longer range jumps and Forster energy transfer. If initially created with random energy within the inhomogeneously broadened DOS, excitons will preferentially migrate to sites of lower energy, i.e. to the tail states of the DOS. Such down-hill migration is naturally associated with a shift of the average emission energy. For every finite temperature, energetic relaxation terminates once thermally assisted jumps become equally probable to non-assisted jumps to lower energy sites. After this relaxation time no further energetic relaxation is expected. The final emission spectrum is red-shifted and well resolved. The process is also dispersive, as each hop reduces the energy of the exciton, it will take longer for the exciton to be able to make a subsequent jump, thus the hopping decelerates giving a non-linear rate for the process. The main problem with measuring the dynamics of singlet exciton migration is the short period over which the exciton lives before decaying, i.e. several 100 picoseconds to a few nanoseconds. This makes it very difficult to make accurate measurements of the full migration dynamics [63]. To this end, we have found that it is much more informative to study these processes on slower time scales afforded by studying triplet exciton migration which is discussed in the next section. However, by way of illustration to the complexity of the dynamics of singlet excitons, Fig. 16 shows the temperature dependence of Forster transfer between PF2/6 and a tetraphenyl porphyrin dopent [64]. Whereas the dipole–dipole coupling mechanism underlying Forster energy transfer [65] is independent of temperature, excitons at finite temperate may hop and so migrate to within the Forster radius of a trap. At room temperature this gives rise to an anomalously large Forster radius which is seen to decrease at low temperatures. From these measurements an estimate of the hopping rates for singlet excitons in polyfluorene can be derived, at 20 K 0.43 ± 0.15 nm^2/ps and at room temperature, 1.44 ± 0.25 nm^2/ps. These values translate into diffusion lengths of 12 nm at 20 K and 22 nm at room temperature. Similar complex migration dynamics are

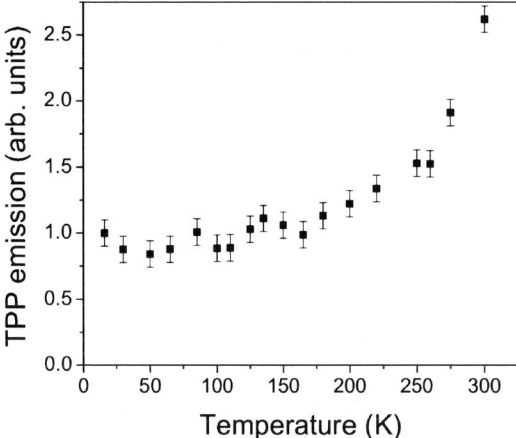

Fig. 16 Temperature dependent emission intensity from TPP dopants in a PF2/6 host. The host is optically excited and the TPP guest is populated by a combination of Forster transfer and exciton migration. Data is corrected for the temperature dependencies of the host and guest photoluminescence quantum yields

also found when studying exciton migration to on-chain keto defects, showing the complex nature of exciton motion in polyfluorenes [66].

2.2
Triplet Diffusion

As we have seen so far, the photophysics of polyfluorenes (and all conjugated polymers) is dominated by their inherent energetic and spatial disorder. This twofold disorder gives rise to time-dependent exciton diffusion i.e. dispersive diffusion so there is no single diffusion constant for the migration of excitons in these polymers. For singlet excitons who have rather short lifetimes, typically a few hundred picoseconds this is hard to show experimentally, but with triplet excitons which have much longer lifetimes this type of migration dynamics can be studied in detail and several earlier theoretical works [62, 67] on non-equilibrium diffusion and energy relaxation in localized state distributions have successfully been applied to triplet diffusion in polyfluorene derivatives [24, 68, 69]. In the framework of these theories the triplet diffusion is treated as a series of incoherent jumps among spatially and energetically localized states. Jumps downhill in energy only depend on the spectral separation of the sites, uphill jumps additionally require thermal activation energy. After pulsed excitation at random energy within the Gaussian DOS the diffusion of the triplet excitons evolve in two fundamentally different migration regimes. Firstly, motion is governed by fast energy relaxation towards low energy sites within the DOS. For the migrating triplet, neighboring sites that allow a next jump (with a lower energy) become fewer after each successful jump. Consequently, successive jumps to even lower energy sites take longer with the elapse of more and more time after excitation. This renders the diffusion time-dependent or dispersive. The exact behavior of the diffusion, $D(t)$, is strongly dependent on the width of the DOS and the available thermal activation energy, i.e. the temperature of the environment, and is not trivial to cast into an explicit analytical expression. Nevertheless, at zero temperature the following expression has been derived [67]:

$$D(t) \sim \frac{1}{t \ln(\nu_0 t)} \tag{1}$$

which can be approximated by $D(t) \sim t^{-1.04}$ in the long time limit [70]. The time scaling factor ν_0 represents the attempt to jump frequency of the triplet. During this dispersive thermalization period the same zero temperature treatment predicts a temperature independent relaxation of the average energy $\varepsilon(t)$ relative to the center of the Gaussian DOS, which in a simplified version yields [67]:

$$\varepsilon(t) \sim -\sigma \left[\ln \left(\ln \nu_0 t \right) \right]^{1/2} \quad \text{as} \quad t \to \infty. \tag{2}$$

Therefore, the relaxation is proportional to the width of the DOS, σ. Furthermore, the time dependent relaxation of the phosphorescence can be measured to actually determine the true width of the triplet DOS of polyfluorene as shown in Fig. 17.

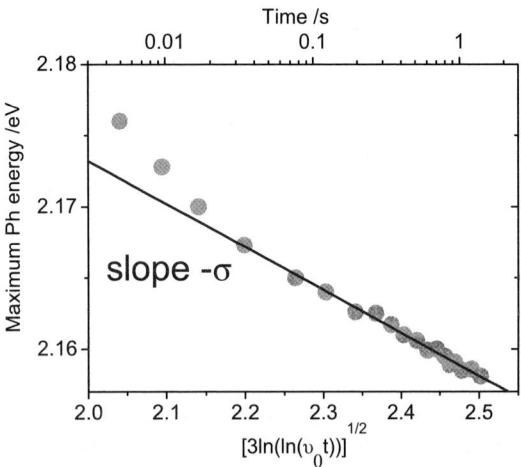

Fig. 17 As the triplet exciton energetically relaxes with increasing time after excitation the Ph spectrum shifts accordingly to lower energies. Here, the peak energies of the phosphorescence spectra of a PF2/6 thin film optically excited at 20 K are plotted according to Eq. 2. The slope yields the true width of the triplet DOS, for polyfluorene ~40 meV. Details are given in [24]

An important implication from Eq. 2 is that the observed phosphorescence spectra are (always) already relaxed within the time resolution of the measurement of phosphorescence spectra. Therefore, the true triplet energy (center of the triplet DOS) is always higher than the apparent phosphorescence spectra, which sets a lower limit only. For PF2/6 the true triplet level is calculated to be 2.26 eV instead of the observed ∼2.15 eV [24] from phosphorescence measurement. This higher value however agrees very well with that measured using the pulse radiolysis energy transfer technique which yields triplet energies in the initial unrelaxed state [71].

Even at $T \neq 0$ the excitons, which at $t = 0$ are excited at random within the DOS, initially thermalize to lower energy sites, accordingly in this time period Eqs. 1 and 2 are still valid. However, for finite temperature the diffusivity will approach an equilibrium value after a certain delay time. At this (temperature dependent) segregation time, t_s, the diffusion due to thermalization within the DOS equalizes the thermally activated hopping; afterwards triplet migration is governed by thermally assisted jumps. Now the non-dispersive, classical, regime is attained, which is described by a time-independent, but still temperature dependent, diffusion constant, D_∞. An analytic expression for the segregation time was calculated as [24]:

$$t_s(T) = t_0 \, e^{\left(\frac{c\delta}{k_B T}\right)^2} \tag{3}$$

with $c = 2/3$ for three-dimensional migration [24] and t_0 is related to v_0 and denotes the dwell time for triplets that migrate through a hypothetical isoenergetic ($\delta = 0$) equivalent structure. Whilst the energetic relaxation as a function of temperature and time of the triplet exciton can be well observed by monitoring phosphorescence spectra, it is more difficult to measure the time or temperature dependent diffusion rate directly. This can only be done indirectly, for example by studying the time dependent emission of emissive triplet acceptors [69]. The triplet diffusion is also the key parameter to understand the dynamics of migration activated triplet–triplet annihilation. For polyfluorene derivatives this issue has been studied in some detail [69] and will be considered in Sect. 3.2.

2.3
Defect Trapping

Time-resolved fluorescence studies of fluorene–fluorenone copolymers in dilute toluene solution, enable us to identify different time regimes in the photoluminescence (PL) decay. Figure 18 shows the results of a maximum entropy method (MEM) analysis of the PL decays of fluorene–fluorenone copolymers [58] collected at the fluorene emission wavelength [66]. The different time regimes of the PL decay are associated with different kinetic species which migrate to the defects, most typically these are the CTS defects

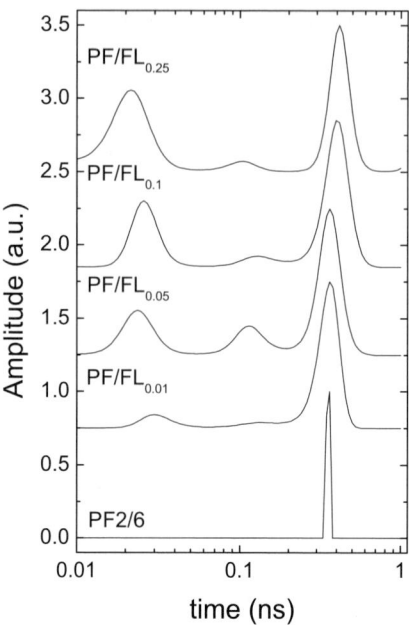

Fig. 18 Maximum entropy method (MEM) analysis of PF2/6 and PF/FLx copolymers. For PF2/6 only a narrow distribution is observed at 360 ps. For PF/FLx copolymers, together with the distribution 60 ps, two additional distributions are observed around 20 and 100 ps

(at keto sites) formed between the fluorenone groups and the nearest neighbor fluorene units, these defects are the acceptors of energy transfer from the polyfluorene donor, alternatively they could be dilute co-monomer units or even intentional dopants. This gives rise to various quenching mechanisms for the fluorene singlet exciton with the quenching occurring in these different time regimes as follows; (1) slow-quenched fluorene singlet excitations (PF_{qslow}), fluorene excitations located far from the traps but free to move along the polymer chains by energy hopping eventually reach a quenching center, (2) unquenched polyfluorene singlet excitations (PF_{unq}), and (3) fast-quenched polyfluorene singlet excitations (PF_{qfast}) where the exciton is created close to a trap and undergoes rapid energy transfer to the trap.

Figure 18, shows the representation of the quenching rate constants associated with both slow (k_1) and fast (k_2) quenching time regimes, also shown is the result from PF2/6 with negligible trap content showing a well-defined single exponential decay time. In the case presented, the slower quenching regime ($k_1 = 1.2 \times 10^{11}$ s^{-1}) is dependent on the fluorenone fraction, such that with increasing Q it becomes more probable that an excitation finds a CTS defect, analogous to what happens in a diffusive quenching process.

The determination of the quenching rate constant associated with the fast regime k_2, suggests that k_2 can qualitatively be fitted with Eq. 4, with

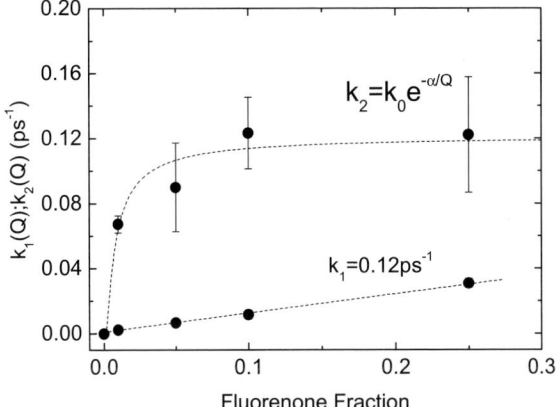

Fig. 19 Determination of k_1 and k_2. The rate constant associated with the slower time regime shows a linear dependence with the fluorenone content. The fast time regime suggests an exponential dependence with fluorenone fraction, compatible with a short range mechanism

$k_0 = 0.12$ ps^{-1} and $\alpha = 0.006$, where k_0 represents the maximum rate constant and α is a parameter without physical meaning, and suggesting the process is described by the Dexter electron exchange mechanism [72]

$$k_2(Q) = k_0 \, e^{-\frac{\alpha}{Q}}. \tag{4}$$

Increasing Q, would decrease the average minimum distance between a self localized PF$_{\text{q-fast}}$ excitation and a possible CTS defect, which suggests that Eq. 4, simply describes the qualitative dependence with distance for the Dexter electron exchange mechanism $k_{ET} = k_0 \, e^{-R}$, where k_0 is the maximum rate constant for energy transfer, occurring when donor and acceptor are at the "collision" distance R_0 and R is the separation between donor and acceptor when they are further apart than R_0.

The fact that fitting k_2 with Eq. 4, leads to k_0 equal to k_1, suggests that what is in fact controlling the quenching process is not the energy migration along the chain but instead the energy transfer from PF units to the CTS defects.

2.4
Amplified Spontaneous Emission

In the literature there are extensive reports on the observation of amplified spontaneous emission (ASE) of polyfluorene [73–76]. ASE occurs at high excitation intensities when it is possible to create a transient excited state population which is greater than the population of the lower lying state to which it radiatively decays, i.e. forming a population inversion. If there is some feedback mechanism of emitted photons, stimulated emission can build

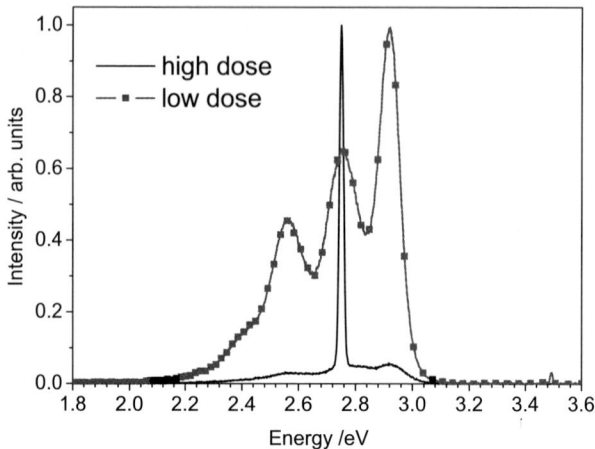

Fig. 20 Normalized emission spectra of a PF2/6 film at 20 K after optical laser excitation with a high and a low pulse intensity of 170 ps duration. The first vibronic overtone, 2.75 eV, of the polyfluorene spectrum undergoes ASE

up and dominate over spontaneous emission. In polymer thin films this is most likely to arise by waveguiding and internal reflection at the substrate interface [76]. However, most of the work has focused on PFO, which is somewhat of a special case and not typical for polyfluorene derivatives, as it has a propensity to form different phases which will be discussed in Sect. 4. As for many other conjugated polymers, typical polyfluorene derivatives, notably PF2/6 and PSBF, ASE exclusively occurs at the energetic position of the first vibronic overtone, i.e. around 2.75 eV at low temperature. A typical example for ASE of PF2/6 is shown in Fig. 20. The 0–0 mode, although of highest intensity, has underlying tail absorption, which quenches the gain, especially over long path lengths through the material typical in a waveguide mode.

3
Exciton–Exciton Interactions

3.1
Singlet–Singlet Annihilation

At high excitation densities in the solid state, the decay of the singlet exciton becomes excitation dependant, bimolecular annihilation of the singlet excitons introduces a fast component to the decay [41, 42, 77, 78], this is shown in Fig. 21. In a number of publications pump-probe spectroscopy has been used to study the phenomena surrounding this accelerated decay. The bimolecular annihilation reaction is effectively energy transfer from one excited singlet to

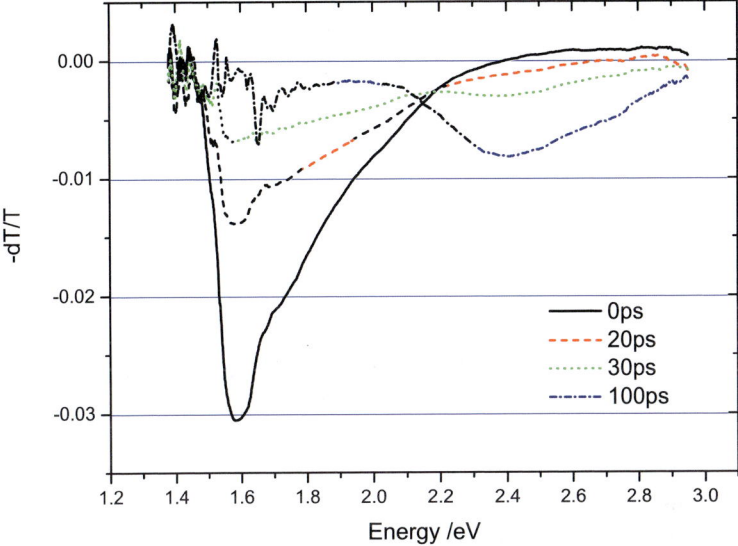

Fig. 21 Pump-probe spectra for the prototypical polyfluorene PF2/6 at high excitation density, showing the photoinduced absorption feature of the singlet population (1.6 eV peak) being rapidly quenched, leading to the formation of charges and characterized by the absorption of the charged state at 2.6 eV

another; thus proceeding in the following way:

$S_1 + S_1 \longrightarrow S_0 + S_n$,
$S_n \longrightarrow (1 - \eta)(S_1 + Q) + \eta(p^+ + p^-)$.

The production of a highly excited singlet state has important consequences; while many of the highly excited S_n states decay non-radiatively to the emissive S_1 state, there is also a possibility for the S_n state to dissociate into a pair of charges. The pump-probe spectra in Fig. 21 shows the quenching of the singlet photoinduced absorption spectrum being replaced with the characteristic absorption of the polyfluorene charged state at 2.4 eV. The stable isobestic point shows that there is a direct donor–acceptor relationship between the decay of the upper singlet state and the generation of the charges. This is the principal mechanism for the photogeneration of charge pairs in the solid state. The excitation density dependence of singlet decay is shown at room temperature and at 10 K in Fig. 22, the figure shows the decay of the singlet photoinduced absorption feature at low and high excitation density, the effect of the accelerated decay of the singlet is clear at high excitation density. This effect is quantified in Fig. 23 which shows that although the initial intensity of the excited state absorption remains linear with excitation fluence, the intensity of the absorption after 100 ps follows a linear law

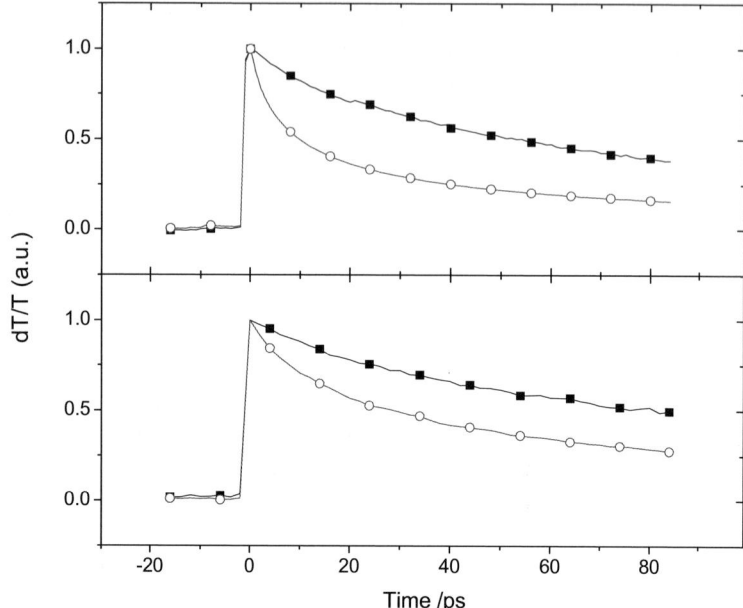

Fig. 22 Decay of the singlet population at 300 K (*top*) and 10 K (*lower*) for high (○) and low (■) excitation density

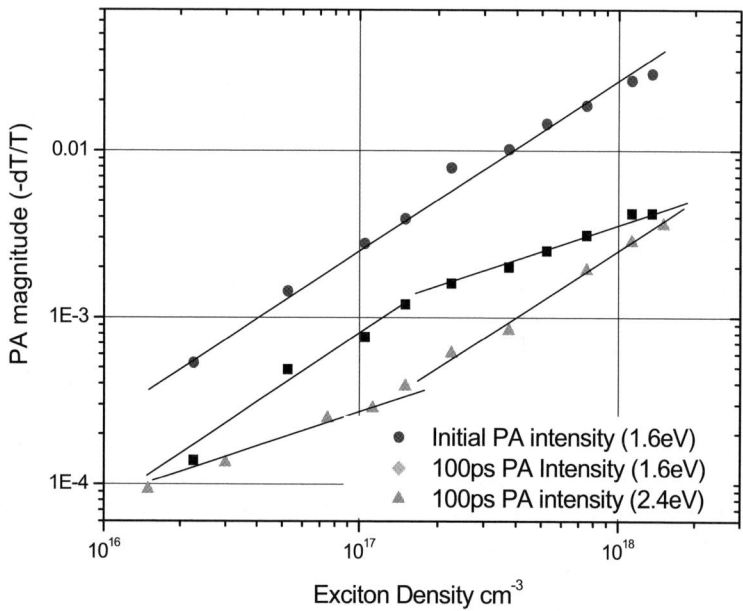

Fig. 23 Excitation density dependence of the photoinduced absorption features in Fig. 21; the threshold for the decay of the singlets being dominated by singlet–singlet annihilation at 1.5×10^{17} cm^{-3} is clear

at low fluence and a quadratic law above a threshold excitation density of 1.5×10^{17} cm^{-3}, where an additional, bimolecular decay process begins to have an effect on the decay of the singlet states. In addition, in the region of the charge state absorption at 2.4 eV there is an increase in the intensity of the absorption at the same point.

The Förster radius for the energy transfer reaction between two excited singlets can be calculated from the overlap of the fluorescence spectrum and the singlet excited state absorption, giving $R_0 = 1.1 \pm 0.5$ nm, which suggests that singlet–singlet annihilation is very inefficient given that the mean separation of the excited states at the observed threshold for the annihilation process is close to 20 nm. This seems to be at variance to the experimental data. The two panels in Fig. 22 showing the excitation dependence of the decay at different temperatures provide the answer; the annihilation becomes possible because at room temperature the singlets are very mobile; exciton diffusion allows the excited states to quickly become close enough to interact and the process remains efficient. At low temperature, the mobility of the singlet excitons drops and the effect of the annihilation becomes less, i.e. singlet–singlet annihilation is also an exciton diffusion controlled process.

3.2
Triplet–Triplet Annihilation and Delayed Fluorescence

With sensitive gated detection an emission contribution, with considerably longer lifetime than the fluorescence lifetime is readily detected in many luminescent polymers, not least polyfluorenes and is termed delayed fluorescence (DF) [79]. It is very frequently observed in studies of polyfluorenes [24, 80] and a typical example is shown in Fig. 24. Using pulse radiolysis, where very high triplet populations can be generated, it is possible to multiply excite single (dilute) chains such that triplet excitons readily migrate and interact on a single chain, annihilate and give strong delayed fluorescence [81]. In general, it may originate from charge carrier recombination (either geminate or non-geminate), inter-system-crossing, or triplet–triplet-annihilation [82]. However, for optical excitation it has been demonstrated that the observed delayed fluorescence of polyfluorene mostly stems from triplet–triplet annihilation [28, 83].

At low temperature, it is reasonable to neglect radiative and non-radiative monomolecular decay and to assume that annihilation is the dominating decay mechanism for the triplets in the time domain much shorter than the triplet lifetime, i.e. for $t < 100$ ms:

$$\text{DF} \sim \frac{dn_T}{dt} = -\gamma_{TT} n_T^2 \tag{5}$$

with n_T being the time-dependent triplet density, γ_{TT} denotes the triplet-triplet-annihilation "constant". The rate limiting step for the annihilation

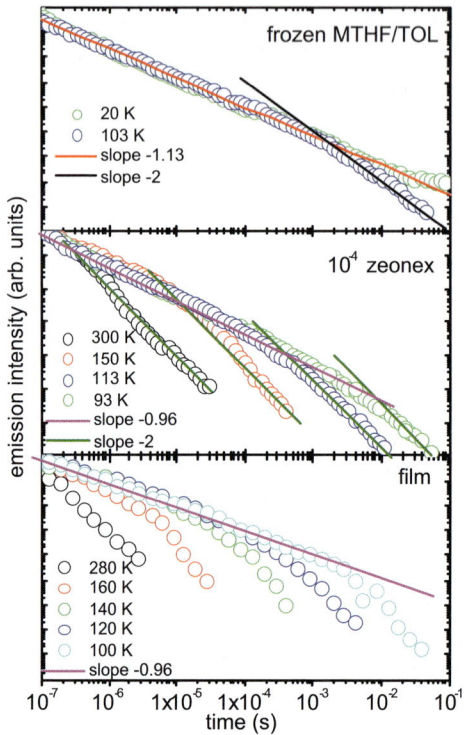

Fig. 24 Delayed fluorescence decays after pulsed optical excitation of PF2/6 as a function of temperature in several media as indicated. Details can be found in [24]

process is triplet exciton migration that brings two triplet excitons within the interaction radius, typically 2 nm. Thus, triplet migration is the key to understanding the dynamics of delayed fluorescence in polyfluorene. If bimolecular triplet annihilation dominates triplet decay then according to Eq. 5 DF and triplet population obey algebraic decay laws with exponents of −2 and −1, respectively. However, according to Eq. 1 at low temperature, during the dispersive time regime the diffusivity of the triplet and thus the triplet annihilation rate slows down with time with a slope of −1. In consequence, the apparent decay of the DF is retarded and features a slope of −1 instead of −2. After the temperature dependent segregation time, Eq. 2, is reached the triplet diffusivity approaches a constant value and so does the triplet annihilation rate. Without the retarding influence from the time dependent triplet annihilation rate, the DF decay proceeds with the classical slope of −2.

Corresponding experimental data for PF2/6 in solid film, frozen solution, and in an inert matrix polymer are shown in Fig. 25. A clear turnover from an initial slope of −1 to a slope of −2 occurs, the turnover occurring earlier for increasing temperature. By analyzing this data according to Eq. 3, such

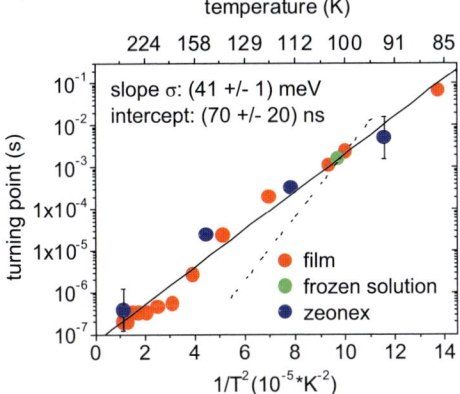

Fig. 25 Presentation of the turn over points between dispersive and non-dispersive triplet diffusion as extracted from Fig. 24 and plotted according to Fig. 3. Details can be found in [24]

as shown in Fig. 25, it is possible to gain the true triplet DOS width and the attempt-to-jump frequency for the triplet exciton, which for PF2/6 works out as 40 meV and 70 ns, respectively [24].

Because of the dispersive nature of the triplet diffusion, the triplet annihilation rate must also be a function of time and temperature. Alternatively, for continuous excitation the annihilation rate becomes a function of excitation dose. Figure 26 shows the saturation of the triplet density during continuous

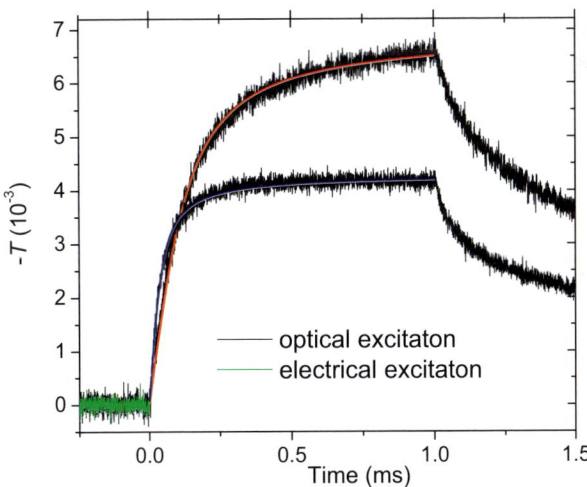

Fig. 26 Transient triplet absorption data of PSBF devices after optical (*black curve*) and electrical (*green curve*) excitation for 1 ms. The *solid lines* are fits according to a model that takes triplet–triplet annihilation into account. Details can be found in [24]

optical and electrical excitation due to triplet–triplet annihilation for a PSBF device. A careful analysis of such data allows an estimation of the triplet annihilation rate at $\gamma_{TT}(I_0) = 1.1 \times 10^{-32} I_0^{0.71}\,\mathrm{m^3\,s^{-1}}$ for PBSF, which is of the order $10^{-15}\,\mathrm{cm^3\,s^{-1}}$ for typical excitation doses [68, 84].

3.3
Singlet–Triplet Annihilation

In a similar way to the annihilation of pairs of singlets or pairs of triplets it is theoretically possible for singlet excitons to annihilate with triplets. As with the singlet–singlet annihilation this is theoretically possible as a Förster transfer between the singlet and the triplet. However, although theoretically possible, once again the Förster radius is very small. In the case of S–T annihilation there have been only a few reports of triplets efficiently quenching singlet states in polyfluorene, either in the solid state and in single molecule studies and the efficiency of the observed process is much lower than singlet–singlet annihilation [85, 86]. The efficiency of S–S annihilation is enhanced by the migration of the singlets, but the migration does not seem to dramatically enhance the efficiency of S–T annihilation. One suggestion for this is that the small spatial extent and perpendicular orientation of the triplet compared to the singlet state prevents the states coming close enough for the reaction to take place [35, 41] on an individual chain, and as the Forster radii are rather small, <10 Å, only intra chain events are likely.

4
The "Beta" Phase of Polyfluorene

4.1
Absorption and Emission

A new "phase" of PFO was first observed by Grell and Bradley et al. [87, 88], who reported the appearance of a shoulder at the onset of the absorption band of films of both neat PFO, PFO dispersed in polystyrene and PFO solutions in poor-solvents (such as cyclohexane). This shoulder could, upon appropriate treatment, evolve into a well-defined peak with a maximum at 437 nm as seen in Fig. 27. In particular, it was observed that this peak becomes more prominent in films exposed to solvents or upon their slow warming from 77 K up to room temperature. On the basis of X-ray fiber diffraction studies, Grell et al. [89] concluded that the lower energy state is associated with particularly extended conformations of PFO chains [90], resulting from the polymer response to some physical stress, either driven by solvent quality or temperature [91]. Furthermore, it has been suggested [91] that octyl side-chains in PFO are particularly effective at inducing and stabilizing such

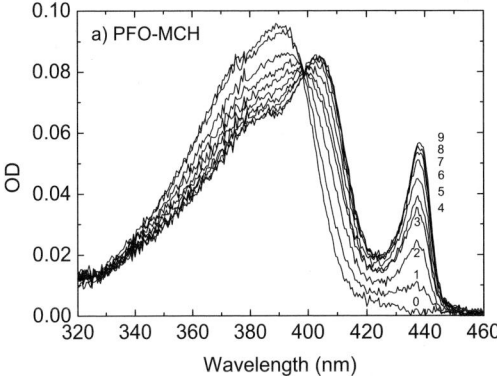

Fig. 27 Growth of a new absorption band below the $\pi-\pi^*$ transition in dilute PFO cooled (1 at 295 K to 9 at 130 K) in a poor solvent, MCH. A clear isobestic point is observed indicative of a simple A → B reaction and is the signature of the formation of a new morphological state of the PFO in which the polymer chains become rigid and planar. This state has been coined the "beta-phase" [88]

intramolecular ordering in solution, when compared to hexyl and dodecyl side-groups. Recently, the concept of side chain-driven planarization upon polymer agglomeration in solvent mixtures due to an increase of the fraction of a poor-solvent was introduced [92, 93].

Optically, the β-phase is very characteristic. In Fig. 27 a new absorption feature, below the $\pi-\pi^*$ transition is seen, typical as a small component but which can be induced to grow especially upon cooling [93]. However, this small band in the absorption spectrum produces profound changes in emission, and in films, emission from the β-phase dominates. β-Phase emission is characterized by a spectrum of very well-defined sharp vibronic replicas with almost no Stokes shift. Indeed the spectrum is very similar to that of a ladder polymer, see Fig. 28. The domination of the β-phase emission comes about through efficient Forster transfer from the "host" PFO to the β-phase inclusions (or guests) in the PFO matrix. Time-resolved energy transfer measurements show this to be a very efficient process [94]. Since the β-phase spectrum is so reminiscent of the ladder polymer emission, it is obvious to conclude that indeed the β-phase is a rigid planar segment of PFO chain, or more likely the β-phase is a fully planar rigid PFO chain.

As with singlet emission, β-phase triplet emission is red shifted compared to that of the PFO to 2.08 eV in thin film at 20 K [23] and the vibronic components are extremely narrow indicting a narrow DOS. In fact, the phosphorescent peaks are ca. 3-times narrower than those measured in a ladder polymer indicating a more homogeneous environment. The ratio of the 0–0 to 0–1 modes in the phosphorescence is very large, with more than 95% of the

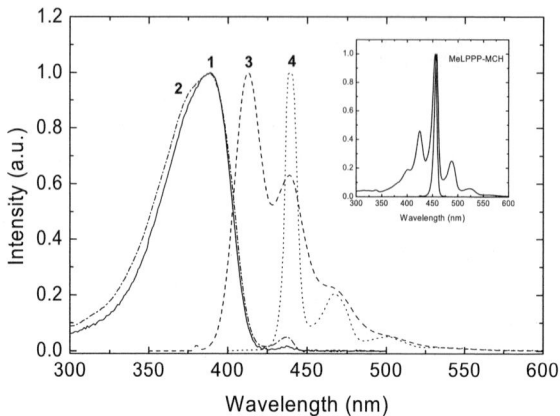

Fig. 28 Emission spectra of cooled and uncooled PFO in MCH. Once the beta phase has formed all emission emanates from this phase. For comparison, the *inset* shows the absorption and emission of rigid MeLPPP

emission being in the 0–0 mode, indicative of a highly planar and rigid backbone conformation. With such a narrow well-defined DOS, both the $T_1 \rightarrow T_n$ and a second transition to a higher lying triplet level can be observed in quasi CW pump probe measurements [23].

4.2
Effect of Alkyl Chain Length

It was originally thought that the β-phase only formed in poly(9,9'-di-*n*-octylfluorene) (PF8) [91, 95], however this is not in fact the case. β-Phase can be induced both in poly(9,9'-di-*n*-septylfluorene) (PF7) and poly(9,9'-di-*n*-nonylfluorene) (PF9) by either cooling solutions (in bad solvent such as MCH) or in cast films [6]. Detailed X-ray and neutron scattering studies on PF7, PF8 and PF9 confirm the presence of 2-D sheet structures associated with the β-phase, in solutions of these polymers [96, 97]. There is also a literature report that a β-phase like state can be induced in poly(9,9'-di-*n*-hexylfluorene) (PF6) [98]. Optical characterization of the β-phase states in PF7 and PF9 confirm that they are very similar to that found in PF8 but they are blue shifted by ca. 7 nm and have different stability over time, with PF8 yielding the most stable β-phase. These findings point to a mechanism of β-phase formation where side chain interactions between adjacent polymer backbones causes interchain "zipping" which has the direct consequence of planarizing the PF backbone, yielding the characteristic β-phase optical properties. This view point is also supported by X-ray results [90].

4.3
Amplified Spontaneous Emission

The β-phase of PFO is probably the most promising conjugated polymer system to achieve true lasing. Here, the emission is concentrated in a very narrow spectral region leading to a large gain at this wavelength. But even more important, the β-phase together with its surrounding non-ordered amorphous polyfluorene matrix constitutes an intrinsic four level system [73]: Singlet exciton excitation in the high energy amorphous PFO host, rapidly leads to excitations being trapped at the low energy β-phase sites [94]. Consequently, the β-phase subsystem is efficiently pumped, and, depending on β-phase concentration within the amorphous phase and excitation dose, excited state

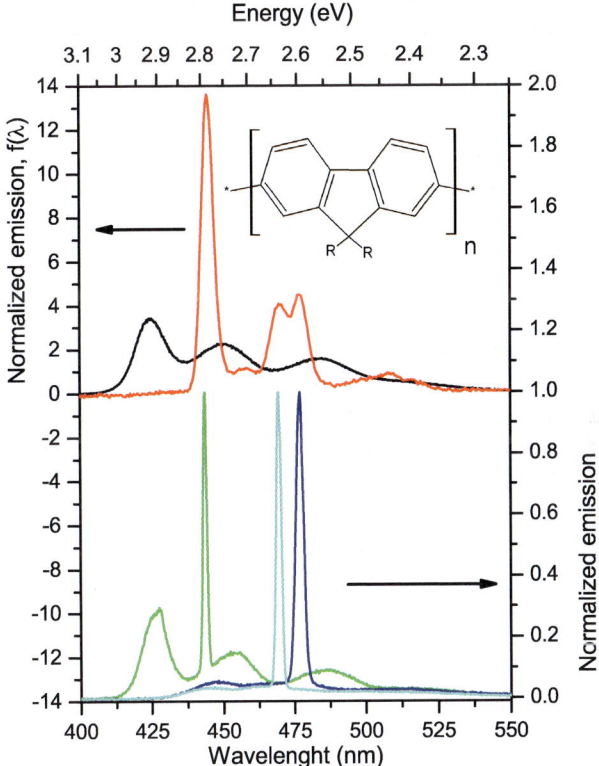

Fig. 29 *Upper panel*: Comparison of amorphous (*black*) and β-phase (*red*) emission spectra taken at 20 K, whereby the integrated emission was normalized. *Lower panel*: Compendium of normalized ASE action spectra of an amorphous film with small amount of β-phase showing ASE to the β-phase ground state (*left narrow peak*) and PFO films with high content of β-phase (*center* and *right narrow peak*) exhibiting lasing to vibronic levels. The chemical repeat unit of PFO is depicted in the inset with R = octyl. Details are given in [73]

inversions have been demonstrated. In this case ASE occurs from the leading 0–0 transition of the β-phase. Given these ideal starting conditions to observe ASE, it is probable that all ASE observed in PFO originates from the β-phase and not intrinsic to amorphous polyfluorene. Figure 29 shows the PL spectra of amorphous and β-phase PFO at low temperature and low excitation dose. For sufficient excitation dose, it is always the β-phase that shows ASE. The ASE wavelength depends on the β-phase concentration within the film, with higher concentrations leading to higher ASE emission wavelength, i.e. from the 0–1 and 0–2 modes [73].

References

1. Kersting R, Lemmer U, Mahrt RF, Leo K, Kurz H, Bassler H, Gobel EO (1993) Femtosecond Energy Relaxation in Pi-Conjugated Polymers. Phys Rev Lett 70(24):3820–3823
2. Meskers SCJ, Hubner J, Oestreich M, Bassler H (2001) Dispersive relaxation dynamics of photoexcitations in a polyfluorene film involving energy transfer: Experiment and Monte Carlo simulations. J Phys Chem B 105(38):9139–9149
3. Grage MML, Wood PW, Ruseckas A, Pullerits T, Mitchell W, Burn PL, Samuel IDW, Sundstrom V (2003) Conformational disorder and energy migration in MEH-PPV with partially broken conjugation. J Chem Phys 118(16):7644–7650
4. Dias FB, Macanita AL, de Melo JS, Burrows HD, Guntner R, Scherf U, Monkman AP (2003) Picosecond conformational relaxation of singlet excited polyfluorene in solution. J Chem Phys 118(15):7119–7126
5. Karabunarliev S, Bittner ER, Baumgarten M (2001) Franck–Condon spectra and electron-libration coupling in para-polyphenyls. J Chem Phys 114(13):5863–5870
6. Bright D, Dias F, Galbrecht F, Scherf U, Monkman A (2008) Alkyl chain length effects of the beta phase in polyfluorene. Adv Func Mat, in press
7. Harrison MG, Moller S, Weiser G, Urbasch G, Mahrt RF, Bassler H, Scherf U (1999) Electro-optical studies of a soluble conjugated polymer with particularly low intrachain disorder. Phys Rev B 60(12):8650–8658
8. Palsson LO, Monkman AP (2002) Measurements of solid-state photoluminescence quantum yields of films using a fluorimeter. Advanced Materials 14(10):757–758
9. Lakowicz JR (1999) Principles of fluorescence spectroscopy. 2nd ed, Kluwer Academic/Plenum, New York, p xxiii, 698
10. Sluch MI, Godt A, Bunz UHF, Berg MA (2001) Excited-state dynamics of oligo(p-phenyleneethynylene): Quadratic coupling and torsional motions. J Am Chem Soc 123(26):6447–6448
11. Karabunarliev S, Baumgarten M, Bittner ER, Mullen K (2000) Rigorous Franck–Condon absorption and emission spectra of conjugated oligomers from quantum chemistry. J Chem Phys 113(24):11372–11381
12. Lieser G, Oda M, Miteva T, Meisel A, Nothofer HG, Scherf U, Neher D (2000) Macromolecules 33:4490
13. Scholes GD, Larsen DS, Fleming GR, Rumbles G, Burn PL (2000) Origin of line broadening in the electronic absorption spectra of conjugated polymers: Three-pulse-echo studies of MEH-PPV in toluene. Phys Rev B 61(20):13670–13678

14. Grage MML, Pullerits T, Ruseckas A, Theander M, Inganas O, Sundstrom V (2001) Conformational disorder of a substituted polythiophene in solution revealed by excitation transfer. Chem Phys Lett 339(1–2):96–102
15. Muller JG, Anni M, Scherf U, Lupton JM, Feldmann J (2004) Vibrational fluorescence spectroscopy of single conjugated polymer molecules. Phys Rev B 70(3):035205
16. Bassler H, Schweitzer B (1999) Site-selective fluorescence spectroscopy of conjugated polymers and oligomers. Accounts Chem Res 32(2):173–182
17. Becker W (2005) Advanced Time-Correlated Single Photon Counting Techniques. Springer Series in Chemical Physics Vol. 81
18. Hintschich SI, Dias FB, Monkman AP (2006) Dynamics of conformational relaxation in photoexcited oligofluorenes and polyfluorene. Phys Rev B 74(4)
19. Di Paolo RE, de Melo JS, Pina J, Burrows HD, Morgado J, Macanita AL (2007) Conformational relaxation of p-phenylenevinylene trimers in solution studied by picosecond time-resolved fluorescence. Chem Phys Chem 8(18):2657–2664
20. Dhoot AS, Ginger DS, Beljonne D, Shuai Z, Greenham NC (2002) Triplet formation and decay in conjugated polymer devices. Chem Phys Lett 360(3–4):195–201
21. Wohlgenannt M, Tandon K, Mazumdar S, Ramasesha S, Vardeny ZV (2001) Formation cross-sections of singlet and triplet excitons in pi-conjugated polymers. Nature 409(6819):494–497
22. Rothe C, King S, Al Attar HA, Monkman AP (2006) Direct measurement of the singlet generation yield in polymer light emitting diodes. Phys Rev Lett 97:076602
23. Rothe C, King SM, Dias F, Monkman AP (2004) Triplet exciton state and related phenomena in the beta-phase of poly(9,9-dioctyl)fluorene. Phys Rev B 70(19):195213
24. Rothe C, Monkman AP (2003) Triplet exciton migration in a conjugated polyfluorene. Phys Rev B 68(7):075208
25. Sinha S, Rothe C, Guntner R, Scherf U, Monkman AP (2003) Electrophosphorescence and delayed electroluminescence from pristine polyfluorene thin-film devices at low temperature. Phys Rev Lett 90(12):127402
26. Monkman AP, Burrows HD, Hartwell LJ, Horsburgh LE, Hamblett I, Navaratnam S (2001) Triplet energies of pi-conjugated polymers. Phys Rev Lett 86(7):1358–1361
27. Monkman AP, Burrows HD, Miguel MD, Hamblett I, Navaratnam S (2001) Triplet state spectroscopy of conjugated polymers studied by pulse radiolysis. Synthetic Metals 116(1–3):75–79
28. Hertel D, Bassler H, Guentner R, Scherf U (2001) Triplet–triplet annihilation in a poly(fluorene)-derivative. J Chem Phys 115(21):10007–10013
29. Rothe C, Brunner K, Bach I, Heun S, Monkman AP (2005) Effects of triplet exciton confinement induced by reduced conjugation length in polyspirobifluorene copolymers. J Chem Phys 122(8):084706
30. King S, Rothe C, Monkman A (2004) Triplet build in and decay of isolated polyspirobifluorene chains in dilute solution. J Chem Phys 121(21):10803–10808
31. Burrows HD, de Melo JS, Serpa C, Arnaut LG, Monkman AP, Hamblett I, Navaratnam S (2001) S-1 similar to >T-1 intersystem crossing in pi-conjugated organic polymers. J Chem Phys 115(20):9601–9606
32. King SM, Rothe C, Dai D, Monkman AP (2006) Femtosecond ground state recovery: Measuring the intersystem crossing yield of polyspirobifluorene. J Chem Phys 124(23):234903
33. Rothe C, King S, Monkman A (2006) Long-range resonantly enhanced triplet formation in luminescent polymers doped with iridium complexes. Nat Mater 5(6):463–466

34. Rothe C, Guentner R, Scherf U, Monkman AP (2001) Trap influenced properties of the delayed luminescence in thin solid films of the conjugated polymer poly(9,9-di(ethylhexyl)fluorene). J Chem Phys 115(20):9557–9562
35. King SM, Vaughan HL, Monkman AP (2007) Orientation of triplet and singlet transition dipole moments in polyfluorene, studied by polarised spectroscopies. Chem Phys Lett 440(4–6):268–272
36. Shank CV, Ippen EP, Fork RL, Migus A, Kobayashi T (1980) Application of Subpicosecond Optical Techniques to Molecular-Dynamics. Philosophical Transactions of the Royal Society of London Series a – Mathematical Physical and Engineering Sciences 298(1439):303–308
37. Tong M, Sheng CX, Vardeny ZV (2007) Nonlinear optical spectroscopy of excited states in polyfluorene. Phys Rev B 75(12)
38. Kraabel B, Klimov VI, Kohlman R, Xu S, Wang HL, McBranch DW (2000) Unified picture of the photoexcitations in phenylene-based conjugated polymers: Universal spectral and dynamical features in subpicosecond transient absorption. Phys Rev B 61(12):8501–8515
39. Virgili T, Cerullo G, Luer L, Lanzani G, Gadermaier C, Bradley DDC (2003) Understanding fundamental processes in poly(9,9-dioctylfluorene) light-emitting diodes via ultrafast electric-field-assisted pump-probe spectroscopy. Phys Rev Lett 90(24):247402
40. Virgili T, Marinotto D, Manzoni C, Cerullo G, Lanzani G (2005) Ultrafast intrachain photoexcitation of polymeric semiconductors. Phys Rev Lett 94(11)
41. King SM, Dai D, Rothe C, Monkman AP (2007) Exciton annihilation in a polyfluorene: Low threshold for singlet–singlet annihilation and the absence of singlet–triplet annihilation. Phys Rev B 76(8)
42. Maniloff ES, Klimov VI, McBranch DW (1997) Intensity-dependent relaxation dynamics and the nature of the excited-state species in solid-state conducting polymers. Phys Rev B 56(4):1876–1881
43. Stevens MA, Silva C, Russell DM, Friend RH (2001) Exciton dissociation mechanisms in the polymeric semiconductors poly(9,9-dioctylfluorene) and poly(9,9-dioctylfluorene-co-benzothiadiazole). Phys Rev B 6316(16):165213
44. King SM, Hintschich SI, Dai D, Rothe C, Monkman AP (2007) Spiroconjugation-enhanced intramolecular charge-transfer state formation in a polyspirobifluorene homopolymer. J Phys Chem C 111:18759–18764
45. Beljonne D, Pourtois G, Silva C, Hennebicq E, Herz LM, Friend RH, Scholes GD, Setayesh S, Mullen K, Bredas JL (2002) Interchain vs. intrachain energy transfer in acceptor-capped conjugated polymers. Proc Natl Acad Sci USA 99(17):10982–10987
46. Oconnor P, Tauc J (1982) Photoinducedmidgap Absorption in Tetrahedrally Bonded Amorphous-Semiconductors. Phys Rev B 25(4):2748–2766
47. Cadby AJ, Lane PA, Mellor H, Martin SJ, Grell M, Giebeler C, Bradley DDC, Wohlgenannt M, An C, Vardeny ZV (2000) Film morphology and photophysics of polyfluorene. Phys Rev B 62(23):15604–15609
48. Ford TA, Avilov I, Beljonne D, Greenham NC (2005) Enhanced triplet exciton generation in polyfluorene blends. Phys Rev B 71(12)
49. Kersting R, Lemmer U, Deussen M, Bakker HJ, Mahrt RF, Kurz H, Arkhipov VI, Bassler H, Gobel EO (1994) Ultrafast Field-Induced Dissociation of Excitons in Conjugated Polymers. Phys Rev Lett 73(10):1440–1443
50. Conwell EM, Mizes HA (1995) Photogeneration of Polaron Pairs in Conducting Polymers. Phys Rev B 51(11):6953–6958
51. Bredas JL, Street GB (1985) Polarons, Bipolarons, and Solitons in Conducting Polymers. Accounts of Chemical Research 18(10):309–315

52. Cabanillas-Gonzalez J, Antognazza MR, Virgili T, Lanzani G, Gadermaier C, Sonntag M, Strohriegl P (2005) Two-step field-induced singlet dissociation in a fluorene trimer. Phys Rev B 71(15)
53. Montilla F, Mallavia R (2007) On the origin of green emission bands in fluorene-based conjugated polymers. Adv Funct Mater 17(1):71–78
54. Bliznyuk VN, Carter SA, Scott JC, Klarner G, Miller RD, Miller DC (1999) Electrical and photoinduced degradation of polyfluorene based films and light-emitting devices. Macromolecules 32(2):361–369
55. Gaal M, List EJW, Scherf U (2003) Excimers or emissive on-chain defects? Macromolecules 36(11):4236–4237
56. Hintschich SI, Rothe C, Sinha S, Monkman AP, de Freitas PS, Scherf U (2003) Population and decay of keto states in conjugated polymers. J Chem Phys 119(22):12017–12022
57. Dias FB, Maiti M, Hintschich SI, Monkman AP (2005) Intramolecular fluorescence quenching in luminescent copolymers containing fluorenone and fluorene units: A direct measurement of intrachain exciton hopping rate. J Chem Phys 122(5):054904
58. de Freitas PS, Scherf U, Collon M, List EJW (2002) (9,9-Dialkylfluorene-*co*-fluorenone) copolymers containing low fluorenone fractions as model systems for degradation-induced changes in polyfluorene-type semiconducting materials. E-Polymers, ARTN 009, 2002
59. Zojer E, Pogantsch A, Hennebicq E, Beljonne D, Bredas JL, de Freitas PS, Scherf U, List EJW (2002) Green emission from poly(fluorene)s: The role of oxidation. J Chem Phys 117(14):6794–6802
60. Dias FB, Pollock S, Hedley G, Palsson LO, Monkman A, Perepichka II, Perepichka IF, Tavasli M, Bryce MR (2006) Intramolecular charge transfer assisted by conformational changes in the excited state of fluorene-dibenzothiophene-*S,S*-dioxide co-oligomers. J Phys Chem B 110(39):19329–19339
61. Meskers SCJ, Hubner J, Oestreich M, Bassler H (2001) Time-resolved fluorescence studies and Monte Carlo simulations of relaxation dynamics of photoexcitations in a polyfluorene film. Chem Phys Lett 339(3–4):223–228
62. Richert R, Bassler H, Ries B, Movaghar B, Grunewald M (1989) Frustrated Energy Relaxation in an Organic Glass. Phil Magazine Lett 59(2):95–102
63. Mollay B, Lemmer U, Kersting R, Mahrt RF, Kurz H, Kauffman HF, Bassler H (1994) Dynamics of Singlet Excitations in Conjugated Polymers – Poly(Phenylenevinylene) and Poly(Phenylphenylenevinylene). Phys Rev B 50(15):10769–10779
64. Lyons BP, Monkman AP (2005) The role of exciton diffusion in energy transfer between polyfluorene and tetraphenyl porphyrin. Phys Rev B 71(23):235201
65. Forster T (1948) Zwischenmolekulare Energiewanderung und Fluoreszenz. Annalen Der Physik 2(1–2):55–75
66. Dias FB, Knaapila M, Monkman AP, Burrows HD (2006) Fast and slow time regimes of fluorescence quenching in conjugated polyfluorene–fluorenone random copolymers: The role of exciton hopping and dexter transfer along the polymer backbone. Macromolecules 39(4):1598–1606
67. Grunewald M, Pohlmann B, Movaghar B, Wurtz D (1984) Theory of Non-Equilibrium Diffusive Transport in Disordered Materials. Philosophical Magazine B – Physics of Condensed Matter Statistical Mechanics Electronic Optical and Magnetic Properties 49(4):341–356
68. Rothe C, Al Attar HA, Monkman AP (2005) Absolute measurements of the triplet-triplet annihilation rate and the charge-carrier recombination layer thickness in

working polymer light-emitting diodes based on polyspirobifluorene. Phys Rev B 72(15):155330
69. Rothe C, King S, Monkman AP (2006) Systematic study of the dynamics of triplet exciton transfer between conjugated host polymers and phosphorescent iridium (III) guest emitters. Phys Rev B 73:245208
70. Karg S, Riess W, Dyakonov V, Schwoerer M (1993) Electrical and Optical Characterization of Poly(Phenylene-Vinylene) Light-Emitting-Diodes. Synthetic Metals 54(1–3):427–433
71. Burrows HD, de Melo JS, Serpa C, Arnaut LG, Miguel MD, Monkman AP, Hamblett I, Navaratnam S (2002) Triplet state dynamics on isolated conjugated polymer chains. Chem Phys 285(1):3–11
72. Dexter DL (1953) A Theory of Sensitized Luminescence in Solids. J Chem Phys 21(5):836–850
73. Rothe C, Galbrecht F, Scherf U, Monkman A (2006) The beta-phase of poly(9,9-dioctylfluorene) as a potential system for electrically pumped organic lasing. Advanced Materials 18(16):2137–2141
74. Takahashi H, Naito H (2005) Amplified spontaneous emission from fluorene-based copolymer wave guides. Thin Solid Films 477(1–2):53–56
75. Long X, Grell M, Malinowski A, Bradley DDC, Inbasekaran M, Woo EP (1998) Spectral narrowing phenomena in the emission from a conjugated polymer. Opt Mater 9 (1–4):70–76
76. Scherf U, Riechel S, Lemmer U, Mahrt RF (2001) Conjugated polymers: lasing and stimulated emission. Curr Opin Solid State Mater Sci 5(2–3):143–154
77. Daniel C, Herz LM, Silva C, Hoeben FJM, Jonkheijm P, Schenning A, Meijer EW (2003) Exciton bimolecular annihilation dynamics in supramolecular nanostructures of conjugated oligomers. Phys Rev B 68(23)
78. Martini IB, Smith AD, Schwartz BJ (2004) Exciton–exciton annihilation and the production of interchain species in conjugated polymer films: Comparing the ultrafast stimulated emission and photoluminescence dynamics of MEH-PPV. Phys Rev B 69(3)
79. Sternlicht H, Robinson GW, Nieman GC (1963) Triplet–Triplet Annihilation and Delayed Fluorescence in Molecular Aggregates. J Chem Phys 38(6):1326–1335
80. Rothe C, Palsson LO, Monkman AP (2002) Singlet and triplet energy transfer in a benzil-doped, light emitting, solid-state conjugated polymer. Chem Phys 285(1):95–101
81. Monkman AP, Burrows HD, Hamblett I, Navaratnam S (2001) Intra-chain triplet–triplet annihilation and delayed fluorescence in soluble conjugated polymers. Chem Phys Lett 340(5–6):467–472
82. Pope M, Swenberg CE (1999) Electronic Processes in Organic Crystals and Polymers. Oxford University Press, Oxford, pp 161–166
83. Rothe C, Monkman A (2002) Dynamics and trap-depth distribution of triplet excited states in thin films of the light-emitting polymer poly(9,9-di(ethylhexyl)fluorene). Phys Rev B 65(7):073201
84. Rothe C, King SM, Monkman AP (2006) Direct measurement of the singlet generation yield in polymer light-emitting diodes. Phys Rev Lett 97(7):076602-1–076602-4
85. Zaushitsyn Y, Jespersen KG, Valkunas L, Sundstrom V, Yartsev A (2007) Ultrafast dynamics of singlet–singlet and singlet–triplet exciton annihilation in poly(3-2′-methoxy-5′-octylphenyl)thiophene films. Phys Rev B 75(19):195201-1–195201-7
86. List EJW, Scherf U, Mullen K, Graupner W, Kim CH, Shinar J (2002) Direct evidence for singlet–triplet exciton annihilation in pi-conjugated polymers. Phys Rev B 66(23):235203-1–235203-5

87. Bradley DDC, Grell M, Long X, Mellor H, Grice A, Inbasekaran M, Woo EP (1997) Influence of aggregation on the optical properties of a polyfluorene. Opt Probes Conjugated Polym 3145:254–259
88. Grell M, Bradley DDC, Long X, Chamberlain T, Inbasekaran M, Woo EP, Soliman M (1998) Chain geometry, solution aggregation and enhanced dichroism in the liquid-crystalline conjugated polymer poly(9,9-dioctylfluorene). Acta Polymerica 49(8):439–444
89. Grell M, Bradley DDC, Ungar G, Hill J, Whitehead KS (1999) Interplay of physical structure and photophysics for a liquid crystalline polyfluorene. Macromolecules 32(18):5810–5817
90. Brinkmann M (2007) Directional epitaxial crystallization and tentative crystal structure of poly (9,9′-di-n-octyl-2,7-fluorene). Macromolecules 40:7532–7541
91. Teetsov J, Fox MA (1999) Photophysical characterization of dilute solutions and ordered thin films of alkyl-substituted polyfluorenes. J Mater Chem 9(9):2117–2122
92. Scherf U, List EJW (2002) Semiconducting polyfluorenes – Towards reliable structure-property relationships. Advanced Materials 14(7):477–487
93. Dias FB, Morgado J, Macanita AL, da Costa FP, Burrows HD, Monkman AP (2006) Kinetics and thermodynamics of poly(9,9-dioctylfluorene) beta-phase formation in dilute solution. Macromolecules 39(17):5854–5864
94. Ariu M, Sims M, Rahn MD, Hill J, Fox AM, Lidzey DG, Oda M, Cabanillas-Gonzalez J, Bradley DDC (2003) Exciton migration in beta-phase poly(9,9-dioctylfluorene). Phys Rev B 67(19):195333
95. Chunwaschirasiri W, Tanto B, Huber DL, Winokur MJ (2005) Chain conformations and photoluminescence of poly(di-n-octylfluorene). Physical Rev Lett 94(10):107402-1–107402-4
96. Knaapila M, Dias FB, Garamus VM, Almasy L, Torkkeli M, Leppanen K, Galbrecht F, Preis E, Burrows HD, Scherf U, Monkman AP (2007) Influence of side chain length on the self-assembly of hairy-rod poly(9,9-dialkylfluorene)s in the poor solvent methylcyclohexane. Macromolecules 40:9398–9405
97. Knaapila M, Garamus VM, Dias FB, Almasy L, Galbrecht F, Charas A, Morgado J, Burrows HD, Scherf U, Monkman AP (2006) Influence of solvent quality on the self-organization of archetypical hairy rods – Branched and linear side chain polyfluorenes: Rodlike chains versus beta-sheets in solution. Macromolecules 39(19):6505–6512
98. Chen SH, Su AC, Su CH, Chen SA (2006) Phase behavior of poly(9,9-di-n-hexyl-2,7-fluorene). J Phys Chem B 110(9):4007–4013

Structure and Morphology of Polyfluorenes in Solutions and the Solid State

Matti Knaapila[1] (✉) · Michael J. Winokur[2] (✉)

[1] Department of Physics, Institute for Energy Technology, P.O. Box 40, 2027 Kjeller, Norway
matti.knaapila@ife.no

[2] Department of Physics, University of Wisconsin, 1150 University Avenue, Madison, WI 53706, USA
mwinokur@wisc.edu

1	Introduction	229
2	Single Molecules	230
3	Solutions	233
3.1	Structural Order of Polyfluorenes in Solution	233
3.2	Water Solutions of Polyfluorenes	237
4	Solid State	240
4.1	Intra- and Intermolecular Structures	240
4.1.1	Branched Side Chain PF2/6	240
4.1.2	Linear Side Chain PF8	246
4.1.3	Other Polyfluorenes	249
4.1.4	PF Oligomers	249
4.2	Macroscopic Alignment of Polyfluorene Chains and Crystallites	251
4.2.1	Alignment	251
4.2.2	Aligned Films of PF2/6	253
4.3	Surface Morphology	259
4.4	Higher Levels of Complexity—Nano and Microscale Assemblies	262
4.4.1	"Bottom-Up" Nanostructures	262
4.4.2	"Top-Down" Nanostructures	264
5	Conclusions and Outlook	267
	References	267

Abstract This account provides a state of the art overview of polyfluorene structure and phase behaviour in solutions and the solid state. This review covers key aspects of the hierarchical intra- and intermolecular self-assembly starting at the molecular level and extentding up to larger length scale structures. This includes crystallization, alignment on surfaces and texture. Many Central ideas are highlighted via structural archetypes. Recent theoretical treatments for understanding these structural properties are discussed and the implications for opto-electronics and photophysics are described.

Keywords Conjugated polymers · Polyfluorenes · Self-organization · Structure · Supramolecules

Abbreviations

$C_{12}E_5$	Pentaethylene glycol monododecyl ether
CP	π-Conjugated polymer
DOS	Density of states
ED	Electron diffraction
F2/6	9,9-Bis(2-ethylhexyl)fluorene
F8	9,9-Dioctylfluorene
F8BT	Poly(9,9-dioctylfluorene-co-benzothiadiazole)
F8T2	Poly(9,9-dioctylfluorene-co-bithiophene)
F8Ox	Poly{2,7-(9,9-dioctylfluorene)-co-1,4-(2,5-bis-(methyl-4'-(6-(3-methyloxetan-3-yl)methoxy)hexyloxy)benzene)}
GIXRD	Grazing-incidence X-ray diffraction
Hex	Hexagonal
HMW	High molecular weight
IXS	Inelastic X-ray scattering
LC	Liquid crystal
LMW	Low molecular weight
MCH	Methylcyclohexane
MEH-PPV	Poly(2-methoxy-5-(2-ethylhexoxy)-1,4-phenylenevinylene)
Nem	Nematic
NEXAFS	Near edge X-ray absorption fine structure
NLO	Nonlinear optics
PANI	Polyaniline
PBS-PFP	Poly{1,4-phenylene-(9,9-bis(4-phenoxy-butylsulfonate))fluorene-2,7-diyl}
PF	Polyfluorene
PF6	Poly(9,9-dihexylfluorene)
PF7	Poly(9,9-diheptylfluorene)
PF8 or PFO	Poly(9,9-dioctylfluorene)
PF9	Poly(9,9-dinonylfluorene)
PF10	Poly(9,9-didecylfluorene)
PF12	Poly(9,9-didodecylfluorene)
PF2/6	Poly(9,9-bis(2-ethylhexyl)-fluorene-2,7-diyl)
PFB	Poly(9,9-dioctylfluorene-co-bis-N,N'-(4-butylphenyl)-bis-N,N'-phenyl-1,4-phenylenediamine)
PDMOF	Poly{2,7-(9,9-bis((S)-3,7-dimethyloctyl))fluorene}
PMMA	Polymethylmethacrylate
PPP	Poly(p-phenylene)
PT	Polythiophene
PTFE	Polytetrafluoroethylene
SAED	Selected area electron diffraction
SANS	Small-angle neutron scattering
SAXS	Small-angle X-ray scattering
ssDNA	Single stranded DNA
SNOM	Scanning near-field optical microscope
TFB	Poly(9,9-dioctylfluorene-co-N-(4-butylphenyl)diphenylamine)
TFT	Tetrahydrofuran
XRD	X-ray diffraction

1
Introduction

In recent years conducting (or conjugated) polymers have become a key building block in the development of device technologies such as flat displays used in mobile phones and televisions. These materials are finding applications in areas as diverse as artificial muscles, electronic noses, plastic solar cells, corrosion inhibition, biosensors, electronic textiles, and nerve cell communications. This rapid evolution is based on the widespread advancements in (1) chemical synthesis of new materials, (2) new optoelectronics applications and (3) novel processing (e.g. ink jet printing). Consequently, the main streams of CP research comprise of synthetic chemistry as well as photo and device physics. However, the complex structures formed by these polymers are not thoroughly understood but this ordering underlies virtually all of the centrally important optical and electrical properties. Whether for enhancing a fundamental understanding of the optoelectronic properties, guiding the development of new device applications or providing new insight into the molecular design of new materials, it is critical to better understand their structural aspects and key structure–property relationships [1–3].

Apart from the emerging applications, the self-assembly of CPs is itself a fascinating subject. From the electronic perspective, these CPs represent a low-dimensional solid with strong covalent bonds along the molecular backbone and much weaker interchain interactions in orthogonal directions. The transport is therefore highly anisotropic and involves self-localized electronic excitations (e.g., solitons, polarons). Alternatively, from the structural perspective, the CPs are soft matter and belong to the class hairy-rod materials with a rigid backbone and flexible side chains [4]. Without the addition of functionalizing side chains the range of potential applications is limited. Linear, unsubstituted CPs (e.g., polyacetylene) exhibit very small increases in conformational entropy on dissolution or melting as well as a strong aggregation tendency. Consequently most of these materials are infusible and insoluble in common solvents. Functionalization via a side chain addition to the backbone forcibly dissolves the polar main chain in the surrounding matrix of a bound solvent and thus effects an attractive interaction between polymer and solvent. This leads to melting point depression and solution processability. In this context, the structural behavior of conjugated hairy-rods is understood in terms of self-organizing block copolymers [5] with rigid backbone and flexible side chains representing distinct blocks and forming microphase-separated domains. From another perspective these hairy-rods can be viewed in terms of thermotropic liquid crystals [6], and this liquid crystallinity stems from the highly anisotropic shape of the molecule.

There is an amazingly rich variety of CPs dependent on the choice of backbone and side chains and the nature of these two constituents has a major

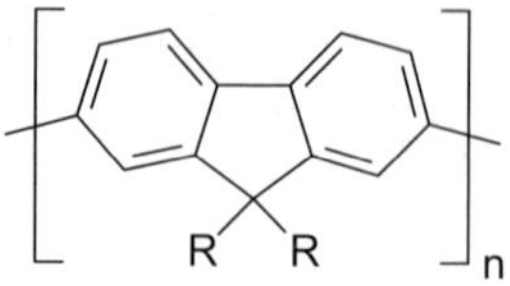

Fig. 1 Chemical structure of PF

influence on the phase behavior and resulting electronic and optical properties. Among the many reported CPs the PF family (Fig. 1) represents a central class of materials [7–10]. This emphasis stems from a combination of desirable qualities including facile synthesis, good environmental stability, excellent processing characteristics and efficient blue emission bands. There have been widespread efforts towards understanding, controlling and manipulating structures of PFs in both bulk and at surfaces. Not surprisingly, these advances have been exploited in a broad array of new polymeric technologies (for these optoelectronic aspects, the reader is referred to other reviews in this volume). These concerted efforts span multiple fields and those focusing on the structure attributes are widely distributed. No broad review of PF structure, assessing both the sold-state and solution, currently exists.

This review article focuses on the many levels of structural hierarchy ranging from the specifics of single PF molecules to those of aggregates in crystallites. The length scales span from Angstroms to tens of nanometers and beyond. The outline of the current review is as follows. First, we introduce the key structural attributes within single PF molecules. Next, we discuss the individual molecules and their aggregates in solution. Then we consider solid state and distinguish intra- and intermolecular structures including the higher levels of structural order in bulk, aligned fibers and aligned thin films. We also deal with similarities and differences between oligomers and polymers. Finally, we summarize and speculate on future studies. Overall this work spans the many landmark contributions in this quest to better understand PF structure.

2
Single Molecules

One of the guiding principles underlying much of the physics in conventional polymers is that many macroscopic macromolecular properties, not least the structure formation, can be satisfactorily understood even when the details of single molecules are ignored [11]. As is discussed later in this review, this idea still remains relevant for the intermolecular self-organization of hairy-rod PFs and for their macroscopic alignment. However, much of the behavior seen in PFs (and virtually all other electronic polymers as well) is an exception to this simplification because all optical and transport phenomena derive from the quantum mechanical properties of the base fluorene monomer and the near-neighbor coupling both along a single chain (i.e., intramolecular) and between chains (i.e., intermolecular).

While both intramolecular and intermolecule properties have relevance, it is often difficult to measure each independently. Two important settings that allow for direct observation of the single chain attributes are when isolated single polymers are dispersed on surfaces [12] or after dissolution of separated, uncorrelated chains in dilute solution. This latter topic is discussed in detail in the next section. The heterogeneous nature of the single chain structure in tandem with an overt sensitivity to the surrounding environment can, at times, make these very challenging systems to study. We expect that, with the advent of X-ray free electron lasers, direct structure studies of single molecules will become possible and this will resolve many key questions that still remain.

Single molecule spectroscopy of polyfluorenes [12] is discussed elsewhere in this volume and therefore we will limit our discussion of this matter. Many articles adopt the view that structure phase behavior and spectroscopic features can be strongly correlated. Clearly if a polymer chain is highly disordered one would expect there to be rather poorer transport properties and rather broad absorption and emission signatures. Parallels can be made to conventional diblock polymers in which highly ordered structures appear at relative large length scales but they still can be amorphous at nanometer distances. On the other hand in highly ordered materials the optoelectronic properties should be reflecting this fact [13, 14]. However, most conjugated polymers, PFs included, are intermediate between these two extremes and so the actual molecular level relationships are often subtle [15].

This subtlety is immediately apparent in direct comparisons of the chemically similar PF derivatives PF2/6 and PF8. PF2/6 (vide infra) is mesomorphic but the overall changes in the observed optical spectroscopy are quite modest. PF8 is also polymorphic but, in this case, there are very striking changes in the optical absorption and emission bands. Some of these may be associated with the formation of semi-crystalline phases and others not.

For both PF2/6 and PF8 the aforementioned main chain characteristics are essentially identical and so any pronounced differences are likely to originate in secondary structural characteristics of the functionalizing side chains. PF8 studies by Bradley and coworkers [16] first identified the unusual spectroscopic emission band now conventionally referred to as the "β phase". The hallmark signature of this peculiar chain structure is a relatively sharp series of emission bands red shifted some 100 meV from those seen when the polymer is prepared in a glassy state. π-Conjugated polymers have strong electron–phonon coupling and so, in addition to the π–π^* emission, there is a manifold of vibronic overtones spaced approximately 180 meV apart and red-shifted from the dominant π–π^* emission band.

The full PL envelope may be modeled using the Franck–Condon expression,

$$\text{PL}(\hbar\omega) \propto [n(\omega)\hbar\omega]^3 \sum_{n_1=0}^{\infty} \cdots \sum_{n_p=0}^{\infty} \prod_{k=1}^{p}(1+c) \times \left[\frac{e^{-S_k} S_k^{n_k}}{n_k!}\right] \Gamma\left\{\delta\left[\hbar\omega - \left(\hbar\omega_0 - \sum_{k=1}^{p} n_k \hbar\omega_k\right)\right]\right\}, \quad (1)$$

where $n(\omega)$ is the refractive index, $\hbar\omega_0$ the π–π^* transition energy, and $\hbar\omega_k$ vibrational mode energies for each mode k with $n_k = 0, 1, 2, \ldots$ overtones. S_k is the Huang–Rhys coefficient (i.e., reflecting the strength of the electron–phonon coupling) and Γ is the lineshape operator (this is a function of the density of states, temperature and oscillator strengths) [17]. A similar expression may be used for photoabsorption but here the effects of heterogeneous broadening are more pronounced.

Modeling of the alkyl side chains in the two respective cases gives distinctly different outcomes (Fig. 2). On average the PF8 main chain adopts a more planar conformation and, additionally, one can differentiate three distinct families (denoted as C_α, C_β, and C_γ) of conformation isomers [17, 18] and correlate them with the variations in the observed PL spectra. In this

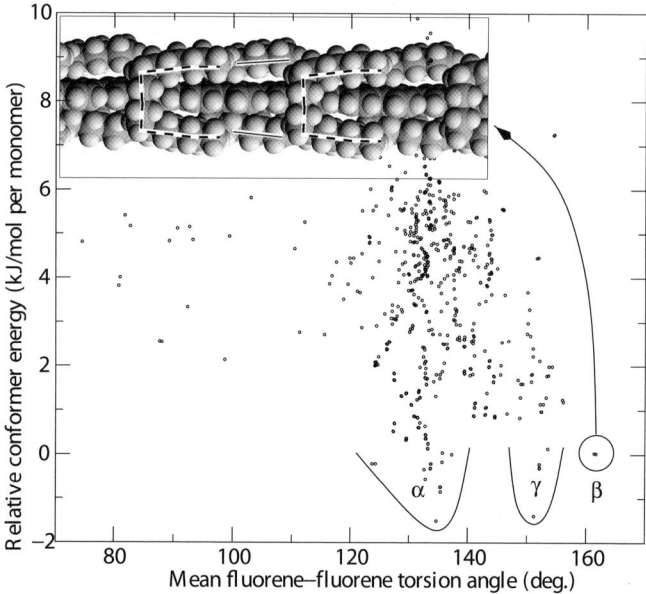

Fig. 2 Low-energy PF8 conformations (referenced to the C_β isomer). Each symbol corresponds to a single, unique tested conformation. C_α, C_β, and C_γ correspond to the conformer families. The *inset* shows the model of the C_β isomer highlighting side chains. Reprinted with permission from [17]. © (2005) by the American Physical Society

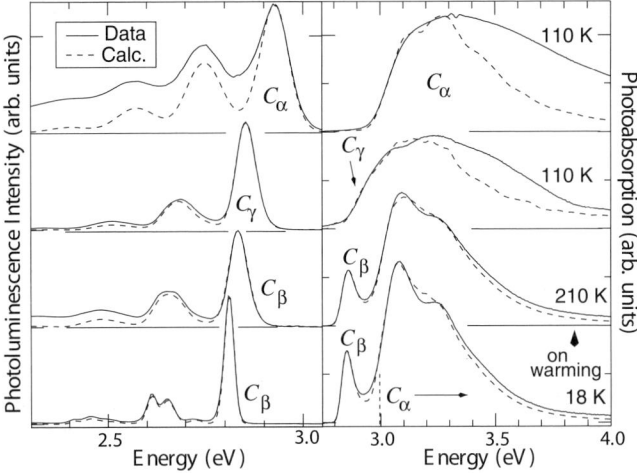

Fig. 3 Examples of PF8 PA and PL spectra from the three conformational isomer families in combination with calculated PL fits to the data using Eq. 1. Reprinted with permission from [17]. © (2005) by the American Physical Society

picture each family corresponds to a different ensemble average fluorene–fluorene torsion angle. As illustrated in Fig. 3 this trifurcation into conformational isomer families fully reproduces the PL experimental data when Eq. 1 is integrated with a tight-binding Frenkel-type exciton model. Guha and coworkers [19, 20] have shown that the differences in the side chain conformations are present in Raman scattering data and correlate well with ab initio quantum chemical calculations. From the perspective of single molecule spectroscopy each polymer chain is independently studied and its structure reflects one specific stochastic ensemble of main chain and side chain configurations in a self-consistent correspondence to the conformational energy surface topology.

3
Solutions

3.1
Structural Order of Polyfluorenes in Solution

PF solution studies are interesting for two chief reasons. First, PFs can be easily processed from solution and any structures adopted in solution will impact those that evolve in the subsequent processing. As might be expected, the solvent plays a major role in establishing the large length-scale morphology of solvent-processed PF thin films [21]. Second, the composition may be continuously varied from isolated single molecules, in the dilute case, all the way to macrophase-separated domains and this provides a broad platform for pursuing studies of the fundamental physics.

Deciphering the main structural aspects of PF solutions is viewed as important but, so far, the published literature remains relatively sparse. Examples include PF8 [22], PF2/6 [23–25], F2/6 oligomers [26], and poly(9,9-dialkylfluorene-co-fluorenone) copolymers [27] in toluene as well as PDMOF in tetrahydrofuran [28].

The solution behavior of PFs may be subdivided in terms of three phenomenological variables—the nature of the solvent, the fraction of polymer, and the nature of the side chain. An additional parameter which becomes dominant in oligomers is the length of the molecule [26]. Once again PF8 can be highlighted as a prime example of a rich and variegated material. Its chain morphology has been detailed in different organic solvents such as chloroform, toluene, TFT and cyclohexane [29] and this work documents the striking structural diversity arising from the quality of the solvent, the first parameter in our discussion. For example, PF8 forms sheet-like particles in 1% ($\sim$ 10 mg/mL) solution in the poor solvent MCH but exhibits an isotropic phase of rod-like polymers at otherwise identical conditions in the better solvent, toluene [25, 30]. In addition, if the second parameter, the polymer fraction is increased from 1% to 3–7% (w/v) in toluene, individual PF8 molecules aggregate and form a large network-like structure [31]. A similar phenomenology is observed for F2/6 oligomers which adopt a rigid rod conformation in dilute toluene solution [26]. However, at concentrations in excess of 30 wt.%, a second dynamic process in the isotropic scattering function relates to clusters of rather spherical overall shape [26]. The effect of side chain length, the third variable, has been also studied for PFs in MCH at concentrations 1–5% (w/v) [32]. It has been found that the aggregation tendency of PFs decreases with increasing side chain length so that at room temperature PF10 with ten side chain beads adopts an isotropic phase. The limit of aggregation can be, once again, controlled by solvent quality. For example, the aggregation is prominent only for PF6 with six side chain beads if a better solvent, such as toluene is utilized [33].

Binary PF mixtures manifest several levels of structural hierarchy. The torsion angle between the monomer units represents the first level of this hierarchy. Wu et al. [28] studied PDMOF in THF and found a structure wherein each monomer unit adopted, at random, four rotational states. These were right- and left-handed torsional rotations of approximately 35 and 144° (consistent with either a 5/1 or 5/2 helices) and modest bond angle fluctuating (c.f., Fig. 4). At much larger distances, Rahman et al. [31] identified, in PF8, extended domain networks. The average distance between these aggregate domains is of the order of 60 nm (c.f., Fig. 5).

When mixed with MCH, PFs functionalized with linear n-alkyl side chains establish even greater levels of hierarchy. At room temperature, PF6, PF7, PF8,

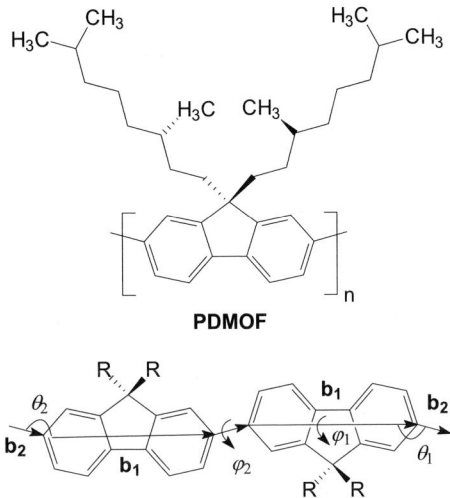

Fig. 4 Chemical structure of PDMOF and geometry of a polyfluorene derivative chain. Reprinted with permission from [28]. © (2004) by the American Chemical Society

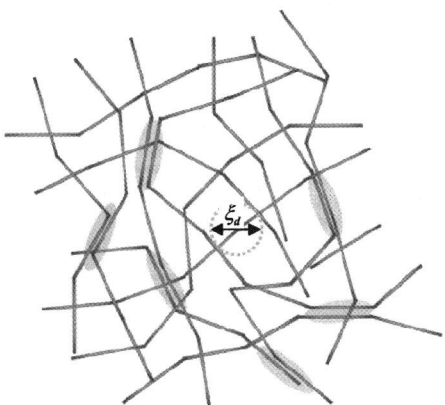

Fig. 5 Schematic representation of the aggregate domains (*shaded area*) tying the PF8 chains to form a cluster. The overlap of the unassociated chains in the cluster as well as in the bulk of the semi-dilute solution generates a dynamic network with the characteristic mesh size of ξ_d. Reprinted with permission from [31]. © (2007) by the American Chemical Society

and PF9 form large (> 10 nm) sheet-like assemblies (thickness of 1–3 nm) which represent solution structure. Interestingly, the larger length scale structures of these sheets display a distinctive odd–even dependence on the side chain length—the PF6 and PF8 sheets are broader and thinner, whereas PF7 and PF9 sheets are thicker with a putative double layer structure. PF10 does not follow this sequence. Only a very small fraction of the polymer assembles into a sheet-like structure and the rest remains dissolved at the molecular level (Fig. 6). Apart from PF10 these PF/MCH mixtures contain, at a shorter length scale, ordered "microcrystalline" structures having an internal period corresponding to the intermolecular periodicity of the solid state β phase of PF8. These domains may also act as nodes for establishing a network-like structure at large distances. The overall idea is illustrated in Fig. 7.

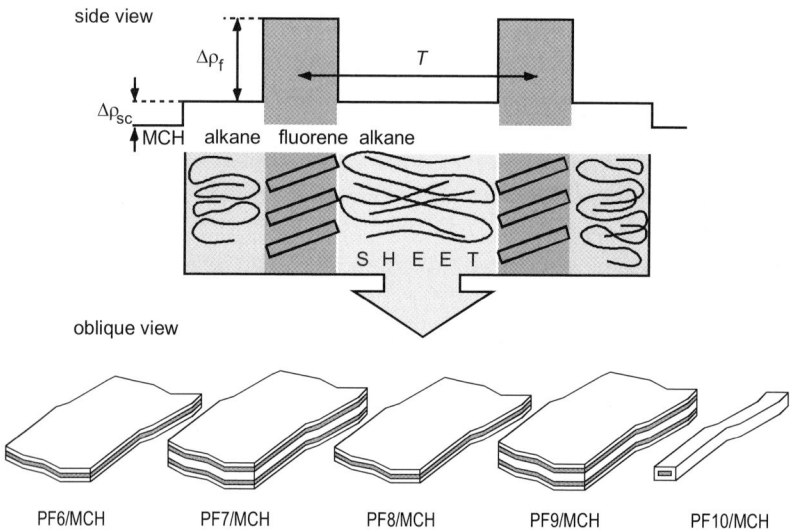

Fig. 6 Schematics of the proposed layer self-organization of PFs in MCH. Reprinted with permission from [32]. © (2007) by the American Chemical Society

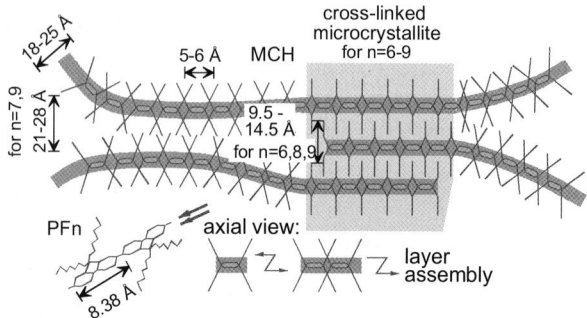

Fig. 7 Schematics of the possible structures of PFn with $n = 6.9$ in MCH. Loose lamellae correspond to those shown in Fig. 6. Reprinted with permission from [32]. © (2007) by the American Chemical Society

3.2
Water Solutions of Polyfluorenes

Water-soluble PFs are a new advance and these materials have opened the path for many new applications. These water-soluble CPs may be used with non-noxious, biocompatible solvents and have uses ranging from chemosensors [34] and biosensors [35] to inkjet processing [36]. Conferring water solubility to PFs is a challenging topic with few examples. One common strategy employed in order to achieve appreciable water solubility is the introduction of neutral or charged hydrophilic functionalities at the terminal position of the PF backbone. Development of this type of PFs has been pioneered by Bazan et al. [37–41] and they have used a number of different charged ammonium groups. This functionalization not only enhances solvent formation but also facilitates molecular recognition of biomolecules such as DNA and peptide nucleic acids. These PFs can be dissolved in high concentrations.

Wang et al. [42] have demonstrated that PFs with charged ammonium groups tethered at the end positions of the side chains, poly(9,9-bis{3′-((N,N-dimethyl)-N-ethylammonium}-propyl)-2,7-fluorene dibromide) are water soluble at concentrations as high as 100 mg/mL, cf., Fig. 8. Co-solvents such as methanol can be employed to ensure the uniformity of the solutions [43]. Another strategy for enhancing the solubility of PFs is to incorporate a surfactant layer separating the polymer from water

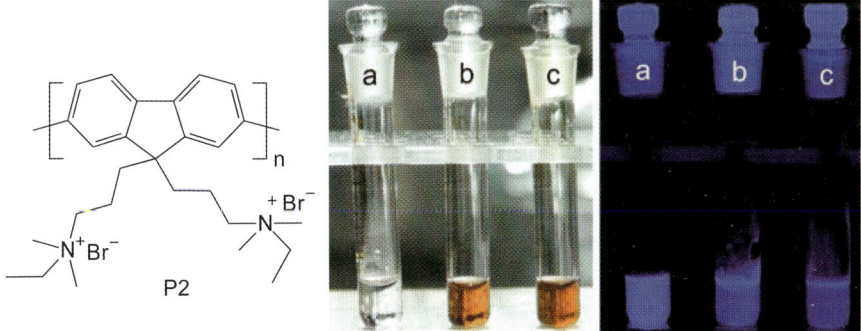

Fig. 8 *Left*: A chemical structure of poly(9,9-bis{3′-((N,N-dimethyl)-N-ethylammonium}-propyl)-2,7-fluorene dibromide) (P2). *Right*: A photograph of P2 with different concentrations in water (*left*). A photograph showing emission of light from P2 with different concentrations in water, the photograph was taken under irradiation with light at 365 nm (*right*). (a) Dilute solution (1×10^{-3} mg/mL), (b) 50 mg/mL, (c) 100 mg/mL. Reprinted with permission from [42]. © (2007) by Wiley-VCH Verlag GmbH & Co. KGaA

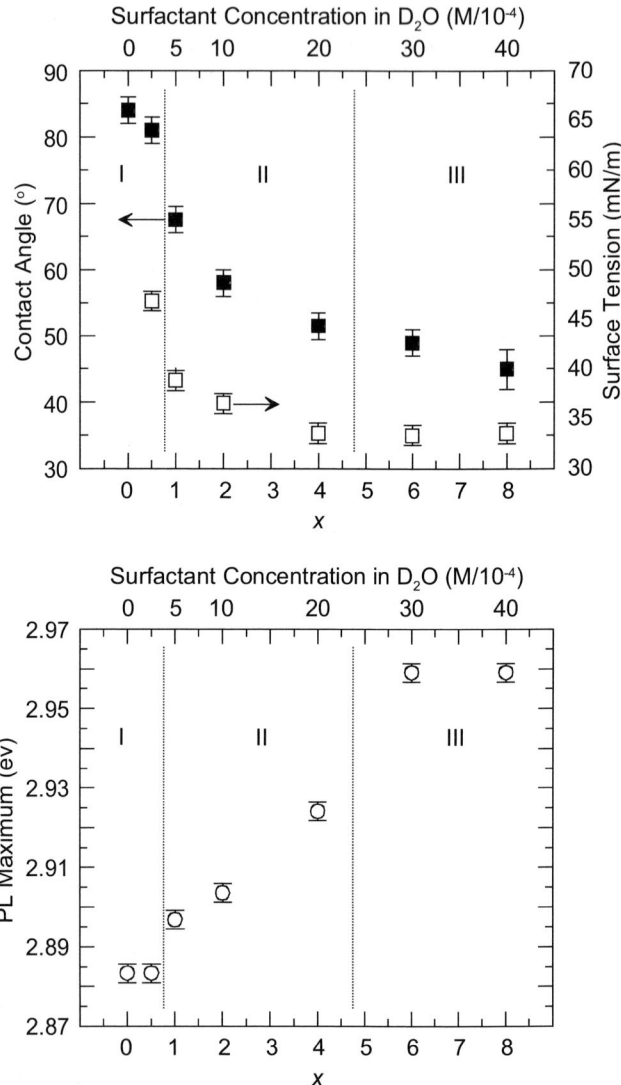

Fig. 9 *Top*: Contact angle (*solid squares*) and surface tension (*open squares*) of PBS-PFP in aqueous $C_{12}E_5$ as a function of molar ratio of surfactant over monomer unit x. The *dotted lines* distinguish tentative phase regimes I–III. $T = 20\,°C$. *Bottom*: Corresponding PL maxima. Monomer concentration was 5×10^{-4} M for all x. The *dotted lines* distinguish phase regimes I–III for which different solution structures are obtained. Reprinted with permission from [48]. © (2006) by the American Chemical Society

molecules [44]. The interaction between polymer and surfactant can arise from hydrophobic–hydrophilic effects [45], charge transfer [46], or molecular recognition [47].

Direct structural studies of these systems are scarce. In one example [48] PBS-PFP was studied in aqueous $C_{12}E_5$ as a function of molar ratio of surfactant to monomer unit, x, when the surfactant concentration was above the critical micelle concentration of aqueous $C_{12}E_5$ under conditions which corresponded to the isotropic liquid L_1 phase regime of the binary water–$C_{12}E_5$ system. Under these conditions, the following phase behavior is found by surface analyses (Fig. 9) and SANS (Fig. 10). The ternary solution is reported to be homogeneous when the polymer concentration is 5×10^{-4} M and the molar ratio of surfactant to monomer, x, is close to 1. This defines the first phase boundary. Elongated objects (of mean length ~ 90 nm) with near circular cross section (~ 3 nm) are observed for $1 < x < 2$. At sufficiently high surfactant concentrations, at $x = 4$, an interference maximum appears at $q \sim 0.015$ Å^{-1} in the SANS spectra and is indicative ordering by the micelles with a characteristic separation distance of 40 nm. This defines the second phase boundary. It is suggested that this ordering is due to electrostatic repulsion rather than steric hindrance.

Interesting analogies can be made with poly{9,9-bis(6-(N,N-trimethylammonium) hexyl)fluorene-co-1,4-phenylene} iodide mixed with aqueous $C_{12}E_5$ [49]. Al Attar and Monkman [50] probed the optical properties of this copolymer in the presence of ssDNA. In the system they found that poly-

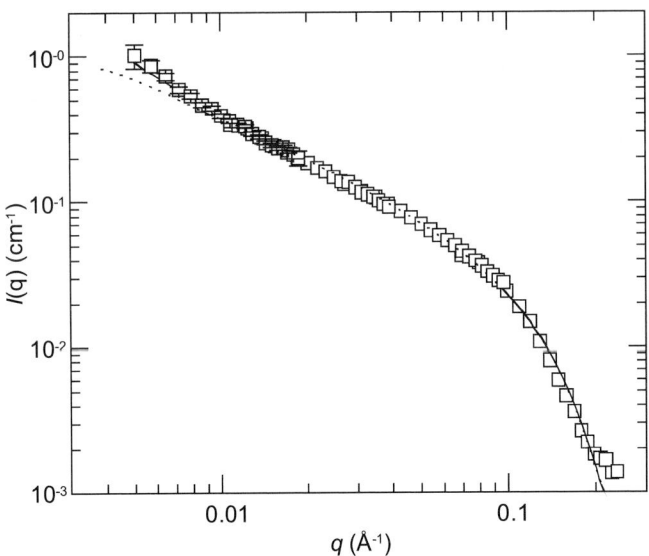

Fig. 10 SANS data of PBS-PFP copolymer with $C_{12}E_5$ surfactant in D2O. The monomer concentration was 5×10^{-4} M and the ratio of surfactant over monomer $x = 2.0$. Also shown are fits by flexible cylinder model (*solid line*) and stiff cylinder (*dashed line*). Reprinted with permission from [48]. © (2006) by the American Chemical Society

mer aggregation occurs only after the surfactant breaks down. They also observed that in solutions with low ssDNA concentrations the surfactant reduces quenching of the complex by preventing charge transfer processes. Charge transfer processes are sensitive to even small distance variations. They suggested that the reduced quenching stems from increasing the distance between the polymer and ssDNA by incorporation of polymer into aggregates. At high ssDNA concentrations the quenching actually *increased* sharply. This was attributed to the increase in the electrostatic force destroying the micelle's structure around the polymer.

4
Solid State

4.1
Intra- and Intermolecular Structures

4.1.1
Branched Side Chain PF2/6

Among the many PF derivatives that have been synthesized in recent years the branched 2-ethylhexyl functionalized polyfluorene, or PF2/6 [24, 51–56], clearly stands out as one of the two most heavily studied PF structural archetypes in both bulk and thin film form. The paper by Lieser et al. [51] is key and this work gives evidence for the formation of 5-fold helices through a combination of XRD, ED and molecular modeling. Additional support for this proposed main chain conformational motif is reported in follow up studies of PF2/6 [52, 53] and F2/6 [18, 57].

The modeling of [51] also emphasizes a very important PF *intra*molecular structural attribute; attention should be placed on the rotation between repeat units. In contrast to ladder-type PPPs [7] in which all phenylene rings are fused, PFs are representative of a "step-ladder" and this allows for a sterically hindered rotation between the repeat units. There is a well-defined competition between π-bonding, which lowers the total energy through planarization of the phenylene rings, and a steric repulsion of the α-hydrogens between adjacent monomers. Apart from a probable exception at very high pressures [58], PF2/6 usually adopts a stiff helical conformation of the main chain marked by a single ensemble average distribution of conformational isomers. Thermal cycling of PF2/6 leads to mesomorphic phase behavior with the presence of nematic, hexagonal and isotropic phases.

Modeling indicates the presence of multiple 2-ethylhexyl conformational isomers with nearly degenerate ground-state energies and a broad range of fluorene–fluorene torsion angles approximating the 144° angle necessary for

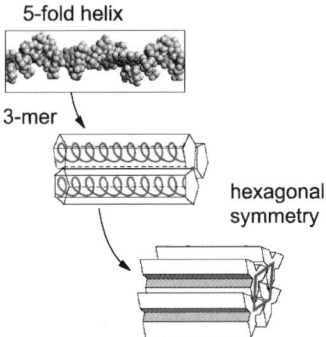

Fig. 11 Intra and intermolecular self-organization of PF2/6 polymer from 5-fold helices to a three chain ensemble and, finally, into an ordered hexagonal array. Adapted from [3]

a 5/2 helix (i.e., two full turns of the helix for every five monomers). Similar results are seen by Marcon et al. [18] in a molecular dynamics study. This heterogeneity leads to a PF material whose photoabsorption and photoluminescence of PF2/6 are, to first order, independent of the specific structural phase and subsequent processing.

In this picture the 5-fold helices constitute a base structure that self-organizes into triangular ensembles of three chains and these subsequently pack onto a hexagonal lattice at sufficiently high PF molecular weight and low enough temperature [24] (Fig. 11). This yield a semi-crystalline Hex phase with equatorial coherence lengths (i.e., the interchain packing) exceeding 20 nm.

An underlying issue with these helices is that the underlying 5-fold symmetry is incommensurate with the observed hexagonal unit cell and must introduce appreciable frustration. This latter effect will impact the crystalline perfection and any deviation from perfect periodicity introduces systematic effects in the Bragg scattering peak widths (and lineshape). Follow-up fiber XRD studies [24] identify a systematic trend in the observed scattering peak width consistent with the presence of paracrystallinity. Moreover, highly oriented PF2/6 fiber XRD data contains reflections which do not quite fit the nominal chain repeat specified by a 5 fold helix. Knaapila et al. [59] suggest that PF2/6 actually forms helices comprised of 21 monomers as opposed to the 5-fold helices, (i.e., in a 4×5 content) (Fig. 12). The 21-mer helical model would better accommodate the 3-fold rotational symmetry of the three chain unit cell and explain previously non-indexed reflections.

In regards to the *inter*molecular self-organization much can be learned from the physics of hairy-rod polymers [60] and PFs, especially PF2/6, are excellent model systems. As is the case in diblock polymers, a deceptively simple model parameterization of the key underlying interactions yields semi-quantitative predictions of the structural phase behavior. These model calculations can be directly compared to the experimental phenomena. The

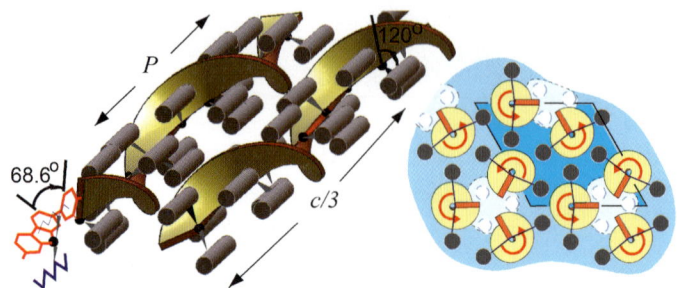

Fig. 12 Schematics of the helix form and side chain packing of PF2/6. In the proposed 21_4 helix the link atom is at the helical axis, around which the main chain and two side chains revolve quite like in a triple helix. The helical pitch P is defined by one complete turn of the backbone. The 21-helicity realizes 3-fold rotation symmetry. The *right* drawing shows possible arrangement of the side chains. The *dashed areas* correspond to sites unoccupied in this particular monomer layer. Reprinted with permission from [59]. © (2007) by the American Chemical Society

most extensive work describes the phase behavior of PF2/6 [24, 54, 55], as functions of both temperature and molecular weight, through a classic mean-field analysis of the free energies.

The main transition of interest in PF2/6 is a crossover from the Hex phase at high molecular weight to a Nem phase at low molecular weight, cf. Fig. 13.

In the mean-field analysis the free energy of the Nem phase is estimated as

$$F_N \approx k_B TVc \ln \frac{f}{e} + k_B TVc \ln \frac{4\pi}{\Omega_N}, \qquad (2)$$

where the first and second terms, respectively, correspond to the translational and orientational entropy. Here $k_B T$ is the Boltzmann factor, V volume of the sample, c concentration of hairy-rod molecules, f the volume fraction of the backbone in the molecule, and e is the Euler number. The quantity Ω_N describes the degree of overall (uniaxial) alignment: The smaller Ω_N is, the more aligned the system.

In contrast, the Hex phase has negligible translational entropy and the interaction between ordered molecules due to the inhomogeneous distribution of side chain ends becomes dominant. The resulting free energy is

$$F_H \approx k_B TVc \ln \frac{4\pi}{\Omega_H} - k_B TV \frac{v}{v_0 l_K^2 l_u}, \qquad (3)$$

where v and l_K are, respectively, the volume and Kuhn length of a side chain, l_u is the distance between two consecutive grafting points (the length of the repeat unit), and v_0 is the volume of one repeat unit of the hairy-rod.

The concentration c is directly related to M_n through $c = M_u/v_0 M_n$ where M_u is the weight per repeat unit. For a given molecular weight and temperature, the phase with the lower free energy is more favorable. Thus, a threshold

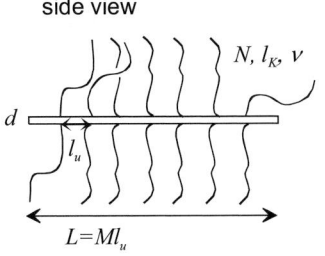

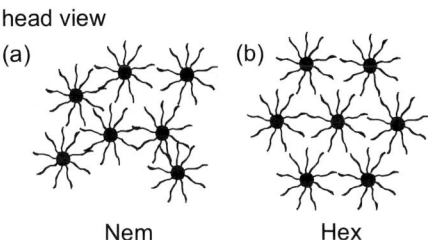

Fig. 13 *Above*: Schematics of a hairy-rod polymer consisting of a stiff backbone and flexible covalently connected side chains. N, l_K, and v are the number of segments, the segment length (the Kuhn length), and the volume of the beds of the side chains, respectively; M, l_u, L, and d are the number of the repeat units, distance between grafting points (the unit length) and the length and diameter of the rod, respectively. *Below*: End-on schematics of (a) nematic (Nem) and (b) hexagonal (Hex) phase of hairy-rod molecules. Adapted from [24]

value, M_n^*, is obtained when the two free energies are equal

$$M_n^* \approx M_u \frac{l_K^2 l_u}{v} \ln \frac{e\Omega_N}{f\Omega_H} \, . \tag{4}$$

Equation 4 implies that for $M_n > M_n^*$, the Hex phase should be observed. Approximating $\Omega_N \approx \Omega_H$ which is justified near the glass transition temperature, the threshold molecular weight is given as

$$M_{n0}^* \approx M_u \frac{l_K^2 l_u}{v} \ln \frac{e}{f} \, . \tag{5}$$

The relation between molecular weight and degree of alignment can be estimated as follows [55]. PF2/6 is regarded as a hairy-rod molecule in which each chain consists of stiff segments of diameter d and length l_K^{HR}, where l_K^{HR} is the Kuhn length of the rod (Fig. 14). The Kuhn segments are assumed to align independently, the total orientational configurational space for the

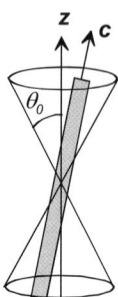

Fig. 14 The illustration of the alignment in 3D consideration. The vectors z and c represent the alignment direction and the backbone of a rigid molecule, respectively. Reprinted with permission from [55]. © (2005) by the American Chemical Society

chain being

$$\Omega_i(t) \approx \left(\frac{d}{l_K^{HR}}(1+C_i t)\right)^{M_n l_u / M_u l_K^{HR}}, \qquad (6)$$

where $t = T - T_g$ is the temperature measured from the glass transition temperature T_g and C_i is a phenomenological coefficient which describes the change in the orientational freedom of a single segment with temperature. Subindices $i = $ N, H refer to "Nem" or "Hex", respectively. When alignment between phases (Ω_i) is considered, the equation for the binodal T vs. M_n can be linearized to yield t as

$$t^* \approx A\left(1 - \frac{M_{n0}^*}{M_n^*}\right), \qquad (7)$$

where A incorporates all the phenomenological constants.

The implications of this theory are as follows: Eq. 4 yields a threshold weight separating a LMW, $M_n < M_n^*$, from HMW, $M_n > M_n^*$, regime. Equation 6 depicts the degree of alignment as a function of T and M_n. Equation 7 gives an approximate expression for the Hex–Nem coexistence line above T_g. When the experimental dimensions of the molecule are introduced into the theory, the Hex–Nem phase transition is predicted as a function of M_n. When $T^* \sim T_g$, Eq. 5 predicts this transition at $M_{n0}^* \sim 10$ kg/mol for PF2/6 [24]. The limit in the case $M_n \gg M_n^*$ is obtained by the constant A extracted from experiment. Finally, the binodal, Eq. 7, is an interpolation between these limiting cases.

The phase diagram plotted in Fig. 15 compiles the theoretical result alongside experimental data for $T > T_g$ [24]. For lower M_n only the Nem phase is possible above T_g. At lower T the Nem–Hex transition as a function of M_n is seen at M_n^* whereas the Hex–Nem transition as a function of temperature is observed for HMW materials. As seen in Fig. 15 the experimental data are in surprisingly good agreement with the theory although the theory only con-

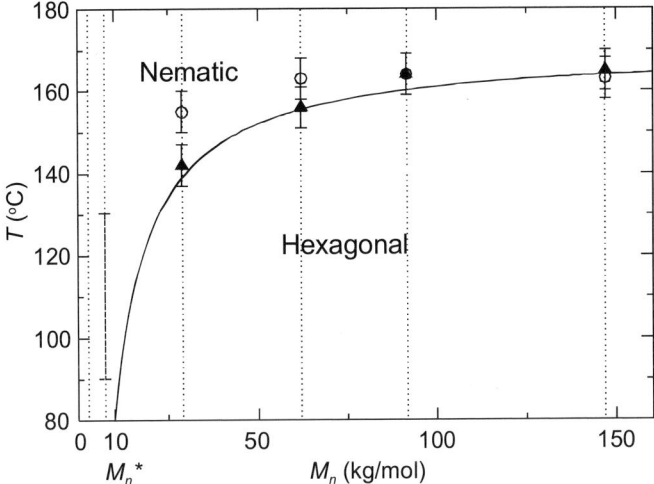

Fig. 15 Theoretical phase diagram of PF2/6 as a function of M_n for $T \geq T_g$ and the compilation of the corresponding experimental results. M_n^* defines low (LMW, $M_n < M_n^*$) and M_n high (HMW, $M_n > M_n^*$) regimes. The *solid line* shows the theoretical Nem–Hex phase transition. The Hex–Nem transitions based on DSC and XRD measurements are marked by *solid triangles* and *open circles*, respectively. The grid-lines correspond to the M_n of experimentally studied materials. The vertical bar at 90–130 °C for $M_n = 8$ kg/mol shows the position of occasionally seen Hex traces. Reprinted with permission from [24]. © (2005) by the American Physical Society

siders monodisperse chains whereas experiment contains polydisperse rods. What is not explicitly considered in this theory is the important fact that the hexagonal unit cell does not contain one but three polymer chains; a feature which stems from the 5-fold symmetry of the chain helical conformation. This suggests that the energy scale relating to the formation of three chains per unit cell is a comparatively small quantity. The mean-field approach itself is quite general and, with appropriate modifications, it is likely other CPs systems can be similarly understood.

Various PFs [61–63] and their pyrrole analogues [64] functionalized with chiral side chains exhibit NLO character and it is not clear how the chirality, helicity and NLO character are related. Because PF2/6 has become a model test system in the study of helical PFs, fully understanding its intramolecular structure can serve as a reference point for further comparisons. The molecular level heterogeneity intrinsic in the single chain structure of PF2/6 insures a high level of frustration and thus it represents a model polymer for paracrystals and paracrystallinity [24].

The near 5-fold helical symmetry of the main chains suggests a near pentagonal shape to the polymer chain and, with respect to interchain self-assembly of the Hex phase, the three-chain per unit cell is the basic packing motif for soft pentagons. This can be contrasted with the small two-fold reconstruction one observes in packing of rigid pentagons (cf. [65]). Finally, we note that the above phase diagram holds only at ambient pressures and this is not the whole picture. Guha et al. [58, 66] have studied PF2/6 at high pressures by spectroscopic methods and found a phase transition involving a potential planarization of the chains at 20 kbar for the bulk and 35 kbar in thin film samples.

4.1.2
Linear Side Chain PF8

Linear side chain PF8 (or PFO) [17, 19, 20, 22, 67–76] represent another heavily investigated PF archetype and can be considered a counterpart to PF2/6. Apparent in all the reported findings to date is the conspicuous physical difference between these two materials despite their close chemical resemblance; they have a similar chemical stoichiometry and structure. The major distinguishing structural characteristics are the chain morphology and phase behavior. PF2/6, as already discussed, appears to be helical with nematic, hexagonal and isotropic phases. Single chain conformational studies indicate a broad distribution of conformational isomers whose inter fluorene–fluorene dihedral twist is approximately 45° [53]. Thus, photoabsorption and photoluminescence of PF2/6 are essentially phase independent. In contrast, as described in the single molecule section, PF8 is both more planar and more diverse, exhibiting upwards of three distinct conformational isomer families (denoted as C_α, C_β, C_γ) each with a different conjugation length and mean torsional angle between the monomers [17]. These intramolecular variations are paralleled by an increased level of interchain structures. PF8 is decidedly polymorphic with a crystalline α phase and meta stable crystalline α' phase, as well as amorphous, glassy, nematic and isotropic phases. Accordingly, the isomers and different "phases" of PF8 show significant photophysical differences despite their identical chemical make up.

In intermolecular (crystallographic) terms, the nematic and isotropic phase can occur at high temperatures (transitions at $\sim 160\,°C$ and $\sim 300\,°C$, respectively) while it is generally accepted that at the room temperature the solid state is dominated by a stable crystalline α phase. This was first proposed to manifest a zigzag conformation with the fiber periodicity of 33.4 Å corresponding to four monomer units [67]. Later on, an orthorhombic lattice of eight chains with the space group $P2_12_12_1$ and lattice parameters $a = 25.6$ Å, $b = 23.4$ Å, and $c = 3.36$ Å (crystallographic **c** axis being along the polymer backbone) and theoretical density 1.041 g/cm^3 was proposed [73] (c.f., Fig. 16). This proposed picture was recently challenged by an alternative model based on a tetraradial construction of the side chains and a unit cell space group $Pnb2_1$ [76]. This model is illustrated in Fig. 17 and employs the following logic. At the onset PF8 was constrained to a planar zigzag conformation with, after accounting for the *n*-octyl sidechains, a four monomer periodicity along the **c** axis. In this model the planar backbone does not preclude the possibility of a left/right

Structure and Morphology of Polyfluorenes in Solutions and the Solid State 247

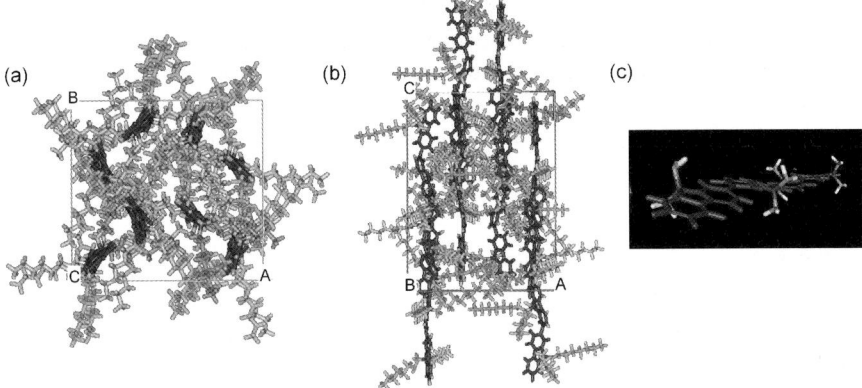

Fig. 16 Proposed model of molecular packing in the crystalline phase of PF8: energy-minimized structure (backbones in *red*) as viewed along **a** the **c** axis and **b** the **b** axis; **c** an expanded view with side chains hidden to demonstrate twisting of the backbone. A few selected pairs of *n*-octyl side chains are highlighted in *green* in (**a**) and (**b**) to allow for easier identification. Reprinted with permission from [73]. © (2004) by the American Chemical Society

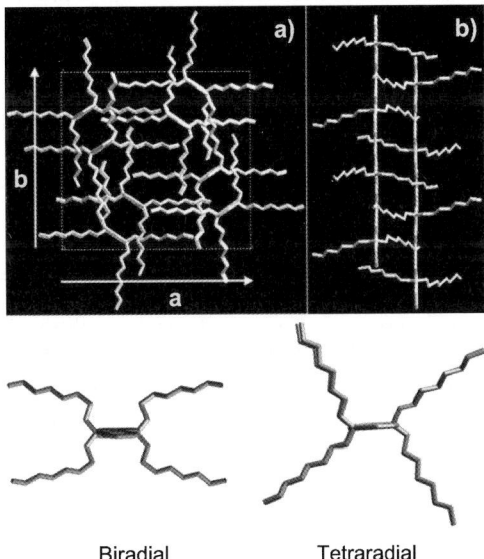

Fig. 17 *Above*: **a** c axis projection of the refined PF8 crystal structure. **b** Projection of the two-chain pattern along the 1–1–0 direction. *Below*: Projections of the PF8 chain conformation along the *c*-axis for the biradial and tetradial configuration of the *n*-octyl side chains. The hydrogen atoms are omitted for clarity. Reprinted with permission from [76]. © (2007) by the American Chemical Society

handedness for each PF8 chain and so, by invoking a systematic sequencing of either biradial and tetraradial side chain conformations [2], an efficient space filling packing of the n-octyl chains could be realized. Brinkmann [76] systematically examined a series of postulated packing schemes and found that the simulated SAED patterns compared most favorably with the tetraradial construction whose refined structure is shown in Fig. 17.

Another crystalline form, metastable α', is comprised of a slightly modified orthogonal lattice along the **b** axis ($b = 23.8$ Å) [75]. Non-crystalline room temperature phases include an (optically isotropic) amorphous phase and a metastable so-called β "bulk" phase (not to be equated with the C_β conformer and β-type optical emission). The overall phase behavior also includes a glassy g phase [12]. Even as complicated as this picture appears it still may be viewed as an oversimplification. A general depiction of the phase behavior is illustrated in Fig. 18.

Some special attention should be placed on the β (bulk) phase as well. Although this form is often reported as being non-crystalline, it gives rise to sharp Bragg reflections commensurate with lamellar order with a long period of 12.3 Å [74] and fiber periodicity of 16.6 Å (which corresponds to two monomer units) [67]. Thus, it differs from "real" crystals in the sense that it is mesomorphic. This phase also includes the presence of absorbed solvent and may be obtained by extended exposure to solvent vapor or solvent (cf., the solvent section). In particular, it has been found to appear as an intermediate step in the transformation from the solvent induced clathrate-like structure to the solvent-free well-ordered α phase [74]. The α and β (bulk) phases may coexist and are closed related to one another but are still structurally incompatible.

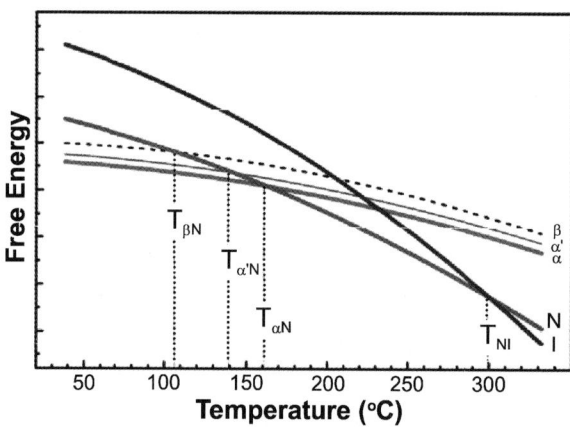

Fig. 18 Schematic free energy diagram, showing relative thermodynamic stabilities of various phases of solid state PF8. Reprinted with permission from [74]. © (2005) by the American Chemical Society

In regards to the intramolecular structure (as discussed in detail in the single molecule section) there may well be inhomogeneities in local structure arising from the side chain conformational isomerism (with either *anti* or *gauche* alkyl chain conformations). The three aforementioned conformational isomer families [17] reflect mean fluorene–fluorene torsion angles between adjacent fluorene monomers of approximately 135, 160, and 150° [17] (or 138, 165, and 155° [19]). Of these, it is the 150° (i.e., C_γ) conformer that is most likely to be incorporated into thermally annealed PF8 polymer samples (and thus be incorporated into the α crystal phase). Brinkmann's [76] planar model has many attractive structural qualities and the scattering data is noteworthy but this work includes no spectroscopic measurements and, without corroborating optical emission data, the backbone conformation in the PF8 crystal cannot be genuinely shown to be truly planar. It is also worth noting that the order–disorder transition observed at 80 °C (cf, Fig. 18) involves the loss of intra *and* interchain structural ordering [70].

4.1.3
Other Polyfluorenes

Apart from the above two examples, PF2/6 and PF8, the molecular structures and phase behavior of other PFs are known in far less detail. Two fluorene–thiophene copolymers, F8BT [77] and F8T2 [78], have been found to form lamellar semi-crystalline structures. An example of GIXRD data from a thin film F8BT sample is depicted in Fig. 19. Elsewhere the nature of the intra- and intermolecular interactions in F8BT have been studied by spectroscopic methods at elevated pressure [79]. As a general trend, by changing the side chain length, the interlayer spacing of a lamellar PF could be systematically changed [80].

Another interesting example is PF6 [81, 82]. Qualitatively PF6 exhibits many similarities to PF8 and the corresponding crystal phases are identified by Su et al. [81]. However, the structure is not identical and, while the α and α' phases of PF8 are both orthorhombic, the α and α' phases of PF6 are, respectively, monoclinic and triclinic. Interestingly, the analogous β "bulk" phase is characterized by a interlamellar periodicity of 14 Å and this is larger than that in PF8. The phase transition temperatures are also shifted and, while the nematic phase of PF8 appear at around 160 °C, the nematic phase of PF6 is not observed until at 250 °C [81].

4.1.4
PF Oligomers

Systematic structural studies of PF monomers and oligomers provide complementary structural information. Single crystal studies by Leclerc et al. [83] of 2,7-dibromo-9,9-dioctylfluorene-chloroform (or just the monomer it-

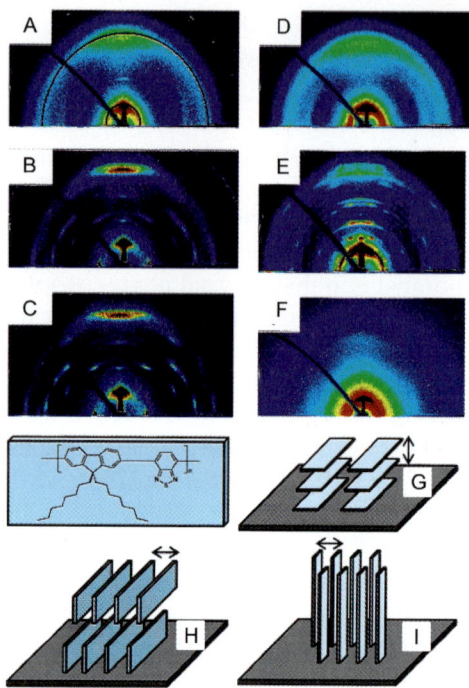

Fig. 19 GIXRD data for F8BT with molecular weight 255 kg/mol (**A–C**) and with molecular weight of 9 kg/mol (**D–F**). (**A** and **D**) Pristine, (**B** and **E**) annealed to T_g and slowly cooled, and (**C** and **F**) annealed to T_m and slowly cooled. The inner and outer rings in (**A**) correspond to the (001) and (004) reflections, respectively. The most likely orientations of the polymers with respect to the substrate is shown in (**G**), with the π-stacking direction indicated by *arrows*. Reprinted with permission from [77]. © (2005) by the American Chemical Society

self [84]), the base monomer of PF8, observe a planar fluorene core with the octyl side chains adopting an all *anti* conformation. A crystallographic study of the thiol-capped F6 monomer, 2,7-di(2-thienyl)-9,9-dihexylfluorene, by Destri et al. [85] yields similar results. This contrasts with the presence of *gauche* conformations in the PF polymers. With oligomers one can employ monodisperse samples and step-wise study the effect of increasing chain length of any number of properties.

Oligofluorenes studies have been performed by Wegner and coworkers [18, 57, 86, 87] and others [88, 89]. The packing frustration evident in the PF2/6 polymer is already present in the chains as short as a trimer [89]. Even at this short length thin films exhibit mesotropic-type phases [89] as opposed to the crystalline monomer samples.

A representative result is plotted in Fig. 20. While the intramolecular characteristics approximate those of high (molecular weight) PF2/6, the intermolecular assemblies of low molecular weight PF2/6 and F2/6 differ. As

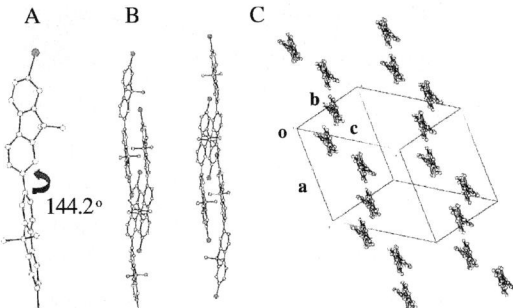

Fig. 20 The single crystal structure of (7,7′-dibromo-9,9,9′,9′-tetrakis(2-ethylhexyl)-2,2′-bifluorene) (or DBBF): **A** the aromatic framework shown without the aliphatic sidechains; **B** the chain packing showing the relation among the different molecules in the unit cell. **C** The 3D packing, projected along the molecular long axis. Note that the conformational angle between the two fluorene moieties in DBBF is 144.2°. Reprinted with permission from [57]. © (2005) by Wiley-VCH Verlag GmbH & Co. KGaA

discussed in the section above, LMW PF2/6 forms a nematic phase [24]. In contrast F2/6 oligomers develop a smectic B phase [57]. This difference arises from the fact that the PF2/6s are polydisperse whereas oligomers are monodisperse.

Another structural attribute that can be achieved in branched side chain PF oligomers are samples containing a single chiral enantiomer [88, 89]. These samples are intrinsically chiral and thus are optically active and can produce circularly polarized emission. Chiral functionalized PFs exhibit analogous properties [62, 63, 90].

Oligomers with n-alkyl functionalization have also received attention. A particularly intriguing question is in regards to the unusual β phase conformational isomer isolated PF8 and its presence in F8 oligomers. Tsoi et al. [91] confirm the presence of the β phase in very short oligomers but, by monitoring the emission energy as a function of chain length, estimate that its conjugation length actually extends up to 30 ± 12 monomers in longer oligomer samples. This observation of β-type emission in short oligomers reinforces the notion that the polymer crystal phases may not be directly correlated with the local conformational state of the polymer backbone.

4.2
Macroscopic Alignment of Polyfluorene Chains and Crystallites

4.2.1
Alignment

Achieving controlled macroscopic alignment of PF films is important for a wealth of reasons. Oriented samples exhibit anisotropic charge transport properties with enhanced mobilities and PF examples include F8BT [92],

PF8 [93], and F2/6 oligomers [94]. Major consequences also appear in the optical properties. In PFs the $\pi-\pi^*$ transition dipole is arrayed nearly parallel to the chain axis and, in oriented samples, it is possible to achieve polarized EL or PL. This was demonstrated very early on in PF2/6 [95] and PF8 [96]. Here ellipsometry is an extremely useful analytical method and the optical constants of spin cast PF2/6 [97, 98] and F8BT [99] thin films have been widely measured. As-spun films exhibit strong optical anisotropy when probed with light having in-plane and out-of-plane polarization. S-Polarized light (i.e., in the plane of the films) undergoes a large absorption with a highly dispersive refractive index. Light polarized normal to the film surface experiences very low absorption and a weakly dispersive refractive index. These observations are consistent with a picture that within the as-cast films the polymer backbones are parallel to the surface but randomly oriented in this plane (i.e., a classic planar-type structure). Thermal annealing of PF films atop rubbed polyimide [97] lead to appreciable uniaxial alignment of the polymers parallel to the rubbing direction and this reorients the optic axis along the rubbing direction. This situation is also manifested in stretch-aligned PF fibers that have been prepared by mechanical drawing at temperatures above the glass transition temperature [24, 59, 67]. An interesting variation of this method employs mechanical stretching of a blend [100] (see the discussion of microfibers in the next section).

Advanced procedures to assess the extent of PF uniaxial alignment have been forwarded by Bradley et al. [101] but the nematic morphology intrinsic to PFs introduces significant secondary complications [102]. Although the majority of PF literature generally focuses on macroscopic alignment of the chain axes, it should be noted that large-scale structural anisotropy of CPs extends to directions [52, 103] orthogonal to the chain axis.

There are notable instances in which improvements in the structural characterization of aligned fibers has led to better understanding of the underlying optical properties. Galambosi et al. [104], for instance, were able to relate fiber XRD of PF2/6 to IXS and thereby present a unified theoretical and experimental treatment and the DOS arising from both the backbone and side chains. With aligned fibers they were able to identify directional components of IXS spectra and DOS.

Although uniaxial alignment within "bulk" fibers often provides a useful setting for more fundamental studies, a more technologically important format is that of a thin film. The presence of multiple interfaces and a large surface to volume ratio often introduces greater structural complexities and these derive from the surface interactions. In thin, uniaxially aligned PF films the stiff PF backbones collectively align locally into a domain and this ensemble, on average, lies in the plane of the film and the director points along a preferred in-plane axis. In the simplest case there is an oriented nematic phase. If, however, the PF chains also include long-range meridional and/or equatorial translational order, there can be aligned smectic or crystalline phases. This leads to opportunities for a secondary alignment and is referred to as biaxiality. Biaxially aligned PFs manifest higher structural hierarchies in their textured crystallites (vide infra).

A common method for achieving uniaxial alignment of PF thin films is based on simply spin-casting of the CP onto a templating substrate (typically a rubbed polyimide layer). In regards to PFs, Miteva et al. [95, 105–109] performed much of the original work. Polyimide is electrically insulating and so if charge transport through the alignment layer is a necessity (i.e., in an LED device) the alignment layer can also be doped with an electroactive material [110]. The presence of a second component in the templating layer modifies the substrate/polymer interface and thus the nature of the uniaxial alignment. The degree of chain alignment is a very strong function of system and process parameters. These include solvent [21] and polymer molecular weight [111].

Alternative methods, including friction transfer [71] and directional epitaxial crystallization [76], have also been used successfully in the alignment of PFs. In the friction transfer method one provides a crystalline templating substrate, (e.g., PTFE) and then applies a CP over layer. As in the case of a rubbed substrate, the CP does not significantly impact the orienting ability of the aligned substrate. This process is commonly referred to as graphoepitaxy.

Directional epitaxial crystallization requires a crystallizable solvent that is heated above its melting point and then exposed to a spatial thermal gradient that sequentially recrystallizes the solvent. As the temperature drops below the melting point the crystallizing solvent front conveniently functions as both a nucleating and orienting surface [76].

Anisotropic orientation of PFs may also have been achieved in LB films and by LB film transfer methods [112] but in this case uniaxial alignment has not as yet been achieved. Alignment can also be achieved using a top down approach rather than a bottom up strategy. Samuel and coworkers [113] used a rubbed polyimide surface coated with an intervening thermotropic liquid crystal layer to produce uniaxial orientation starting at the top surface of a PF2/6 LED device.

4.2.2
Aligned Films of PF2/6

PF2/6 films, depending on the explicit PF2/6 material and environmental conditions, can manifest uniaxial [54, 114] or biaxial alignment [52, 55, 115]. Figure 21 plots schematically an example of the uniaxial chain alignment (parallel to the substrate surface) and equatorial patterning (perpendicular to the surface) by LMW and HMW PF2/6 thin films atop templating rubbed polyimide substrates. In keeping with the terminology introduced above, the experimental observations are as follows.

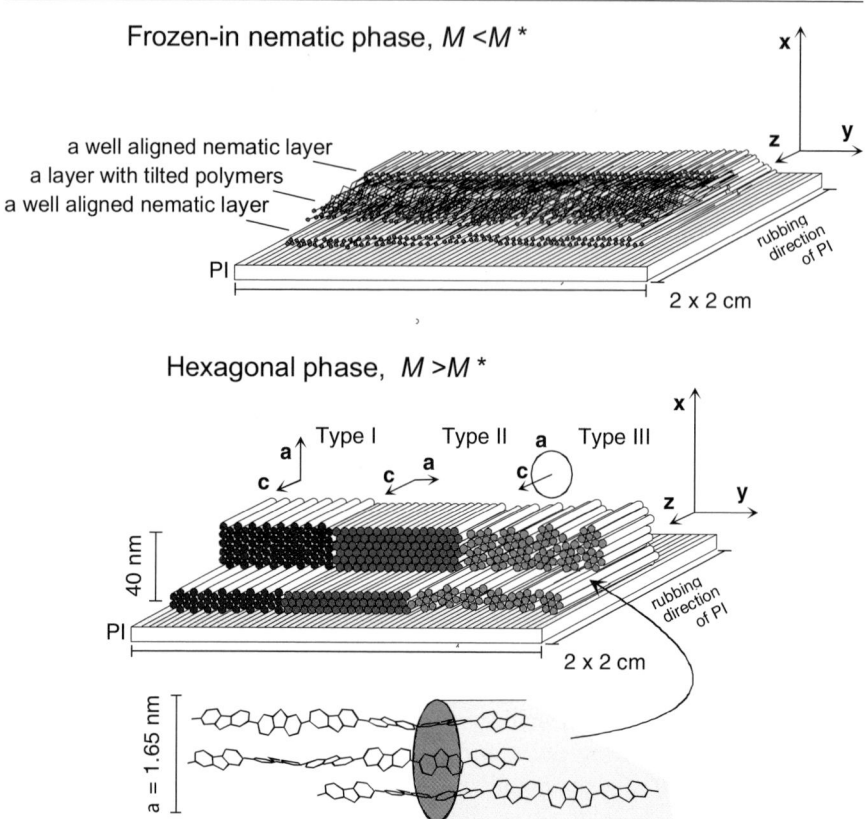

Fig. 21 Experimental geometry and schematics of the aligned PF2/6 films. *Above*: Uniaxially aligned frozen-in nematic PF2/6 microstructures. *Below*: Biaxially aligned hexagonal PF2/6 microstructures with crystallite types I–III. Assuming chain alignment (i.e., the c axis) along the rubbing direction, then the equatorial and meridional directions may be defined by the (*xy*0) plane and *z*-axis, respectively. See [114, 115] for details

Case 1: $M_n < M_n^*$. The structure of uniaxially aligned LMW PF2/6 has been recently studied by combination of optical spectroscopy, NEXAFS and GIXRD, Fig. 22 exhibits a representative NEXAFS data and the subsequent analysis. These data support a structural model in which there is a graded morphology such that the top and bottom surfaces exhibit extensive planar, uniaxial alignment while the film interior is less well oriented and includes both planar and tilted (i.e., non-planar) PF2/6 chains [114].

Case 2: $M_n > M_n^*$. Once the molecular weight exceeds a threshold value, M_n^* [24], PF2/6 forms hexagonal unit cells [51]. In this case the structure displays secondary effects due to the thin film geometry and the presence of interfacial forces. Here, there is a pronounced equatorial anisotropy with a measurable contraction of the lattice constants in the out-of-plane direction [52]. These surface interactions also give rise to a distinctive biaxial

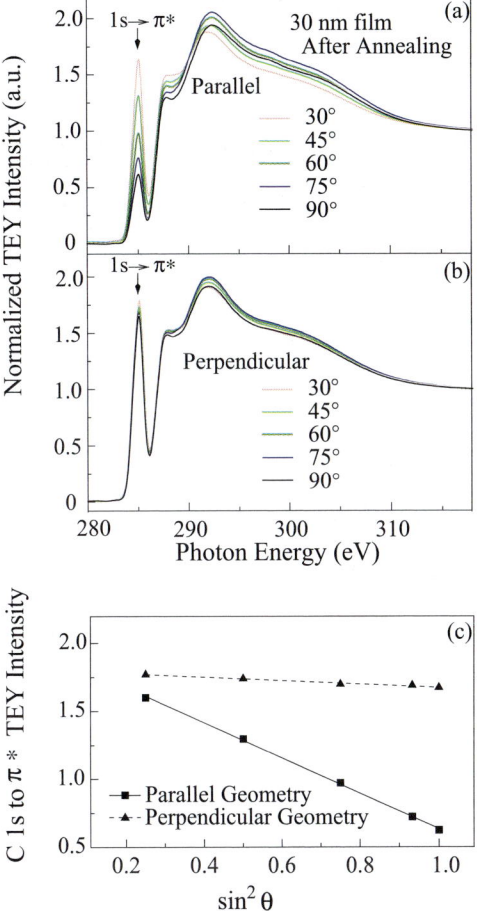

Fig. 22 Example NEXAFS data from an aligned PF2/6 film spin cast on rubbed PI. "Parallel Geometry" refers to E-field parallel to the rubbing direction. θ is the angle between surface normal and electric field vector of the incident light in conjunction with the rubbing direction of the polyimide substrate. See [114] for details

alignment marked by uniaxial (meridional) chain alignment and a multimodal equatorial anisotropy. In this situation the **c** axis is defined as the direction along the rod-like backbone (see GIXRD data in Fig. 23). The two dominant equatorial crystallite orientations, type I and II, have their respective **a** axes parallel and perpendicular to the surface normal [52]. These orientations form a mosaic texture and, in thin films, the crystallites extend through the entire thickness of the film [55]. These two equatorial orientations (i.e., type I and II crystallites) are also found to exhibit almost identical meridional orientation distributions [115] (Fig. 24). The equatorial ordering includes similar paracrystalline attributes in both the type I and II crystal-

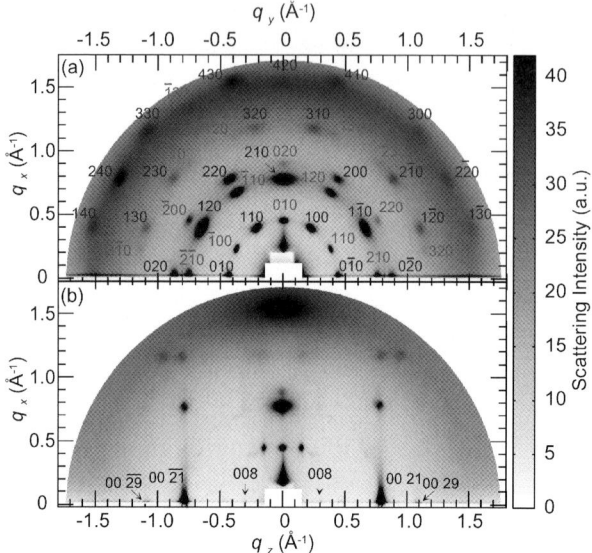

Fig. 23 GIXRD images from the PF2/6 film studied. **a** (xy0) plane, $\phi = 0°$ and **b** (x0z) plane, $\phi = 90°$. The GIXRD patterns were measured with the incident beam along the z- and y-axes, respectively. *Blue* and *red* indices show the primary reflections of the types I and II, respectively. Reprinted with permission from [115]. © (2007) by the American Chemical Society

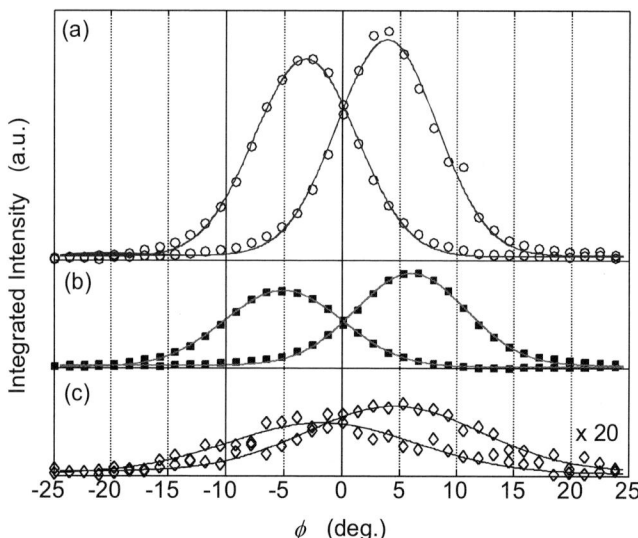

Fig. 24 Integrated intensities of the second hexagonal GIXRD reflections of aligned PF2/6 film corresponding to the sample of which the data is shown in Fig. 23. **a** Type I. **b** Type II. **c** Type III. *Solid lines* are corresponding Gaussian fits. Reprinted with permission from [115]. © (2007) by the American Chemical Society

lites. A small fraction of equatorial scattering is superimposed on a background of hexagonal phase polymer with a cylindrically isotropic orientation, denoted as type III [115]. This third type is most prominent in doped aligned films [116]. The overall model is thus summarized in Fig. 21.

The *uniaxial* PF2/6 alignment can be quantified in terms of the mean-field theory discussed previously. The degree of alignment in equilibrium (Ω) is a function of the number-averaged molecular weight (M_n) as described by Eq. 6. This prediction has been studied by photoabsorption in [55] and there it has been shown that the solid angle Ω is expressed in terms of the dichroic ratio in absorption (R) as

$$\frac{\Omega}{4\pi} \approx \frac{2}{R} + O\left(R^{-2}\right), \qquad (8)$$

where R is defined as

$$R = \left|\overline{E}_{\parallel}/\overline{E}_{\perp}\right|, \qquad (9)$$

where $\overline{E}_{\parallel}$ and $\overline{E}_{\perp}$ are the maximum values of the absorbance for light polarized parallel and perpendicular to the z axis (cf., Figs 14 and 25). In a perfectly aligned sample the z axis is parallel to the molecular c axis.

Equations 6, 8, and 9 illustrate the relationship between phase behavior and Ω as a function of M_n. In particular, they imply that if Ω increases exponentially with M_n then, correspondingly, R decreases exponentially.

Figure 26 plots R from the PF2/6 films as a function of M_n when the uniaxial alignment has been allowed to reach saturation (at elevated temperature) in the Nem phase regime (cf., Fig. 15). The overall picture for the alignment under similar conditions is that the degree of alignment first increases (R linearly) and then drops (R exponentially) with M_n. So LMW PF2/6 behaves quite differently than the HMW material, not only in terms of self-assembly but also in terms of the overall alignment.

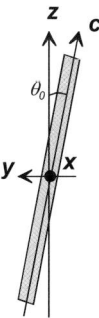

Fig. 25 The illustration of the alignment in a 2D consideration. The vectors z and c represent the alignment direction and the backbone of a rigid molecule, respectively. Reprinted with permission from [115]. © (2007) by the American Chemical Society

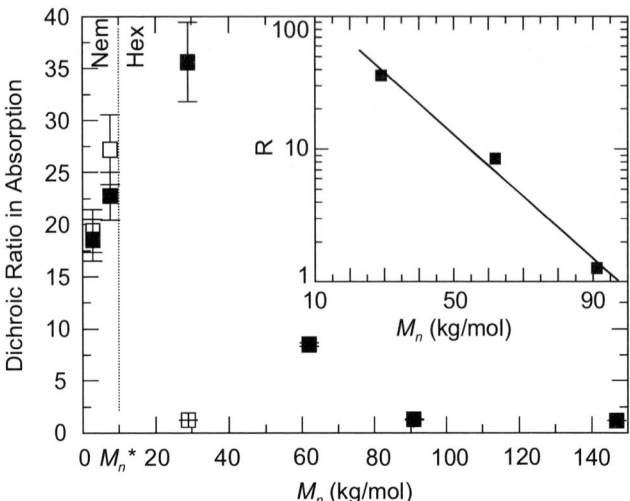

Fig. 26 Dichroic ratios in absorption (R) as a function of M_n as measured at 380 nm. The *open squares* correspond to the samples annealed at 80 °C for 10 minutes and the *solid squares* those at 180 °C for 18 h. A *dotted line* shows M_n^*, the Nem–Hex transition point of PF2/6. An *inset* shows the region of an exponential drop between $M_n = 10^4 - 10^5$ g/mol and a linear fit. Reprinted with permission from [55]. © (2005) by the American Chemical Society

The maximum degree of alignment as a function of M_n is achieved at the boundary of these regimes, at around $M_n^* \sim 10$ kg/mol. These regimes have been characterized as follows [55].

- $M_n < M_n^*$: The small LC regime, regime 1. Because the lengths of molecules in the regime 1 are approximately twice the persistence length [23], flexibility is not dominant and therefore the alignment increases with M_n [117].
- $M_n > M_n^*$: Chain flexibility now becomes important. If inflexible, the molecules in the melt would align perfectly apart from thermal fluctuations. As predicted by Eq. 6, the flexibility changes the Ω scaling as $\Omega \sim (\text{const.})^{M_n}$. An exponential increase in Ω implies an exponential decay in R. This is experimentally confirmed between 10–100 kg/mol in the inset of Fig. 26.
- $M_n \gg M_n^*$: R reaches its minimum (i.e., it cannot decrease further from unity).

Although the alignment is discussed under the assumption of a Nem phase, the same progression holds for the Hex phase with one exception. There is no regime 1 because the Hex phase does not exist when $M_n < M_n^*$.

These above concepts do not differentiate between which crystallite types actually exist in PF2/6 films (see Fig. 21). The connection between uniaxial and biaxial orientation of crystallites has been described in [115]. In this example R is first related to the 2D orientational order parameter s with

$$s = \frac{R-1}{R+2}. \tag{10}$$

The order parameter s links R to the mosaic distribution of the azimuthal rotation angle about the surface normal (ϕ). The former is measured using optical absorption spectroscopy whereas the latter is measured separately for each crystallite types using GIXRD. In this task it has been assumed that the rod-like molecules are always parallel to the (0yz) plane (i.e., perfectly planar alignment) and a two-dimensional order parameter can be given as

$$s = \langle 2\cos^2\theta - 1 \rangle = \int f(\theta) \cos 2\theta \, d\theta. \tag{11}$$

At a high degree of orientation, this may be simplified by $\cos\alpha \simeq 1 - \alpha^2/2$

$$s \simeq 1 - 2\theta_0^2, \tag{12}$$

where $\theta_0 = \sqrt{\langle\theta^2\rangle}$ may be physically interpreted as the angle accessible for the rotational motion of a molecule (see Fig. 25).

In one experimental test case [115] R corresponded to measured values $\theta_0 = 11°-15°$. These numbers were then compared to those obtained by GIXRD. For the three crystallite types I–III respective values of $\theta_0^I = 8.8 \pm 0.2°$, $\theta_0^{II} = 9.9 \pm 0.2°$, and $\theta_0^{III} = 15 \pm 1°$ were observed. The difference between the GIXRD and optical measurements stems from the fact that GIXRD is preferentially sensitive to crystalline material. To reconcile this difference it was necessary to include a non-crystalline volume fraction of $g_{nc} \simeq 0.06...0.12$; a value which was comparable to the value estimated from diffuse scattering ($\sim 10\%$).

4.3
Surface Morphology

The surface and interfacial morphologies within CP films and their blends are clearly an extremely important aspect of many device applications in which charge generation and energy transfer are key (e.g., see [118–120]). The morphology of many PFs is well-known and PF8 surfaces, for example, often consist of nanometer-sized crystalline grains [73, 76]. Ample studies have shown that the surface quality of the PF films is a strong function of both the processing conditions and the explicit functionalization. The molecular level attributes already discussed, such as chain length, side chain branching, and molecular weight, are equally important in establishing the interfacial structure.

Teetsov and Vanden Bout [121–123] have used SNOM to study linear side chain PFs as a function of side chain length. These authors found polymer-

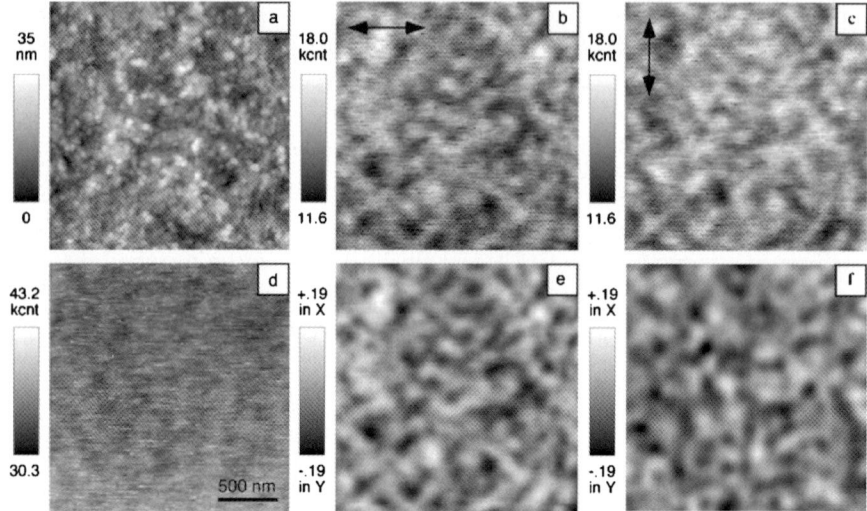

Fig. 27 2 × 2 μm SNOM images of annealed films of PF6 (**a** and **d**), PF8 (**b** and **e**), and PF12 (**c** and **f**): topography (**a–c**) and SNOM fluorescence anisotropy (**d–f**). Reprinted with permission from [122]. © (2002) by the American Chemical Society

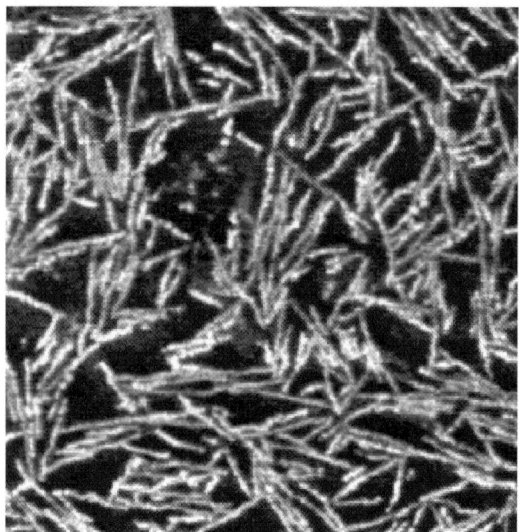

Fig. 28 Tapping mode AFM phase images of a thin PF8 deposit on mica. The area is 1.0 × 1.0 μm. Reprinted with permission from [124]. © (2002) by the American Chemical Society

specific ribbon-like domains 15–30 nm in size. An example is shown in Fig. 27. As a result they concluded that it was the molecular scale interchain interactions that were ultimately responsible for these systematic variations in the morphology.

Surin et al. [124] studied several PF films on mica and reported that the microscopic morphology is also strongly correlated with the molecular architecture. PFs with branched side chains revealed a smooth featureless surface down to the nanometer scale whereas PFs, with linear side chains, formed networks of fibrillar structures in which the chains are closely packed. An example of PF8 fibrilles is shown in Fig. 28.

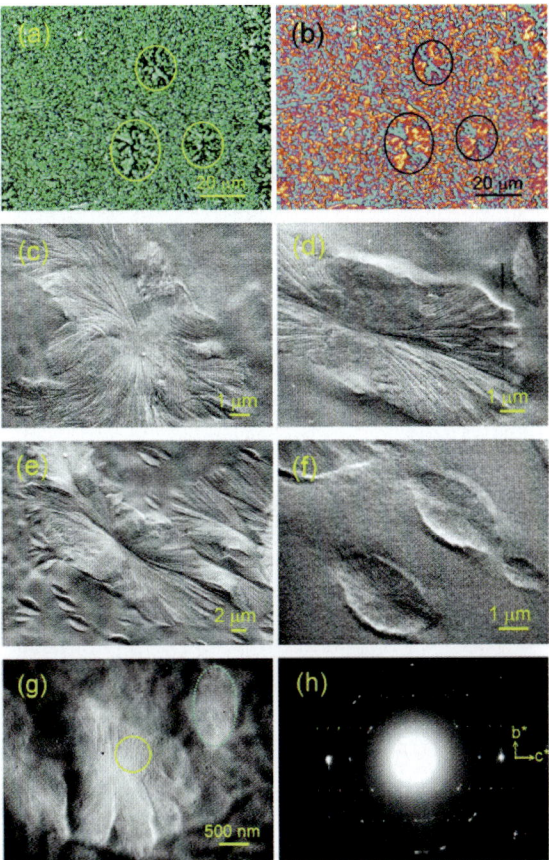

Fig. 29 Axialitic morphology of PF8 films melt-crystallized for 6 min at 145 °C, followed by a rapid cooling: **a,b** Polarized micrographs without or with gypsum plate inserted, **c–f** Secondary electron images at several locations under different magnifications, and **g** a representative BFI and **h** the corresponding SAED pattern. The axialites first appear as leaf-like entities (**f**), which grow and splay mainly in the axial direction (**d**) and become less anisotropic with transversely nucleated branches (**c**). A representative BFI image (**g**) and the corresponding SAED pattern (**h**, taken from the axialite at the central view, slightly rotated to align the b^*-axis with the meridian) indicate that molecular chains run transverse to the long axis of fibrillar features that correspond to slender edge-on crystalline lamellae. Note the presence of a junior-axialite in the *upper right* corner of (**g**). Reprinted with permission from [125]. © (2006) by the American Chemical Society

The surface morphology of aligned F8BT [111] and PF2/6 [56] films have been studied as a function of molecular weight. In the case of PF2/6 there is a crossover in the surface morphology between smooth (Nem phase with $M_n < M_n^*$) and rough (Hex phase, $M_n > M_n^*$) at the threshold molecular weight M_n^* [56]. The domains at the PF2/6 surface appear as aligned furrows and ridges with dimensions comparable to the crystallite size as obtained by XRD.

The examples reported in [56, 111] were of samples achieving a high level of uniaxial alignment. These films were annealed for several hours over which it was assumed that equilibrium had achieved. Far more variegated features are seen to develop during thermal annealing. Chen et al. [125] studied PF8 films from rapid quenching up to some minutes of thermal annealing. They identified a rich interplay of asymmetrically growing spherulites or axialites. These objects were composed of slender edge-on crystalline lamellas with the preference of growth in the axial direction (see the richly textured images in Fig. 29).

4.4
Higher Levels of Complexity—Nano and Microscale Assemblies

4.4.1
"Bottom-Up" Nanostructures

There is a wide range of so-called bottom-up methods for achieving complex structural hierarchies within polyfluorene nanostructures. Synthesis of block

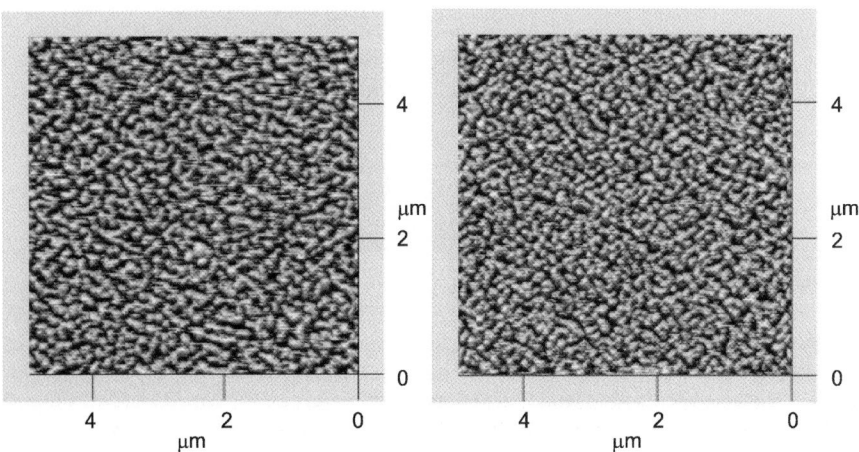

Fig. 30 AFM images (tapping mode) of thin films of the PF-*b*-PANI triblock copolymers. PF is equipped with 3,7,11-trimethyldodecyl and PANI with undecyl side groups, which have been spin-coated from two different solvents: (*left*) chloroform and (*right*) toluene. Reprinted with permission from [126]. © (2005) by Elsevier

copolymers represents one obvious approach for introducing additional competing interactions at the molecular level. Examples of microphase-separated structures of PANI/PF/PANI and PT/PF/PT triblock copolymers have been reported by Scherf and coworkers [126]. The polymers formed segregated domains of sizes ranging from 50 to 300 nm. One example is shown in Fig. 30. In these films the segregation can be controlled by the solvent. This allows selective tuning of the photoluminescence quenching [127]. Elsewhere, Jenekhe and coworkers [128] have combined PF2/6 with poly(γ-benzyl-L-glutamate). These three articles represent just a small slice of the types of materials that can be synthesized and self-assembled.

PF blends provide another clear route for controlling the nanoscale domain architecture. Unlike block copolymers, in which phase separation is constrained by chemical bonds, with blends one can adjust the conditions all the way from complete miscibility to macrophase separation. Examples of dir-

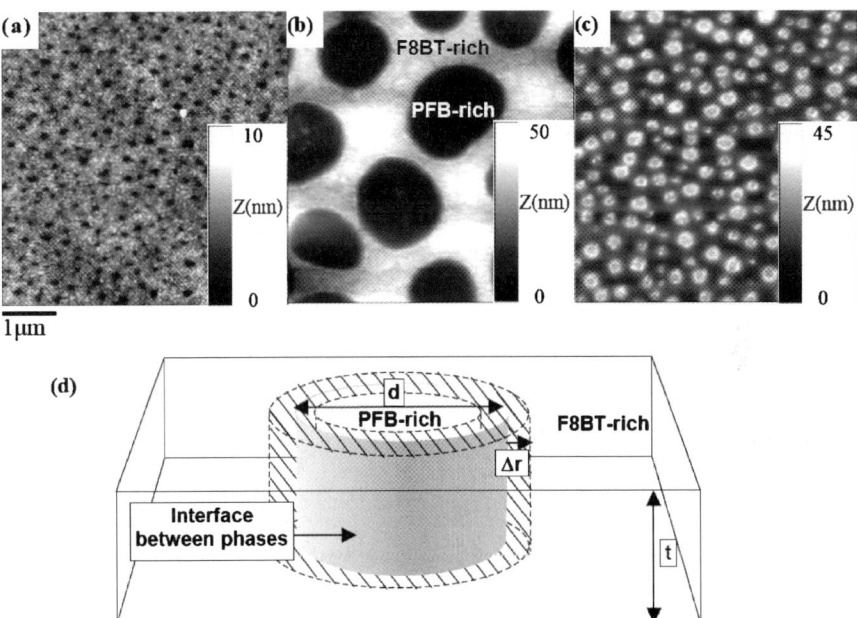

Fig. 31 AFM images for thin films of PFB/F8BT blends with ratios: **a** 1 : 5, **b** 1 : 1, and **c** 5 : 1. **d** 3D schematic representation of blended polymer film. The PFB-rich phase is represented by the volume within the *dark gray cylinder* and the F8BT rich phase is represented by the *hollow box* on the outside of the *dark gray cylinder*. Photogenerated charges within a thin cylindrical shell (*light gray region*) about the interface between the two mesoscale phases (*dark gray cylinder*), become collected charge at electrodes. t is the film thickness, d is the average diameter of the circular phase, and Δr is the distance over which charge can migrate, within the minor phase, to reach the interface. Reprinted with permission from [119]. © (2002) by the American Chemical Society

ect imaging of this segregation include a blend of PF8 and polystyrene [129], PF8 and PMMA [130], F8BT and PFB [131], F8BT, PFB, and TFB [132] or PF8 and F8BT [133–135]. A prominent example of the technological implications is shown in work by Friend et al. [119]. Here they analyzed the impact of varying composition and segregation in a PFB/F8BT blend in a series of photovoltaic devices. Examples of the observed morphological structures are shown in Fig. 31. The PFB and F8BT, respectively, function as hole- and electron-acceptors in the photovoltaic device and the quantum yield is strongly composition dependent. Moreover, Friend and coworkers showed that charge-transport was the main factor limiting the device performance and, additionally, demonstrated that this can be optimized by controlling the microphase separation.

There are examples demonstrating higher level structural order in PF networks through solution chemistry and gel formation (c.f., Fig. 5). In these cases the assemblies are based on cross-linked microcrystallites and physical bonding [31]. Covalently cross-linked PFs have also been introduced in the solid state [136–138]. Cross-linking, by means of photopolymerization of F3 or F7-containing LC oligomers, has been used to control charge carrier mobilities [137]. A cross-linkable F8Ox copolymer has been used in the fabrication of multilayer LEDs [136]. In this example the cross-linkable layer was utilized in multilayer fabrication to separate poly(3,4-ethylenedioxythiophene) and F8BT. Although extensive structural studies are not yet reported, we believe this is certainly another topic area in which in-depth structure studies would be beneficial.

4.4.2
"Top-Down" Nanostructures

Complementing the bottom-up methods which rely on competing molecular interactions are a number of "top-down" approaches for producing PF nanostructures. An intriguing microporous template-based method has been developed by Redmond and coworkers using either F8T2 [139] or PF8 [140–142] to prepare nanowires. In this method the bulk polymer, PF for example, is deposited onto the surface of a porous anodic alumina membrane having a typical pore size of a few hundred nanometers. Then the system is thermally annealed, often with applied pressure, to induce pore filling. Excess polymer is removed mechanically after it solidifies on cooling. Thereafter, the PF nanowires are released by soaking the template (typically in aqueous NaOH which etches away the template). Finally, the wire residue is washed with water and suspended in decane. Examples of these nanowires are shown in Fig. 32. Optical emission studies can identify polarized emission indicating that the PF chains tend to align parallel to the wire axis. This method can allow also for preparation of hollow PF nanotubes [142] (c.f., Fig. 33). These kinds of nanowires can be subsequently aligned on films. These authors have

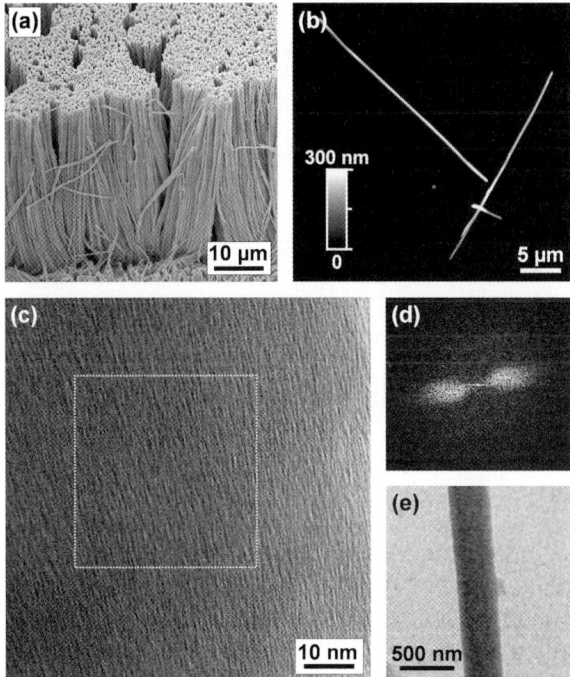

Fig. 32 a SEM image of a melt wetted PF8 nanowire array following template removal. **b** Tapping-mode AFM image of nanowires on a glass substrate. **c** High-resolution bright-field TEM image of a nanowire region. **d** 2D FFT of the region of the image indicated by the *white rectangle* in (**c**). **e** Lower magnification TEM image of the wire. Reprinted with permission from [140]. © (2007) by Elsevier

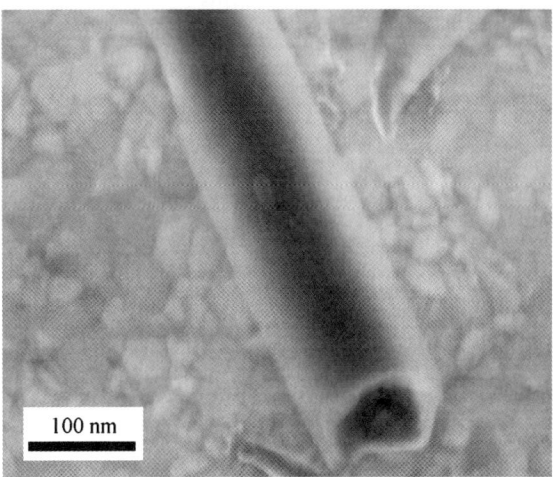

Fig. 33 SEM image of a PF8 nanotube. Reprinted with permission from [142]. © (2008) by the American Chemical Society

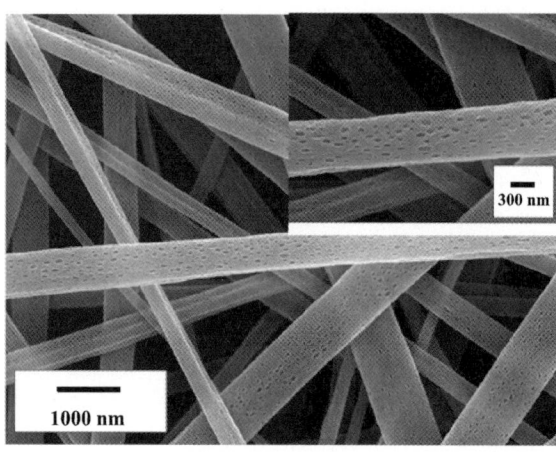

Fig. 34 SEM images of PF8/PMMA blend electrospun nanofibers with **a** 10 and **b** 50 wt. % of PF8. Reprinted with permission from [145]. © (2007) by the American Chemical Society

also shown the nanowires can exhibit a large fraction of β-type phase when PF8 is used [141]. It may well be that by placing PF8 into confined geometries one can achieve some degree of additional control over the phase that forms.

PF nanofibers are a second example of a top-down nanostructure. These fibers are conveniently prepared by electrospinning [143, 144], a technique applied to PFs in work by Chen et al. [145] and Jenekhe et al. [146]. An example of PF8 nanofibers is shown in Fig. 34. In this study various different PFs were blended with PMMA and depending on the molar ratio and mother solution used, either uniform or core-shell structures were obtained. PF aggregation in the electrospun fibers appears to be much smaller than that in the spin-coated films due to the geometrical confinement of the electrospinning process and this leads to higher luminescence efficiency. The fibers with PMMA demonstrate a full light-emitting spectrum as a function of composition [145]. Moreover, a blend of PF8 and MEH-PPV is utilized in the fabrication of field-effect transistors [146].

5
Conclusions and Outlook

PFs, and conducting polymers in general, exhibit an enormously complex range of structural behavior with a polymer specific hierarchal self-assembly spanning many different length scales. In a few archetypal systems, primarily PF2/6 and PF8, it has been possible to provide both a theoretical or modeling framework for understanding the origin of some of these structural forms and the phase behavior. At times it is necessary to analyze the structure starting at the molecular level but, in some cases, a coarser grain analyses has proven effective in providing appreciable insight. Ideally one would like a generalizable and self-consistent methodology that could scale from microscopic to macroscopic length scales given that much CP behavior originates at nanometer distances. Moreover such methods would greatly improve the strategies available for efficient molecular design and engineering. At the larger length scales PFs are necessarily influenced by both the device environment and operation. Thus, there should be an increased emphasis on in-situ investigations. The current range of structural studies has already given critical input for many areas of PF research, synthetic chemistry in particular has benefited. These studies also continue to have major implications for optoelectronic applications. Examples include photovoltaic cells, LEDs, FETs, and biosensors. As already noted, the many reports in this review represent only the tip of the proverbial iceberg in terms of what has and what can be achieved. We expect that structure studies will continue to play a key role in the years to come.

References

1. Hoeben FJM, Jonkheijm P, Meijer EW, Schennig APHJ (2005) Chem Rev 105:1491
2. Winokur MJ, Chunwachirasiri W (2003) J Polym Sci B 41:2630
3. Winokur MJ (2007) In: Skotheim TA, Reynolds JR (eds) Handbook of Conducting Polymers. CRC Press LLC, Boca Raton, FL, p 1
4. Menzel H (1996) In: Salamone JC (ed) Polymer Materials Encyclopedia. CRC Press, Boca Raton, FL, p 2916
5. Grosberg AY, Khokhlov AR (1994) Statistical Physics of Macromolecules. American Institute of Physics, Woodbury, NY

6. de Gennes PG, Prost J (1998) The Physics of Liquid Crystals. Oxford University Press, Oxford
7. Grimsdale AC, Müllen K (2006) Adv Polym Sci 199:1
8. Knaapila M, Stepanyan R, Lyons BP, Torkkeli M, Monkman AP (2006) Adv Funct Mater 16:599
9. Neher D (2001) Macromol Rapid Commun 22:1365
10. Scherf U, List EJW (2002) Adv Mater 14:477
11. de Gennes P-G (1979) Scaling Concepts in Polymer Physics. Cornell University Press, Ithaca
12. Becker K, Lupton JM (2005) J Am Chem Soc 127:7306
13. Dubin F, Melet R, Barisien T, Grousson R, Legrand L, Schott M, Voliotis V (2006) Nat Phys 2:32
14. Guillet T, Berréhar J, Grousson R, Kovensky J, Lapersonne-Meyer C, Schott M, Voliotis V (2001) Phys Rev Lett 87:087401
15. Clark J, Silva C, Friend RH, Spano FC (2007) Phys Rev Lett 97:206406
16. Ariu M, Lidzey DG, Sims M, Cadby AJ, Lane PA, Bradley DDC (2002) J Phys Condens Matter 14:9975
17. Chunwaschirasiri W, Tanto B, Huber DL, Winokur MJ (2005) Phys Rev Lett 94:107402
18. Marcon V, van der Vegt N, Wegner G, Raos G (2006) J Phys Chem B 110:5253
19. Arif M, Volz C, Guha S (2006) Phys Rev Lett 96:025503
20. Volz C, Arif M, Guha S (2007) J Chem Phys 126:064905
21. Banach MJ, Friend RH, Sirringhaus H (2004) Macromolecules 37:6079
22. Grell M, Bradley DDC, Long X, Chamberlain T, Inbasekaran M, Woo EP, Soliman M (1998) Acta Polym 49:439
23. Fytas G, Nothofer HG, Scherf U, Vlassopoulos D, Meier G (2002) Macromolecules 35:481
24. Knaapila M, Stepanyan R, Torkkeli M, Lyons BP, Ikonen TP, Almásy L, Foreman JP, Serimaa R, Güntner R, Scherf U, Monkman AP (2005) Phys Rev E 71:041802
25. Knaapila M, Garamus VM, Dias FB, Almásy L, Galbrecht F, Charas A, Morgado J, Burrows HD, Scherf U, Monkman AP (2006) Macromolecules 39:6505
26. Somma E, Loppinet B, Chi C, Fytas G, Wegner G (2006) Phys Chem Chem Phys 8:2773
27. Dias FB, Knaapila M, Monkman AP, Burrows HD (2006) Macromolecules 39:1598
28. Wu L, Sato T, Tang H-Z, Fujiki M (2004) Macromolecules 37:6183
29. Kitts CC, Vanden Bout DA (2007) Polymer 48:2322
30. Dias FB, Morgado J, Macanita AL, da Costa FP, Burrows HD, Monkman AP (2006) Macromolecules 39:5854
31. Rahman MH, Chen C-Y, Liao S-C, Chen H-L, Tsao C-S, Chen J-H, Liao J-L, Ivanov VA, Chen S-A (2007) Macromolecules 40:6572
32. Knaapila M, Dias FB, Garamus VM, Almásy L, Torkkeli M, Leppänen K, Galbrecht F, Preis E, Burrows HD, Scherf U, Monkman AP (2007) Macromolecules 40:9398
33. Knaapila M, Almásy L, Garamus VM, Ramos ML, Justino LLG, Galbrecht F, Preis E, Schref U, Burrows HD, Monkman AP (2008) Polymer 49:2033
34. Levitsky IA, Kim J, Swager TM (2001) Macromolecules 34:2315
35. Chen L, McBranch DW, Wang H-L, Helgeson R, Wudl F, Whitten DG (1999) Proc Natl Acad Sci USA 96:12287
36. Sirringhaus H (2005) Adv Mater 17:2411
37. Gaylord BS, Heeger AJ, Bazan GC (2002) Proc Natl Acad Sci USA 99:10954
38. Chi C, Mikhailovsky A, Bazan GC (2007) J Am Chem Soc 129:11134

39. Baker ES, Hong JW, Gaylord BS, Bazan GC, Bowers MT (2006) J Am Chem Soc 128:8484
40. Xu Q-H, Gaylord BS, Wang S, Bazan GC, Moses D, Heeger AJ (2004) Proc Natl Acad Sci USA 101:11634
41. Liu B, Bazan GC (2005) Proc Natl Acad Sci USA 102:589
42. Wang H, Lu P, Wang B, Qiu S, Liu M, Hanif M, Cheng G, Liu S, Ma Y (2007) Macromol Rapid Commun 28:1645
43. Ma W, Iyer PK, Gong X, Liu B, Moses D, Bazan GC, Heeger AJ (2005) Adv Mater 17:274
44. Burrows HD, Lobo VMM, Pina J, Ramos ML, Seixas de Melo J, Valente AJM, Tapia MJ, Pradhan S, Scherf U (2004) Macromolecules 37:7425
45. Tapia MJ, Burrows HD, Knaapila M, Monkman AP, Arroyo A, Pradhan S, Scherf U, Pinazo A, Pérez L, Móran C (2006) Langmuir 22:10170
46. Tapia MJ, Burrows HD, Valente AJM, Pradhan S, Scherf U, Lobo VMM, Pina J, Seixas de Melo J (2005) J Phys Chem B 109:19108
47. Burrows HD, Tapia MJ, Silva CL, Pais AACC, Fonseca SM, Pina J, Seixas de Melo J, Wang Y, Marques EF, Knaapila M, Monkman AP, Garamus VM, Pradhan S, Scherf U (2007) J Phys Chem B 111:4401
48. Knaapila M, Almásy L, Garamus VM, Pearson C, Pradhan S, Petty MC, Scherf U, Burrows HD, Monkman AP (2006) J Phys Chem B 110:10248
49. Burrows HD, Knaapila M, Monkman AP, Tapia MJ, Fonseca SM, Ramos ML, Pyckhout-Hintzen W, Pradhan S, Scherf U (2008) J Phys Condens Matter 20:104210
50. Al Attar HA, Monkman AP (2007) J Phys Chem B 111:12418
51. Lieser G, Oda M, Miteva T, Meisel A, Nothofer H-G, Scherf U, Neher D (2000) Macromolecules 33:4490
52. Knaapila M, Lyons BP, Kisko K, Foreman JP, Vainio U, Mihaylova M, Seeck OH, Pålsson L-O, Serimaa R, Torkkeli M, Monkman AP (2003) J Phys Chem B 107:12425
53. Tanto B, Guha S, Martin CM, Scherf U, Winokur MJ (2004) Macromolecules 37:9438
54. Knaapila M, Kisko K, Lyons BP, Stepanyan R, Foreman JP, Seeck OH, Vainio U, Pålsson L-O, Serimaa R, Torkkeli M, Monkman AP (2004) J Phys Chem B 108:10711
55. Knaapila M, Stepanyan R, Lyons BP, Torkkeli M, Hase TPA, Serimaa R, Güntner R, Seeck OH, Scherf U, Monkman AP (2005) Macromolecules 38:2744
56. Knaapila M, Lyons BP, Hase TPA, Pearson C, Petty MC, Bouchenoire L, Thompson P, Serimaa R, Torkkeli M, Monkman AP (2005) Adv Funct Mater 15:1517
57. Chi C, Lieser G, Enkelmann V, Wegner G (2005) Macromol Chem Phys 206:1597
58. Guha S, Chandrasekhar M (2004) Phys Stat Sol 241:3318
59. Knaapila M, Torkkeli M, Monkman AP (2007) Macromolecules 40:3610
60. Stepanyan R, Subbotin A, Knaapila M, Ikkala O, ten Brinke G (2003) Macromolecules 36:3758
61. Blondin P, Bouchard J, Beaupré S, Belletête M, Durocher G, Leclerc M (2000) Macromolecules 33:5874
62. Craig MR, Jonkheijm P, Meskers SCJ, Schenning APHJ, Meijer EW (2003) Adv Mater 15:1435
63. Lakhwani G, Meskers SCJ, Janssen RAJ (2007) J Phys Chem B 111:5124
64. Koeckelberghs G, De Cremer L, Persoons A, Verbiest T (2007) Macromolecules 40:4173
65. Schilling T, Pronk S, Mulder B, Frenkel D (2005) Phys Rev E 71:036138
66. Martin CM, Guha S, Chandrasekhar M, Chandrasekhar HR, Guentner R, Scanduicci de Freitas P, Scherf U (2003) Phys Rev B 68:115203
67. Grell M, Bradley DDC, Ungar G, Hill J, Whitehead KS (1999) Macromolecules 32:5810

68. Ariu A, Lidzey DG, Bradley DDC (2000) Synth Met 111/112:607
69. Kawana S, Durrell M, Lu J, Macdonald JE, Grell M, Bradley DDC, Jukes PC, Jones RAL, Bennett SL (2002) Polymer 43:1907
70. Winokur MJ, Slinker J, Huber DL (2003) Phys Rev B 67:184106
71. Misaki M, Ueda Y, Nagamatsu S, Yoshida Y, Tanigaki N, Yase K (2004) Macromolecules 37:6926
72. Rothe C, King SM, Dias F, Monkman AP (2004) Phys Rev B 70:195213
73. Chen SH, Chou HL, Su AC, Chen SA (2004) Macromolecules 37:6833
74. Chen SH, Su AC, Chen SA (2005) J Phys Chem B 109:10067
75. Chen SH, Su AC, Su CH, Chen SA (2005) Macromolecules 38:379
76. Brinkmann M (2007) Macromolecules 40:7532
77. Donley CL, Zaumseil J, Andreasen JW, Nielsen MM, Sirringhaus H, Friend RH, Kim J-S (2005) J Am Chem Soc 127:12890
78. Kinder L, Kanicki J, Swensen J, Petroff P (2003) Proc SPIE Int Soc Opt Eng 5217:35
79. Schmidtke JP, Kim JS, Gerschner J, Silva C, Friend RH (2007) Phys Rev Lett 99:167401
80. Yang G-Z, Wang W-Z, Wang M, Liu T (2007) J Phys Chem B 111:7747
81. Chen SH, Su AC, Su CH, Chen SA (2006) J Phys Chem B 110:4007
82. Vamvounis G, Nyström D, Antoni P, Lindgren M, Holdcroft S, Hult A (2006) Langmuir 22:3959
83. Leclerc M, Ranger M, Bélanger-Gariépy F (1998) Acta Cryst C 54:799
84. McFarlene S, McDonald R, Veinot JGC (2006) Acta Cryst E 62:o859
85. Destri S, Pasini M, Botta C, Porzio W, Bertini F, Marchiò L (2002) J Mater Chem 12:924
86. Chi C, Im C, Enkelmann V, Ziegler A, Lieser G, Wegner G (2005) Chem Eur J 11:6833
87. Jo JH, Chi CY, Hoger S, Wegner G, Yoon DY (2004) Chem Eur J 10:2681
88. Geng Y, Trajkovska A, Katsis D, Ou JJ, Culligan SW, Chen SH (2002) J Am Chem Soc 124:8337
89. Güntner R, Farrell T, Scherf U, Miteva T, Yasuda A, Nelles G (2004) J Mater Chem 14:2622
90. Oda M, Nothofer H-G, Lieser G, Scherf U, Meskers SCJ, Neher D (2000) Adv Mater 12:362
91. Tsoi WC, Charas A, Cadby AJ, Khalil G, Adawi AM, Iragi A, Hunt B, Morgado J, Lidzey DG (2008) Adv Funct Mater 18:600
92. Sirringhaus H, Wilson RJ, Friend RH, Inbasekaran M, Wu W, Woo EP, Grell M, Bradley DDC (2000) Appl Phys Lett 77:406
93. Redecker M, Bradley DDC, Inbasekaran M, Woo EP (1999) Appl Phys Lett 74:1400
94. Yasuda T, Fujita K, Tsutsui T, Geng Y, Culligan SW, Chen SH (2005) Chem Mater 17:264
95. Grell M, Knoll W, Lupo D, Meisel A, Miteva T, Neher D, Nothofer H-G, Scherf U, Yasuda A (1999) Adv Mater 11:671
96. Grell M, Bradley DDC, Inbasekaran M, Woo EP (1997) Adv Mater 9:798
97. Lyons BP, Monkman AP (2004) J Appl Phys 96:4735
98. Tammer M, Monkman AP (2002) Adv Mater 14:210
99. Ramsdale CM, Greenham NC (2002) Adv Mater 14:212
100. King SM, Vaughan HL, Monkman AP (2007) Chem Phys Lett 440:268
101. Gather MC, Bradley DDC (2007) Adv Funct Mater 17:749
102. Winokur MJ, Cheun H, Knaapila M, Monkman A, Scherf U (2007) Phys Rev B 75:113202

103. Sirringhaus H, Brown PJ, Friend RH, Nielsen NM, Bechgaard K, Langeveld-Voss BMW, Spiering AJH, Janssen RAJ, Meijer EW, Herwig P, de Leeuw DMR (1999) Nature 401:685
104. Galambosi S, Knaapila M, Soininen AJ, Nygård K, Huotari S, Galbrecht F, Scherf U, Monkman AP, Hämäläinen K (2006) Macromolecules 39:9261
105. Grell M, Knoll W, Lupo D, Meisel A, Miteva T, Neher D, Nothofer H-G, Scherf U, Yasuda A (1999) Adv Mater 11:671
106. Miteva T, Meisel A, Nothofer H-G, Scherf U, Knoll W, Neher D, Grell M, Lupo D, Yasuda A (1999) Proc SPIE Int Soc Opt Eng 3797:231
107. Miteva T, Meisel A, Grell M, Nothofer HG, Lupo D, Yasuda A, Knoll W, Kloppenburg L, Bunz UHF, Scherf U, Neher D (2000) Synth Met 111-112:173
108. Nothofer H-G, Meisel A, Miteva T, Neher D, Forster M, Oda M, Lieser G, Sainova D, Yasuda A, Lupo D, Knoll W, Scherf U (2000) Macromol Symp 154:139
109. Lupo D, Yasuda A, Grell M, Neher D, Miteva T (2000) Polyimide Layer Comprising Functional Material, Device Employing the Polyimide Layer, Manufacturing the Device, Eur Pat Appl. Sony International (Europe) GmbH 10785 Berlin (DE), Max Planck Institut für Polymerforschung 55128 Mainz (DE), EP 1 011 154, p 29
110. Meisel A, Miteva T, Glaser G, Scheumann V, Neher D (2002) Polymer 43:5235
111. Banach MJ, Friend RH, Sirringhaus H (2003) Macromolecules 36:2838
112. Worsfold O, Hill J, Heriot SY, Fox AM, Bradley DDC, Richardson TH (2003) Mater Sci Eng C 23:541
113. Godbert N, Burn PL, Gilmour S, Markham JPJ, Samuel IDW (2003) Appl Phys Lett 89:5347
114. Cheun H, Liu X, Himpsel FJ, Knaapila M, Scherf U, Torkkeli M, Winokur MJ (2008) Macromolecules (submitted)
115. Knaapila M, Hase TPA, Torkkeli M, Stepanyan R, Bouchenoire L, Cheun H-S, Winokur MJ, Monkman AP (2007) Cryst Growth Des 7:1706
116. Knaapila M, Torkkeli M, Lyons BP, Hunt MRC, Hase TPA, Seeck OH, Bouchenoire L, Serimaa R, Monkman AP (2006) Phys Rev B 74:214203
117. Khokhlov AR (1991) Theories based on the Onsager Approach. In: Ciferri A (ed) Liquid Crystallinity in Polymers. VCH Publishers, New York, p 97
118. Rozanski LJ, Cone CW, Ostrowski DP, Vanden Bout DA (2007) Macromolecules 40:4524
119. Snaith HJ, Arias AC, Morteani AC, Silva C, Friend RH (2002) Nano Lett 2:1353
120. Snaith HJ, Friend RH (2004) Thin Solid Films 451/452:567
121. Teetsov J, Vanden Bout DA (2000) J Phys Chem B 104:9378
122. Teetsov J, Vanden Bout DA (2002) Langmuir 18:897
123. Teetsov JA, Vanden Bout DA (2001) J Am Chem Soc 123:3605
124. Surin M, Hennebicq E, Ego C, Marsitzky D, Grimsdale AC, Müllen K, Brédas J-L, Lazzaroni R, Leclère P (2004) Chem Mater 16:994
125. Chen S-H, Su A-C, Chen S-A (2006) Macromolecules 39:9143
126. Asawapirom U, Güntner R, Forster M, Scherf U (2005) Thin Solid Films 477:48
127. Tu G, Li H, Forster M, Heiderhoff R, Balk LJ, Sigel R, Scherf U (2007) Small 3:1001
128. Kong X, Jenekhe SA (2004) Macromolecules 37:8180
129. Kulkarni AP, Jenekhe SA (2003) Macromolecules 36:5285
130. Biagioni P, Celebrano M, Zavelani-Rossi M, Polli D, Labardi M, Lanzani G, Cerullo G, Finazzi M, Duo L (2007) Appl Phys Lett 91:191118
131. Arias AC, MacKenzie JD, Stevenson R, Halls JJM, Inbasekaran M, Woo EP, Richards D, Friend RH (2001) Macromolecules 34:6005

132. Xia Y, Friend RH (2005) Macromolecules 38:6466
133. Cadby AJ, Dean R, Elliott C, Jones RAL, Fox AM, Lidzey DG (2007) Adv Mater 19:107
134. Higgins AM, Martin SJ, Goghegan M, Heriot SY, Thompson RL, Cubitt R, Dalgliesh RM, Grizzi I, Jones RAL (2006) Macromolecules 39:6699
135. Morgado J, Moons E, Friend RH, Cacialli F (2000) Adv Mater 13:810
136. Charas A, Alves H, Alcácer L, Morgado J (2006) Appl Phys Lett 89:143519
137. Farrar SR, Contoret AEA, O'Neill M, Nicholls JE, Richards GJ, Kelly SM (2002) Phys Rev B 66:125107
138. Inaoka S, Roitman DB, Advincula RC (2005) Chem Mater 17:6781
139. O'Brien GA, Quinn AJ, Tanner DA, Redmond G (2006) Adv Mater 18:2379
140. O'Carroll D, Irwin J, Tanner DA, Redmond G (2007) Mater Sci Eng B 147:298
141. O'Carroll D, Iacopino D, O'Riordan A, Lovera P, O'Connor É, O'Brien GA, Redmond G (2008) Adv Mater 20:42
142. Moynihan S, Iacopino D, O'Carroll D, Lovera D, Redmond G (2008) Chem Mater 20:996
143. Pinto NJ, Johnson AT Jr, MacDiarmid AG, Mueller CH, Theofylaktos N, Robinson DC, Miranda FA (2003) Appl Phys Lett 83:4244
144. Reneker DH, Chun I (1996) Nanotechnology 7:216
145. Kuo C-C, Lin CH, Chen W-C (2007) Macromolecules 40:6959
146. Babel A, Li D, Xia Y, Jenkhe SA (2005) Macromolecules 38:4705

Optically Active Chemical Defects in Polyfluorene-Type Polymers and Devices

Stefan Kappaun[1] · Christian Slugovc[2] · Emil J. W. List[1,3] (✉)

[1]NanoTecCenter, Weiz Forschungsgesellschaft mbH, Franz-Pichler-Straße 32, 8160 Weiz, Austria

[2]Institute for Chemistry and Technology of Materials, Graz University of Technology, Stremayrgasse 16, 8010 Graz, Austria

[3]Institute of Solid State Physics, Graz University of Technology, Petersgasse 16, 8010 Graz, Austria
e.list@tugraz.at

1	Introduction	274
2	Synthesis of Polyfluorenes and Polyfluorene-Based Materials	275
3	Defect Sites Generated During Polymer Synthesis	278
4	Enhancement of the Green Defect Emission in the Solid State Upon Thermal Stress	284
5	Green Defect Emission Emerging Under Device Operation	286
6	Approaches for Realizing Stable Blue Emitter Materials	288
7	Conclusion	289
	References	290

Abstract Polyfluorenes and polyfluorene-type polymers have emerged as a promising class of emitter materials for realizing blue polymeric light-emitting devices. However, during device operation material degradation leads to the occurrence of an unwanted green emission band at 2.2–2.3 eV. This chapter reviews the latest scientific investigations on the origin of this low energy emission band, putting special focus on chemical (i.e., keto defect sites, hydroxy-terminated polyfluorenes, etc.) and interface defects. Along this line, the formation of defect sites during polymer synthesis, their enhancement in the solid state upon, e.g., thermal stress, and the defect emission under device operation are discussed. Finally, novel approaches for realizing stable blue emitter materials are included, demonstrating that a thorough consideration of the herein presented results in materials design paves the way to polyfluorene-type polymers suitable for practical applications.

Keywords Chemical defects · Electroluminescence · Light-emitting diodes · Photoluminescence · Polyfluorene-type polymers

1
Introduction

Since the discovery of conducting polymers by Shirakawa, MacDiarmid and Heeger in the 1970s, research on these materials has evolved as a lively field of activity. The observation that hydrocarbon polymers (e.g., polyacetylenes) exhibit electrical conductivity upon doping (i.e., oxidation or reduction of the corresponding material) resulted in a tremendous scientific effort, particularly boosted by the huge number of possible applications [1]. The field of research on conducting polymers has been up to now a very interdisciplinary one, falling at the intersection of chemistry, physics, and engineering. The progress made in each of these disciplines in the last three decades has provided competitive materials for industrial applications that exhibit a unique combination of properties, namely the electronic and optical properties of metals and semiconductors combined with the processing advantages and mechanical characteristics of polymers [2]. Concerning their applications, conducting polymers have emerged as a promising class of active materials in electronic and optical devices, among them polymeric light-emitting devices (PLEDs), photodetectors, photovoltaic cells, sensors, field effect transistors (FETs), and optical lasers [3]. However, due to various difficulties and problems encountered during application of the corresponding polymers in electronic and optical devices (e.g., material degradation) there is considerable skepticism that these materials will ever reach the levels of purity and stability required for long-lifetime commercial devices [1–3].

In this context, polyfluorenes (PFs) and polyfluorene-based materials have received considerable attention, especially because of their intense blue photo- and electroluminescence but also for their liquid crystalline properties [4–8]. Although improvements concerning color purity and device stability of polyfluorene-based PLEDs have recently been made, the strict requirements for commercialization, demanding tens of thousands of hours of operation, are still unreached goals for blue light-emitting PLEDs.

While the electroluminescence spectra of unstressed polyfluorene-based PLEDs show very similar bands to those observed in photoluminescence, prolonged operation of the devices frequently leads to the appearance of low energy emission bands. As exemplarily depicted in Fig. 1, the photophysical properties of polyfluorenes are significantly altered upon degradation so that instead of the blue emission an undesired greenish emission color is obtained. Despite intense scientific efforts, the origin of this low-energy emission band is still controversial and is discussed in the literature. Mainly, aggregates or excimers [9–15] on the one hand, or direct emission from chemical defects ("keto defects") on the other hand are thought to be responsible for the observed optical effect [16–18]. The latter scenario is strongly supported by the results from Lupton and coworkers showing that on-chain fluorenone defect emission from single polyfluorene molecules can be found

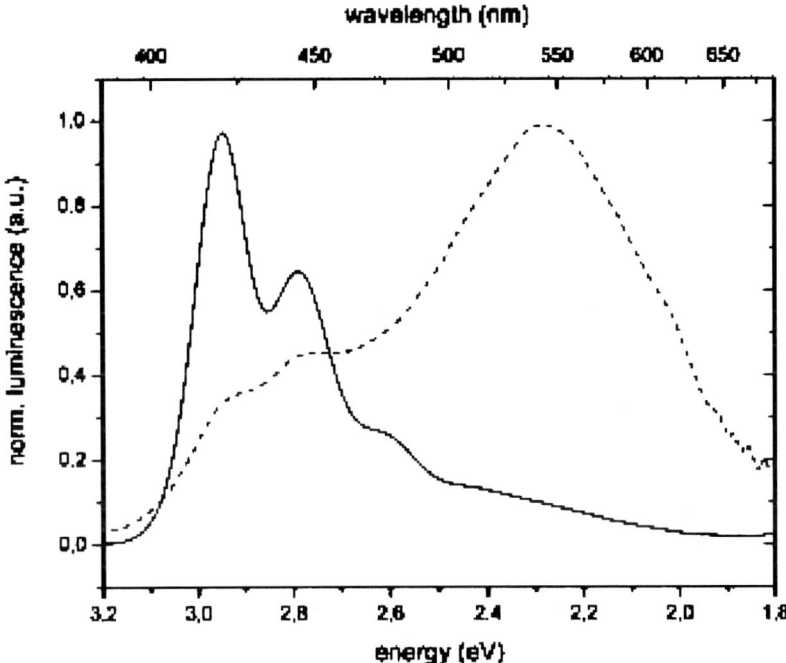

Fig. 1 Normalized photoluminescence spectra of pristine polyfluorene (*solid line*) and after degradation under intense UV irradiation (*dashed line*; modified from [18])

in the absence of intermolecular interactions using single molecule spectroscopy [19].

However, all authors working on the topic agree on the fact that there is a relationship between the occurrence of the green emission and an ongoing chemical degradation process. Therefore, it becomes clear that a fundamental understanding of the degradation processes is crucial for the rational design of novel conjugated blue-emitting polymers exhibiting adequate material stabilities. To that aim, we herein review the latest scientific work on the different chemical defect sites giving rise to green emission bands in polyfluorenes, putting special emphasis on keto defect sites as a possible explanation. Finally, highly promising approaches for realizing stable blue emitter materials are included.

2
Synthesis of Polyfluorenes and Polyfluorene-Based Materials

Because polyfluorenes have undoubtedly emerged as a very attractive class of blue-emitting polymers, not surprisingly, a number of synthesis methods have been applied for the preparation of polyfluorenes and polyarylenes [20], especially reductive coupling reactions and transition-metal-mediated cross-coupling reactions [21].

The reductive coupling of fluorene derivatives and other arylenes, which in the literature is often referred to as Yamamoto coupling (see Scheme 1), is one of the most versatile methodologies for the preparation of conjugated polymers. Starting from dihalide precursor materials (usually dibromo species) and nickel(0) reagents, this reaction can be efficiently applied for the synthesis of, e.g., polyfluorenes, giving polymers with high molecular weights in good to excellent yields [22]. Here, the outcome of the coupling reaction is strongly dependent on the nickel reagent. While good results can be obtained with the commonly applied bis(cyclooctadiene)nickel(0) ($Ni(COD)_2$) in combination with 2,2'-bipyridyl, the use of cheaper nickel(II) derivatives with reducing agents has been shown to be a less effective protocol [23]. Yamamoto coupling can be considered to be an efficient and convenient method for the preparation of polyfluorenes and other polyarylenes; however, its scope is restricted to the formation of homopolymers and statistical copolymers. Strictly alternating copolymers are not accessible with the Yamamoto coupling but can be obtained from transition-metal-mediated cross-coupling reactions.

Transition-metal-mediated cross-coupling reactions have evolved as important methods in modern organic chemistry and provide a versatile toolbox for the preparation of conjugated polymers [4, 6]. Among the different transition-metal-mediated cross-coupling reactions, Suzuki coupling (or the Suzuki–Miyaura reaction) is especially referred to as a very general and selective coupling reaction, which also gives polyfluorenes in good yields (Scheme 1) [4, 6, 24, 25]. Along this line, Suzuki coupling has received considerable attention due to a wide range of tolerated functional groups, mild reaction conditions, the low toxicity of the starting materials, and the fact that dry solvents are generally not required [26–30]. For the synthesis of conjugated polymers such as polyfluorenes, the corresponding precursor materials (Scheme 1) bearing halides and boronic acids or boronic acid esters have to be reacted in the presence of a palladium catalyst (e.g., tetrakis(triphenylphosphine)palladium(0), $Pd(PPh_3)_4$). In an "AA–BB type coupling reaction", the exact 1 : 1 stoichiometry of the dihalo and diborono compound is an important parameter in order to obtain high molecular weight polymers. An exact 1 : 1 stoichiometry is, however, experimentally hard to realize, so starting materials containing the halo and borono group in the same precursor molecule have been successfully used for polymerization reactions in a so-called AB-type coupling reaction (Scheme 1) [31]. While the latter approach circumvents the stoichiometry criterion, its drawbacks are the cumbersome preparation of the precursor materials and that alternating polymers are not accessible. In contrast, the AA–BB type coupling reaction permits the facile synthesis of strictly alternating copolymers and can, therefore, be considered a valuable amendment to the Yamamoto coupling.

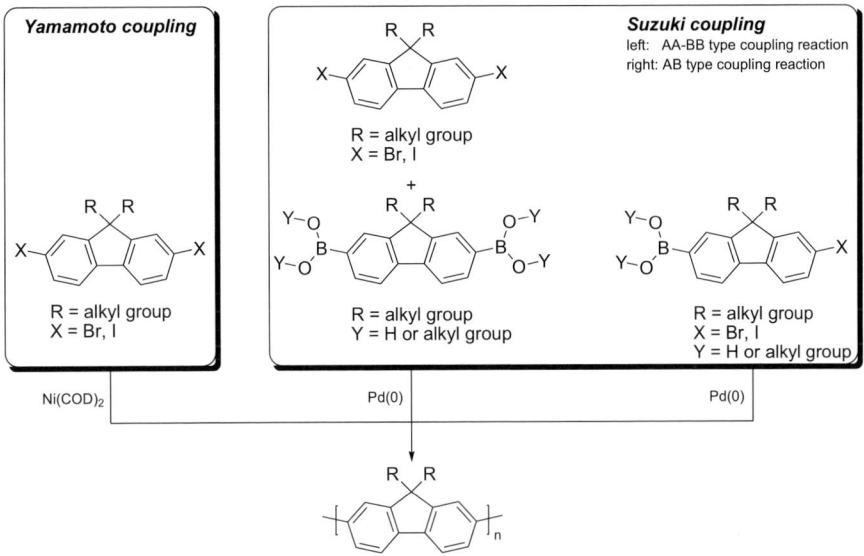

Scheme 1 Synthesis of polyfluorenes via Yamamoto and Suzuki coupling. With Suzuki coupling (AA–BB type coupling reaction), the synthesis of strictly alternating copolymers is possible

The Yamamoto and Suzuki coupling reactions are definitely the most frequently used methods for the preparation of polyfluorene systems, nevertheless, it should be noted that other transition-metal-mediated cross-coupling reactions (e.g., Stille coupling) are occasionally applied to the synthesis of polyfluorene-based materials [32]. In the context of possible preparation methods, the $FeCl_3$-mediated oxidative coupling reactions of fluorenes will also be mentioned. This reaction type, from which the first processable polyfluorene systems were obtained in the late 1980s [33, 34], is nowadays hardly used, particularly because of its unsatisfying regioselectivity and the low molecular weights of the synthesized polymers [16].

A clear advantage of organic conjugated polymers over their inorganic counterparts is their solubility in organic solvents. In order to improve the solubility of the precursor molecules for polymer synthesis, as well as to increase the solubility of conjugated polymers for permitting solution processing techniques, substitution of the monomer molecules with, e.g., alkyl chains is an important part of all described synthesis routes [5]. Consequently, fluorene monomers are also typically substituted at their CH_2 bridge connection (C-9) by treating the corresponding unsubstituted molecules with a strong base, followed by the addition of an alkyl halide (see Scheme 2) [35]. Furthermore, this versatile approach provides the possibility of covalent attachment of other functional groups such as heteroaromatic compounds and

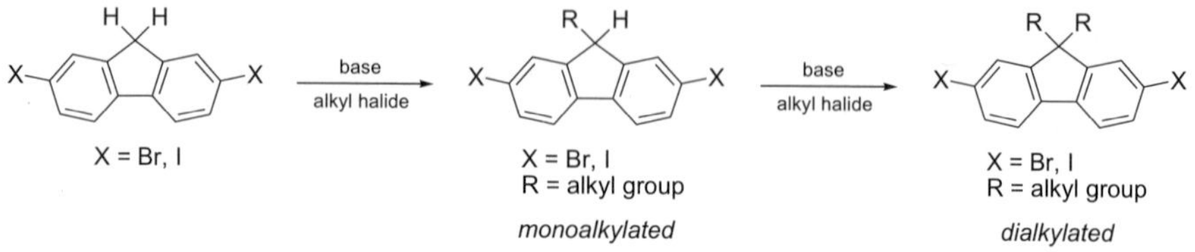

Scheme 2 The alkylation reaction of, e.g., 2,7-dihalofluorene gives mono- and dialkylated products. The complete removal of non- and monoalkylated fluorene derivatives is important for increasing material stability

is, thus, frequently used to prepare functional materials with tailor-made material properties [36].

Even though the alkylation of fluorenes and, generally speaking, arylenes is an important tool for making conjugated polymers processable, incomplete alkylation and insufficient product purification contribute significantly to stability problems encountered during practical application of the corresponding materials in different device architectures, such as PLEDs. Studying the origin and consequences of these stability problems as well as elaborating strategies to overcome material degradation have resulted in tremendous scientific efforts, which are reviewed in subsequent sections.

3
Defect Sites Generated During Polymer Synthesis

Polyfluorenes are considered to be among the most promising classes of conjugated materials for realizing PLEDs, especially because of their high luminescence quantum yields, thermal stability, good solubility, and charge carrier mobility [37]. However, PLED fabrication from polyfluorenes suffers from degradation processes under device operation, resulting in the formation of a low-energy emission band at 2.2–2.3 eV that turns the desired blue emission color into an undesired greenish-blue emission [16].

The origin of this green emission band is still controversial. Initially, a spontaneous formation of aggregates in the ground state and/or excimer formation [10–15] in the photoexcited state were anticipated to give this undesired green emission. Conversely, keto defect sites ("keto defects") were identified as a potential origin of the low-energy emission band [16, 17], which is also in full accordance with recent experimental results [19, 38] and quantum chemical calculations [39].

From the chemist's point of view, the keto defect sites can be formed during polymer synthesis as a consequence of incomplete monomer alkylation, as well as a result of photo-, electro-, or thermooxidative degradation processes occurring after polymer synthesis. Acting as low-energy trapping sites for singlet excitons being populated by an excitation energy transfer from the polyfluorene chain, these keto defects can be made responsible for the low-energy emission band at 2.2–2.3 eV (see Sect. 4) [16, 17].

To support the hypothesis of the formation of keto defects during polymer synthesis, a simple but very impressive experiment has been conducted by the Scherf group. They prepared polyfluorenes starting from 9-monoalkylated and 9,9-dialkylated 2,7-dibromofluorene derivatives utilizing a standard Yamamoto protocol giving, indeed, polyfluorenes with different polymer characteristics. In this experiment, the monoalkylated monomers provided polymers with significantly lower molecular weights and broader molecular weight distributions, clearly indicating that side reactions had occurred during polymer synthesis. To identify the formed by-products from side reactions, IR spectroscopy proved suitable for the detection of keto defect sites (IR band around 1721 cm^{-1} stemming from the carbonyl stretching mode; see Fig. 2) in polymers obtained from monoalkylated starting materials. The keto defects were not present in pristine polyfluorenes synthesized from dialkylated monomers. Besides the generation of keto defects during polymer synthesis, defect sites can also result from photo-, electro-, or thermooxidative degradation processes. As can be concluded from Fig. 2, polyfluorenes prepared from thoroughly purified monomers without non- or monoalkylated impurities do not exhibit the characteristic keto band around 1721 cm^{-1} in the corresponding IR spectra. However, oxidative stress leads to the formations of keto defect sites, which can be readily monitored via IR spectroscopy (see Fig. 2) [16, 17].

In addition to this experimental evidence for the existence of keto defects, a significant impact on the photophysical properties of the corresponding polymers could be identified (see Fig. 3). While the monoalkylated polymers exhibited a strong contribution of the green emission band in solution and, in particular, in the solid state after polymer synthesis, the corresponding greenish emission was not found in the pristine dialkylated polyfluorene. In the latter case, the low-energy emission band was only observed after photo-, electro-, or thermooxidative stress.

The formation of keto defects during polymer synthesis can be readily explained by a mechanism proposed by the Scherf group (see Scheme 3). In this mechanism, the initial step is the reduction of monoalkylated fluorene species by the nickel catalyst giving fluorenyl anions, which subsequently react with atmospheric oxygen during work-up. Finally, the resulting hydroperoxide anions undergo a rearrangement reaction resulting in the formation of fluorenone moieties [16, 17, 40]. Therefore, the thorough removal of non- or monoalkylated fluorene derivatives from the dialkylated monomers can generally be considered a crucial point in the synthesis of stable polyfluorenes.

It is worth noting that various peroxide species can also be formed in palladium-catalyzed Suzuki cross-coupling reactions in the presence of oxy-

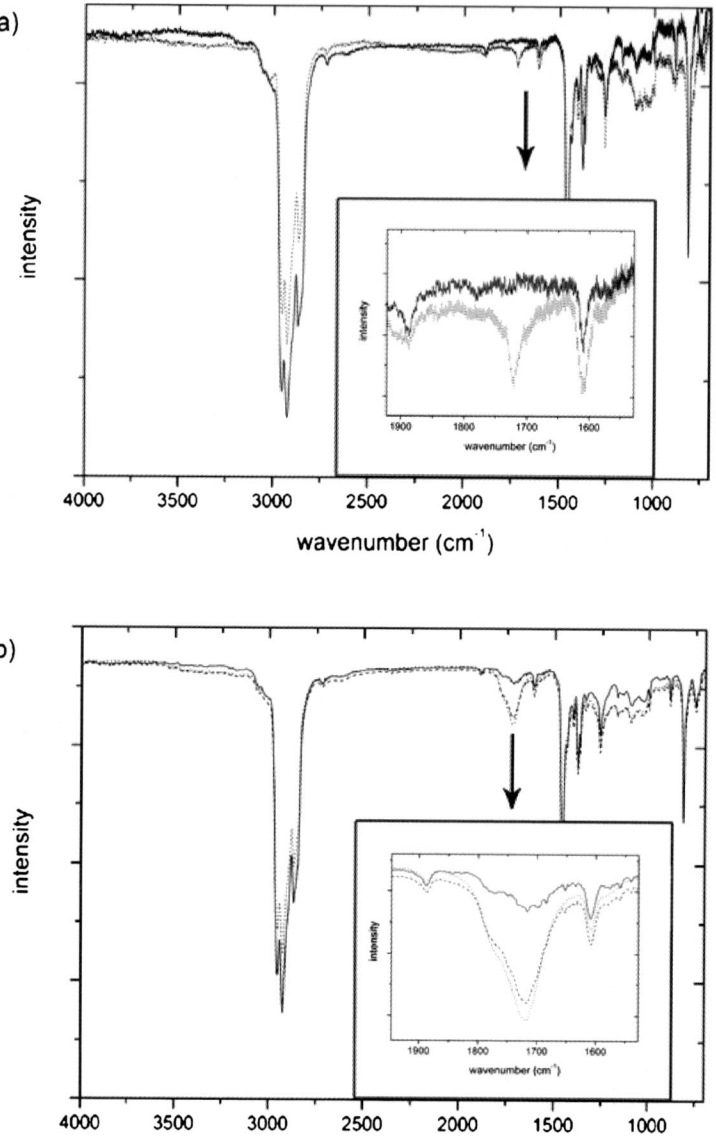

Fig. 2 a Infrared spectra of monoalkylated (*dotted line*) and dialkylated polyfluorenes (*solid line*) show that in the case of monoalkylated polymers keto defects are already present after polymer synthesis. **b** Besides polymer synthesis, keto defects can also be generated by, e.g., photooxidative degradation. Here, the corresponding infrared spectra after photooxidative degradation of dialkylated polyfluorenes are shown (modified from [16, 17])

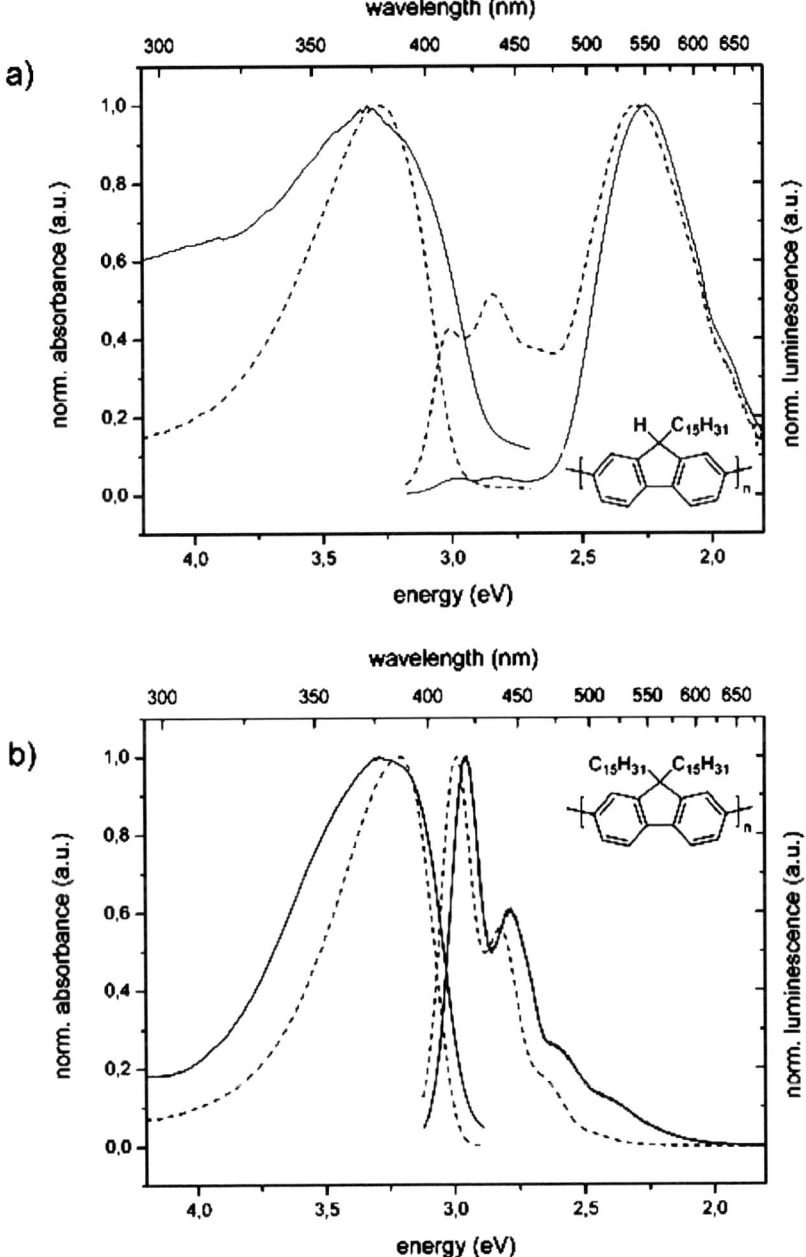

Fig. 3 a Absorption and emission spectra of monoalkylated polyfluorenes in solution (*dashed line*) and in the solid state (*solid line*). **b** Absorption and emission spectra of dialkylated polyfluorenes in solution (*dashed line*) and in the solid state (*solid line*) (modified from [16, 17])

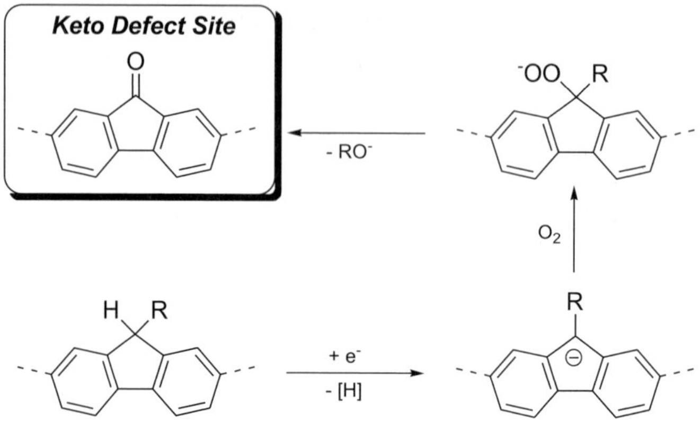

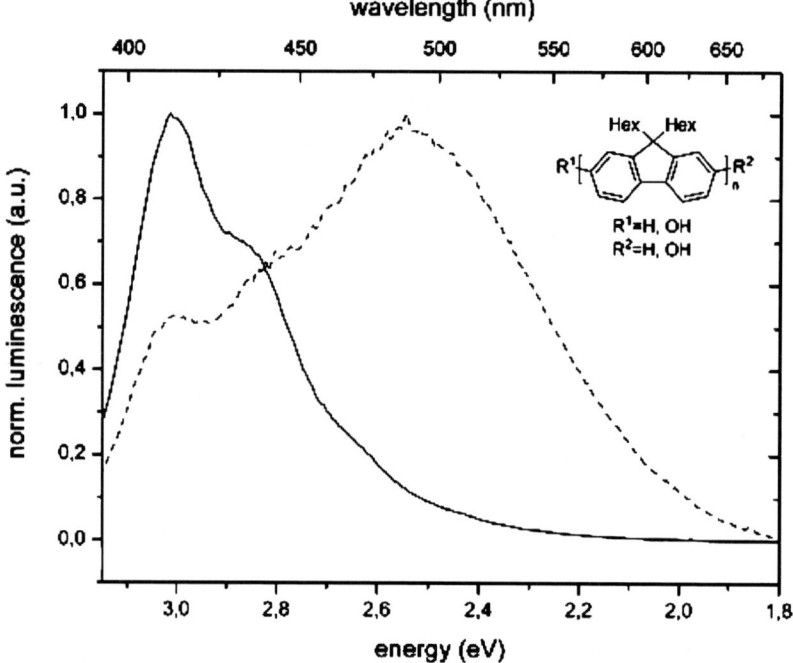

Scheme 3 Proposed mechanism for the generation of keto defect sites during the synthesis of polyfluorenes (modified from [16, 17, 40])

Fig. 4 Emission spectra of hydroxy-terminated polyfluorenes in the solid state before (*solid line*) and after (*dashed line*) the addition of base. The detailed study of the effect of deprotonation and its contribution to the low-energy emission band in, e.g., PLEDs is part of our current research (Kappaun S and coworkers, unpublished results)

gen, leading to similar effects [25]. In addition to this, the generation of peroxides in Suzuki cross-coupling reactions also paves the way to other chemical defects, especially to hydroxy-terminated polyfluorenes [41]. In this

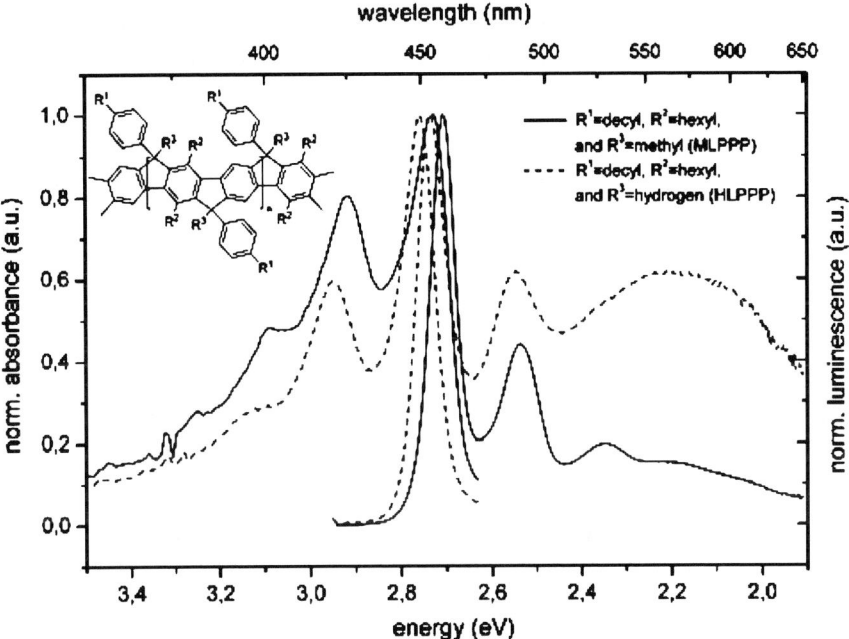

Fig. 5 Absorption and emission spectra of hydrogen- (HLPPP, *dashed line*) and methyl-substituted (MLPPP, *solid line*) ladder-type polyparaphenylenes. Again, full alkylation leads to more stable materials with regard to the photophysical properties (modified from [43])

context it has been shown that hydroxy end groups, obtained from protiodeboronation reactions of boronic acids or boronic acid esters [28], are capable of undergoing deprotonation reactions accompanied by a drastic change of the photophysical properties (i.e., the occurrence of a green emission band; see Fig. 4) (Kappaun S and coworkers, unpublished results). The detailed study of this effect of deprotonation, which has also been observed in other conjugated materials [42], and its contribution to the low-energy emission band of polyfluorene-based devices under operating conditions is part of our ongoing research activities.

From the above results it can be concluded that keto defect sites are preferably formed during polymer synthesis when non- or monoalkylated fluorene species are present in the reaction mixture. This points to the necessity of avoiding even small amounts of these components in order to provide polyfluorenes and polyfluorene-based materials without such "centers of degradation" and to realize high molecular weight polymers. This prerequisite for polyfluorenes with increased stability can also be transferred to other blue emitter materials as shown, e.g., by Romaner et al. [43] for ladder-type polyparaphenylenes. In this study, again, full alkylation was identified as an important parameter for highly stable materials as it was derived from comparison of hydrogen (HLPPP) and methyl-substituted (MLPPP) ladder-type polyparaphenylenes (Fig. 5).

Having discussed different aspects of the chemical origin of keto defects and its general impact on photophysical properties from the synthesis point of view, the keto defect and its impact on the solid state physics and device performance will now be considered.

4
Enhancement of the Green Defect Emission in the Solid State Upon Thermal Stress

To further clarify the formation of keto-type defects, especially in defect-free polyfluorenes, thermal degradation measurements have been found to be the method of choice. This method has the advantage of being less dependent upon the particular film thickness than is photooxidation (where preferably the top layer of the film is affected) or device degradation experiments. Since such experiments can be easily carried out in vacuum, air, or any other atmosphere they allow the simple investigation of the influence of atmospheric components such as oxygen or water, as well as of structural changes upon heating above the glass transition temperature (T_g) of the polymer. In particular, this testing method permits a fast screening of the active material with clear implication on the device performance where thermal stress on the material is an inherent issue.

Figure 6 shows the typical behavior of a pristine polyfluorene film in PL emission heated in air for 1 h at 66, 100, 133, 166 and 200 °C, respectively. While the blue emission at 2.96 and 2.78 eV initially increases, which can be attributed to the removal of residual solvent or ordering effects, it decreases significantly after heating the sample to higher temperatures (166 and 200 °C, respectively). At the same time the broad, low energy emission band around 2.3 eV evolves. As shown in the insert of Fig. 6, the IR feature at 1720 cm^{-1} increases simultaneously, which indicates that chemical defects are created upon material degradation, as discussed above, e.g., for the photooxidative degradation processes. In contrast, heating the polymer in vacuum in the absence of oxygen and water does not lead to any change in the PL emission spectrum nor to a strong alteration of the PL quantum yield. These results make it clear that thermal stress in a device in the presence of oxygen and water will lead to the formation of keto-type defects at the 9-position of polyfluorenes, accompanied by the appearance of the unwanted green emission. Yet, as reported by Müllen et al., arylation [58] of the bridging positions can dramatically improve the stability of the polymer (see Sect. 5).

The experimental results shown hitherto do not directly concern the nature of the green emission band around 2.2–2.3 eV but show the clear correlation between the observation of the low energy emission band and the

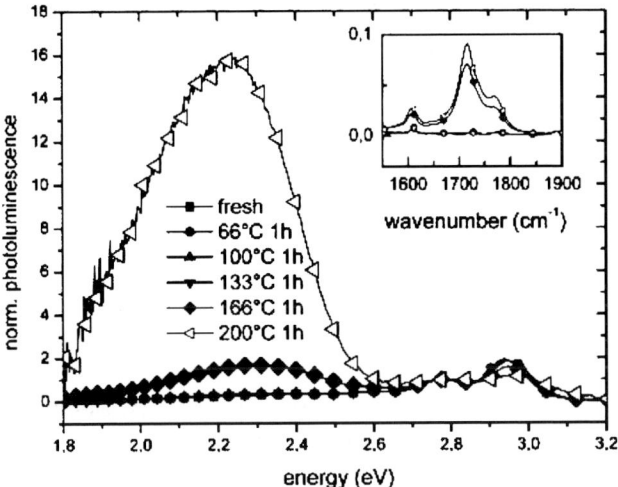

Fig. 6 Photoluminescence spectra of a pristine polyfluorene film thermally degraded in air (1 h at given temperature; excitation wavelength 390 nm). The *insert* shows the corresponding IR spectrum (modified from [18])

emerging of correlated IR bands. For a conscientious spectroscopic study of the origin of the green emission, Scherf et al. synthesized and studied model molecules for degraded polyfluorenes in the form of copolymers of 9,9-difarnesylfluorene and 9-fluorenone moieties. These copolymers exhibit exactly the same properties as degraded polyfluorenes and bear, additionally, the advantage that the 9-fluorenone content can be exactly controlled. In particular, these model polymers have been invoked in a number of experiments for determination of whether the excited species in polyfluorenes emitting in the green spectral region is an excimer, an aggregate, or an on-chain defect [38]. It was found experimentally that:

1. The low energy emission band can be directly excited
2. The intensity ratio of the fluorescence and the low energy emission band does not show any significant concentration dependence
3. It shows a distinct behavior upon solvent polarity as predicted by theory [39]
4. The low energy emission band displays a pronounced vibronic structure at low temperatures
5. Regarding excitation energy migration, it behaves like a guest molecule in a guest–host blend system

All these findings are direct evidence against excimer or aggregate formation as the primary source for the low energy emission band in polyfluorene-type materials. The observation of the low energy emission in the most dilute solutions is taken as additional evidence against any physical dimerization (i.e.

aggregation) effects. Moreover, Lupton et al. used the same model polymers to observe an on-chain fluorenone defect emission from single molecules in the absence of intermolecular interactions using single molecule spectroscopy [19]. All these findings make it very likely that the origin of the green emission band stems from an on-chain emission from the 9-fluorenone moieties, which acts as a chemically formed defect.

5
Green Defect Emission Emerging Under Device Operation

A common mode of device degradation in polyfluorene-type PLEDs is the formation of keto defects during device operation and the related change of the emission spectrum with a broad peak emerging around 2.3 eV (Fig. 7).

Typically, the intensity of the defect emission increases with ongoing operation time as depicted in Fig. 8. Especially for polymers containing an intrinsic amount of keto defects, the relative spectral contribution of the keto emission in electroluminescence is considerably stronger than in photoluminescence. This was initially attributed to the fact that the keto defects act as a trap for charge carriers, thus leading to increased exciton localization at the defect sites. On the other hand, ongoing investigations, including device operation at very low temperatures, have shown that the photoluminescence emission at 2.3 eV vanishes when the temperature is strongly reduced. This behavior is not straightforwardly explained when charge trapping is held responsible for the enhanced defect emission in polyfluorene PLEDs (compared with photoluminescence results). In fact, this explanation would now require the assumption that charge trapping on the keto sites occurs over an activation barrier that can only be overcome by thermal activation. These ongoing experiments lead to the conclusion that the high current densities in PLEDs result in an increased temperature in the polymer film and a subsequent enhanced excitation energy migration within the polyfluorene conjugation segments and, therefore, to an enhancement of the low energy emission, in addition to charge carrier trapping.

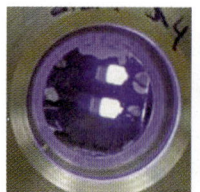 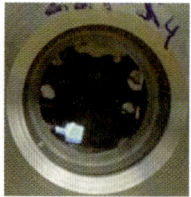

Fig. 7 Electroluminescence of typical polyfluorene-type PLEDs. The pristine emission (*left*) and the emission after electrical stress (standard device operation, *right*) are shown

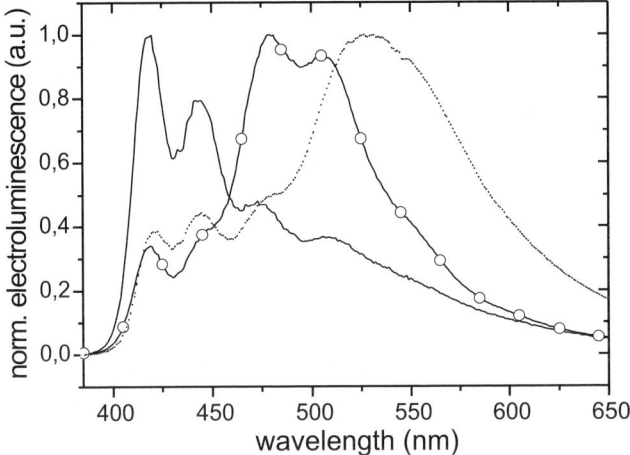

Fig. 8 Electroluminescence spectrum of an ITO/PF/Al device (*solid line*), of the same device after continuous operation under air showing changes due to electrooxidative degradation (*dotted line*), and under operation of a different device in argon showing changes due to interface degradation (*line with symbols*)

In addition to the above mentioned defect emission band at 2.3 eV upon device operation, one also finds an additional individual emission band that is located at 2.45–2.6 eV (energetically located between the "regular" blue emission bands at 2.9 eV and the broad keto-related emission at 2.3 eV). These emission features appear most strongly for devices with calcium electrodes. It was found that this spectral feature is located close to the cathode of the device and is most probably related to a chemical degradation reaction caused by the low work function metal electrodes [44].

Several experiments with calcium/aluminium electrodes using different polyfluorene-type polymers revealed that the metal deposition parameters (such as evaporation rate or base pressure) determine the intensity of the defect emission band, indicating that the residual oxygen may play a critical role. After slow evaporation (< 1 Å s^{-1}) of very thin calcium layers at a base pressure of approximately 2×10^{-6} mbar, which leads to a higher oxygen concentration in the interface region, the defect emission is frequently observed while the effect is not found if the electrode is deposited at a higher rate (> 10 Å s^{-1}) where less oxygen is incorporated into the interface region. This leads to the conclusion that the emissive defects located at the polymer–cathode interface are created in polyfluorene-based PLEDs during electrode deposition. The interface defects emitting at 2.45–2.6 eV are related to a degradation of the polymer in interplay with residual atmospheric species, most likely oxygen or water, and the low work function electrode. According to their spectral signature, the interface defects are not directly related to the above described keto-type bulk defects. Thus, having identified a second independent defect, a proper interface control, optimized electrode deposition parameters, and the use of suitable electron transport, hole blocking, or protection layers may in future help to improve the spectral stability and enhance the lifetime of blue light-emitting polyfluorene-based PLEDs.

6
Approaches for Realizing Stable Blue Emitter Materials

Although polyfluorenes and polyfluorene-based materials are of highest interest for practical applications, the most striking drawback of this class of materials is the appearance of the green emission band during device operation. The primary assignment of the green emission to aggregate and/or excimer formation from adjacent polyfluorene chains led to various synthetic efforts, such as copolymer generation, attachment of dendronic substituents, endcapping, spiro-linking, incorporation of kinked structures or oxygen scavengers as promising alternatives for realizing stable blue emitter materials [45–50]. However, all these approaches provided more or less unsatisfying results concerning the long-term stability of the corresponding materials, also indicating that substantial chemical modifications are necessary to provide adequately stable materials.

In this context, a very promising approach has been presented by the Holmes group substituting the CH_2 bridge (C-9) of fluorenes with silicon. Utilizing a Suzuki cross-coupling reaction (Scheme 4), they prepared poly-9,9-dihexyl-2,7-dibenzosiloles, which exhibit emission properties hardly affected by oxidative stress [51]. Simultaneously, other groups focused on the synthesis and properties of similar silafluorenes providing a set of novel blue emitter materials [52–54].

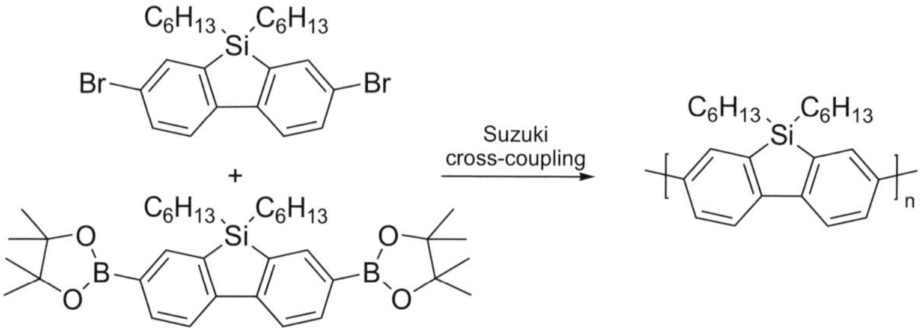

Scheme 4 Synthesis of poly-9,9-dihexyl-2,7-dibenzosiloles [51]

Besides these impressive results, excellent examples of highly stable blue emitter materials have been reported by Müllen et al. [55–59] and Scherf et al. [43, 60]. These authors concentrated on bridged polyparaphenylenes, polyindenofluorenes, polyphenanthrylenes, and other ladder-type analogues

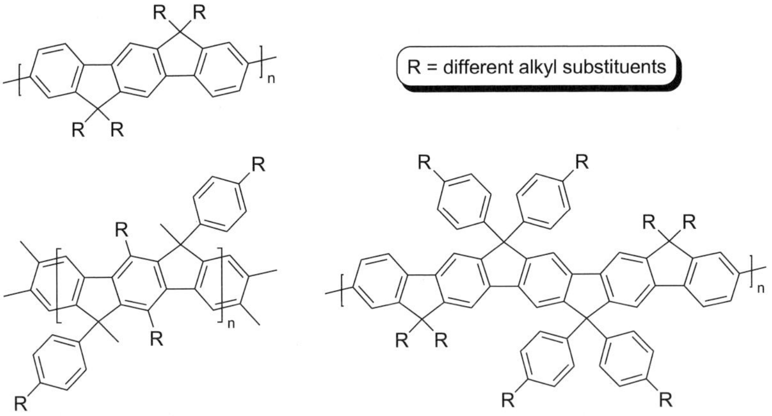

Fig. 9 Representative examples of bridged polyparaphenylene derivatives [55–60]

(some examples are depicted in Fig. 9), making materials with adequate material stability available [55–60]. In these ladder-type materials, methine bridges keep the polymer backbone planar, which also influences the photophysical characteristics of the corresponding polymers so that red-shifted emission maxima with increasing ladder quality are obtained [21]. Even though these polymers permit suppression of the undesired low-energy emission band, substitution of all bridgeheads has been shown to be crucial for realizing improved color purities and long-term stabilities. Consequently, quantitative alkylation or arylation of the bridging positions, as well as the substitution of other "reactive sites" capable of undergoing undesired side reactions, have to be addressed in order to provide sufficient material stability for applications in, e.g., PLEDs [21, 55–60].

7
Conclusion

Although polyfluorenes have emerged as a promising class of materials for realizing blue polymeric light-emitting devices, their degradation under device operation accompanied by a change in the photophysical properties can be considered a significant shortcoming. Large scientific efforts have been devoted to study and to overcome this effect. It has been shown that, in particular, thermal stress in a device in the presence of oxygen and water can lead to the formation of keto-type defects at the 9-position of polyfluorenes, accompanied by the appearance of the undesired green emission band.

In this work we have reviewed the latest scientific results, putting special focus on chemical defects (keto defects, fluorenoles, etc.) as potential explanation for the low energy emission band of polyfluorenes. Along this line, we have discussed the defect chemistry of polyfluorenes and polyfluorene-based materials from a chemist's point of view, summarizing possible side reactions occurring during polymer synthesis and have identified the absence of non- or monoalkylated monomers as an important prerequisite for the preparation of stable blue emitter materials. Moreover, the impact of chemical defects on the solid state physics and device performance has been presented, also demonstrating that polymer–electrode interface defects can contribute significantly to the undesired green emission of polyfluorenes.

With these results in mind, the rational design of novel and, more importantly, stable blue emitter materials becomes possible. Therefore, we have included the latest synthesis efforts of the Müllen, Holmes, and Scherf group, who have shown that a thorough consideration of the degradation processes in materials design indeed provides the possibility of realizing stable blue emitter materials tailored towards particular needs.

Acknowledgements The authors would like to thank the team of coworkers and collaborators that have been involved in the interdisciplinary research projects on semiconducting polyfluorenes over the past years, especially Stefan Gamerith, Josemon Jacobs, Lorenz Romaner, Martin Gaal, Heinz-Georg Nothofer, Michael Graf, Roland Güntner, Michael Forster, Patricia Scandiucci de Freitas, Günther Lieser, Akio Yasuda, Gabi Nelles, Dieter Neher, Tzenka Miteva, Dessislava Sainova, Alexander F. Pogantsch, Franz P. Wenzl, Andrew C. Grimsdale, Egbert Zojer, and Jean Luc Brédas. We also wish to thank Klaus Müllen (Mainz) and Ullrich Scherf (Wuppertal) for their fruitful cooperation as well as for their continuous and generous support of our investigations.

References

1. Hadziioannou G, Malliaras GG (2007) Preface. In: Hadziioannou G, Malliaras GG (eds) Semiconducting polymers. Wiley-VCH, Weinheim
2. Heeger AJ (2007) Foreword. In: Hadziioannou G, Malliaras GG (eds) Semiconducting Polymers. Wiley-VCH, Weinheim
3. Gong X, Moses D, Heeger AJ (2006) Polymer-based light-emitting diodes (PLEDs) and displays fabricated from arrays of PLEDs. In: Müllen K, Scherf U (eds) Organic light-emitting devices. Wiley-VCH, Weinheim, p 151
4. Schlüter AD (2001) J Polym Sci A 39:1533
5. Bolognesi A, Pasini MC (2007) Synthetic methods for semiconducting polymers. In: Hadziioannou G, Malliaras GG (eds) Semiconducting polymers. Wiley-VCH, Weinheim, p 1
6. Leclerc M (2001) J Polym Sci A 39:2867
7. Sović T, Kappaun S, Koppitz A, Zojer E, Saf R, Bartl K, Fodor-Csorba K, Vajda A, Diele S, Pelzl G, Slugovc C, Stelzer F (2007) Macromol Chem Phys 208:1458
8. Neher D (2001) Macromol Rapid Commun 22:1365
9. Lee JI, Klärner G, Miller RD (1999) Chem Mater 11:1083
10. Weinfurter K-H, Fujikawa H, Tokito S, Taga Y (2000) Appl Phys Lett 76:2502
11. Bliznyuk VN, Carter S, Scott JC, Klärner G, Miller RD (1999) Macromolecules 32:361
12. Lemmer U, Heun S, Mahrt RF, Scherf U, Hopmeier M, Siegner U, Göbel EO, Müllen K, Bässler H (1995) Chem Phys Lett 240:373

13. Conwell E (1997) Trends Polym Sci 5:218
14. Grell M, Bradley DDC, Ungar G, Hill J, Whitehead KS (1999) Macromolecules 32:5810
15. Cimrová V, Scherf U, Neher D (1996) Appl Phys Lett 69:608
16. List EJW, Guentner R, Scandiucci de Freitas P, Scherf U (2002) Adv Mater 14:374
17. Scherf U, List EJW (2002) Adv Mater 14:477
18. Gamerith S, Gadermaier C, Scherf U, List EJW (2004) Phys Status Solidi A 201:1132
19. Becker K, Lupton JM, Feldmann J, Nehls BS, Galbrecht F, Gao D, Scherf U (2006) Adv Funct Mater 16:364
20. Scherf U (1999) Topics Curr Chem 201:163
21. Grimsdale AC (2006) The synthesis of electroluminescent polymers. In: Müllen K, Scherf U (eds) Organic light-emitting devices. Wiley-VCH, Weinheim, p 215
22. Grell M, Knoll W, Lupo D, Meisel A, Miteva T, Neher D, Nothofer H-G, Scherf U, Yasuda A (1999) Adv Funct Mater 11:671
23. Yamamoto T (1992) Prog Polym Sci 17:1153
24. Corbet J-P, Mignani G (2006) Chem Rev 106:2651
25. Echavarren AM, Cardenas DJ (2004) Mechanistic aspects of metal-catalyzed C,C- and C,X-bond-forming reactions. In: De Meijere A, Diederich F (eds) Metal-catalyzed cross-coupling reactions. Wiley-VCH, Weinheim, p 1
26. Miyaura N, Yamada K, Suzuki A (1979) Tetrahedron Lett 20:3437
27. Miyaura N, Suzuki A (1979) Chem Commun, p 866
28. Miyaura N (2004) Metal-catalyzed cross-coupling reactions of organoboron compounds with organic halides. In: De Meijere A, Diederich F (eds) Metal-catalyzed cross-coupling reactions. Wiley-VCH, Weinheim, p 41
29. Suzuki A (2005) Chem Commun, p 4759
30. Felpin F-X, Ayad T, Mitra S (2006) Eur J Org Chem 2679
31. Yokoyama A, Suzuki H, Kubota Y, Ohuchi K, Higashimura H, Yokozawa T (2007) J Am Chem Soc 129:7236
32. Asawapirom U, Günther R, Forster M, Farrell T, Scherf U (2002) Synthesis 9:1136
33. Fukada M, Sawada K, Yoshino K (1989) Jpn J Appl Phys 28:L1433
34. Fukada M, Sawada K, Yoshino K (1993) J Polym Sci A 31:2465
35. Dudek SP, Pouderoijen M, Abbel R, Schenning APHJ, Meijer EW (2005) J Am Chem Soc 127:11763
36. Evans NR, Devi LS, Mak CSK, Watkins SE, Pascu SI, Köhler A, Friend RH, Williams CK, Holmes AB (2006) J Am Chem Soc 128:6647
37. Su HJ, Wu FI, Tseng Y-H, Shu C-F (2005) Adv Funct Mater 15:1209
38. Romaner L, Pogantsch A, Scandiucci de Freitas P, Scherf U, Gaal M, Zojer E, List EJW (2003) Adv Funct Mater 13:597
39. Zojer E, Pogantsch A, Hennebicq E, Beljonne D, Brédas J-L, List EJW (2002) J Chem Phys 117:6794
40. List EJW, Gaal M, Guentner R, Scandiucci de Freitas P, Scherf U (2003) Synth Met 139:759
41. Kappaun S, Zelzer M, Bartl K, Saf R, Stelzer F, Slugovc C (2006) J Polym Sci A 44:2130
42. Zelzer M, Kappaun S, Zojer E, Slugovc C (2007) Monatsh Chem 138:453
43. Romaner L, Heimel G, Wiesenhofer H, Scandiucci de Freitas P, Scherf U, Brédas J-L, Zojer E, List EJW (2004) Chem Mater 16:4667
44. Gamerith S, Nothofer H-G, Scherf U, List EJW (2004) Jpn J Appl Phys 2 43:L891
45. Setayesh S, Grimsdale AC, Weil T, Enkelmann V, Müllen K, Meghdadi F, List EJW, Leising G (2001) J Am Chem Soc 123:946
46. Lupton JM, Schouwink P, Keivanidis PE, Grimsdale AC, Müllen K (2003) Adv Funct Mat 13:154

47. Pogantsch A, Wenzl FP, List EJW, Leising G, Grimsdale AC, Müllen K (2002) Adv Mater 14:1061
48. Nakazawa YK, Carter SA, Nothofer H-G, Scherf U, Lee VY, Miller RD, Scott JC (2002) Appl Phys Lett 80:3832
49. Ritchie J, Crayston JA, Markham JPJ, Samuel IDW (2006) J Mater Chem 16:1651
50. Li JY, Ziegler A, Wegner G (2005) Chem Eur J 11:4450
51. Chan KL, McKiernan MJ, Towns CR, Holmes AB (2005) J Am Chem Soc 127:7662
52. Chen RF, Fan Q-L, Liu S-J, Zhu R, Pu K-Y, Huang W (2006) Synth Met 156:1161
53. Wang E, Li C, Mo Y, Zhang Y, Ma G, Shi W, Peng J, Yang W, Cao Y (2006) J Mater Chem 16:4133
54. Mo Y, Tian R, Shi W, Cao Y (2005) Chem Commun, p 4925
55. Jacob J, Zhang J, Grimsdale AC, Müllen K, Gaal M, List EJW (2003) Macromolecules 36:8240
56. Yang C, Scheiber H, List EJW, Jacob J, Müllen K (2006) Macromolecules 39:5213
57. Mishra AK, Graf M, Grasse F, Jacob J, List EJW, Müllen K (2006) Chem Mater 18:2879
58. Jacob J, Sax S, Gaal M, List EJW, Grimsdale AC, Müllen K (2005) Macromolecules 38:9933
59. Jacob J, Sax S, Piok T, List EJW, Grimsdale AC, Müllen K (2004) J Am Chem Soc 126:6987
60. Scherf U (1999) J Mater Chem 9:1853

Single Molecule Spectroscopy of Polyfluorenes

Enrico Da Como[1] (✉) · Klaus Becker[1] · John M. Lupton[2]

[1]Photonics and Optoelectronics Group, Department of Physics and CeNS, Ludwig-Maximilians-Universität, Amalienstrasse 54, 80799 Munich, Germany
enrico.dacomo@physik.uni-muenchen.de

[2]Department of Physics, University of Utah, 115 South 1400 East, Salt Lake City, UT 84112, USA

1	Introduction	294
2	Single Molecule Spectroscopy: Experimental Methods	298
3	How Shape Controls Function: The Single Chain β Phase	300
3.1	Influence of Chain Conformation on β-Phase Formation	300
3.2	Influence of Chain Planarity on the Photophysical Stability	304
3.3	The β Phase: Strain-Induced Depolarisation in Extended π-Electron Systems	306
4	Identification of Single Keto Defects on PF Chains	311
5	Conclusions	315
	References	316

Abstract Single molecule spectroscopy resolves some of the crucial issues in the photophysics of polyfluorenes. After an introduction on single molecule spectroscopy we present the ability of this technique in revealing the intramolecular nature of the photophysical phenomena encountered in the different phases of polyfluorene and in fluorenone–fluorene copolymers. First, we correlate the peculiar emission properties of the phases to their chain extension, probed by using polarisation sensitive studies. Moreover, by looking at single chromophores in the β phase, we demonstrate how a slight bending in this one-dimensional structure does not disrupt the π conjugation while impacting on the exciton linewidth, a crucial parameter for the comprehension of energy and charge transfer. As a definitive answer to the colour degradation, we report the single chain spectroscopy of fluorenone–fluorene copolymers, demonstrating the monomolecular origin of the green emission. More generally, the presented experiments illustrate the unique possibilities offered by single molecule spectroscopy in correlating the structure and the resulting electronic properties in conjugated polymers.

Keywords β Phase · Keto defect · π Conjugation · Single molecules · Polyfluorene

1
Introduction

In the last decade single molecule spectroscopy (SMS) has emerged as one of the most powerful experimental techniques in revealing correlations between macroscopic observables and their molecular nature. The unique potential of probing the electronic structure of a single molecule, avoiding inhomogeneous broadening, gave remarkable advances in the comprehension of how light-harvesting complexes [1, 2], proteins [3], defect centres [4], semiconductor nanocrystals [5] and conjugated polymers [6, 7] arrive at certain functions according to their electronic structure.

SMS has the potential to remove the inhomogeneous disorder and at the same time allows for the collection of statistics of different physical properties. In this way, properties which are obtained with an ensemble averaged value can be reconstructed accounting for the contribution of each single molecule. Such an approach becomes particularly interesting for strongly heterogeneous materials such as conjugated polymers, where every individual single molecule can have very different electronic properties. In these macromolecular semiconductors the interplay between intermolecular and intramolecular disorder has a huge impact on the physical description of charge and energy transfer [8–10]. For example, the crossover between incoherent hopping and the recently observed ballistic transport phenomena [11, 12] requires a microscopic understanding of how each single molecule contributes to the collective macroscopic effect. Additionally, SMS revealed many new physical effects which were initially predicted only theoretically and then subsequently observed with the possibility of looking at a single quantum object [13, 14].

The requirements for performing SMS are the isolation between the molecules and the reliable detection of the optical response coming from them. A general approach to realise such an experiment consists in the study of highly diluted samples, where molecules are separated by several microns in solid or liquid solutions. Single molecules can be subsequently addressed using optical microscopy. Both near-field and far-field techniques have been used successfully. Far-field techniques facilitate the implementation of spectroscopic techniques, while near-field probes take advantage of a highly localised field to enhance the optical signals [15]. Because of the substantial advantages in studying statistical correlations between different single molecules by considering a large number of single molecules simultaneously, far-field imaging techniques have been established as a reliable way to probe many single molecules with reproducible experimental settings. As described in detail in the following section, the measurements presented in this chapter are all performed with a far-field microscope.

SMS techniques can be further distinguished according to the detected spectroscopic observable. Clearly, a particular material under consideration for achieving reliable and fast single molecule fluorescence detection may not always be easy to identify. Because of the necessity of studying materials with a low intrinsic fluorescence, various other light–matter interaction processes have been recently considered for SMS. For example, Raman [16] and Rayleigh scattering [17, 18] techniques have shown great potential, particularly when combined with each other or with the probing of a second optical response [19]. In the study of single organic molecules, fluorescence spectroscopy has evolved to consider more and more complex systems such as multichromophoric supramolecules [20] and conjugated polymers [7, 21]. Single molecule studies of conjugated polymers have given remarkable immediate breakthroughs in revealing the multichromophoric nature of these complex materials. Because of the inhomogeneity and disorder characteristics of such plastic semiconductors [22], ensemble fluorescence spectroscopy can give only a very limited picture of the fundamental processes that take place at the nanometre scale. In conjugated polymers and more generally in molecular materials, physical phenomena are determined by both the intramolecular structure and dynamics (such as energy transfer) and the intermolecular interactions [23]. The physical phenomena underlying device operation, i.e. electronic delocalisation, energy and charge transfer, are influenced by these two entangled aspects, inter- and intramolecular interactions. In this context, SMS isolation and observation of the intrinsic properties of the individual molecules provides a unique tool to reveal the intramolecular nature of phenomena observed in macroscopic devices. In this chapter we will summarise a body of work performed on the SMS of several types of polyfluorenes (PFs) [24–28].

PFs are a remarkable class of π-conjugated materials characterised by a fluorene repeat unit which can be functionalised with different side groups [29] or copolymerised with other conjugated units [30]. They show interesting light-emitting properties in the deep-blue part of the spectrum [31–33] and charge mobilities of up to $\sim 10^{-3}$ cm^2/Vs for both electrons and holes [34]. Together with the well-known applications in organic light-emitting diodes (OLEDs), they can be used as active materials in organic lasers due to their low gain threshold [35] and the negligible overlap of the fluorescence emission with polaron absorption [36]. When copolymerised with electron or hole acceptors they have shown remarkable characteristics in terms of photoconductivity, opening new potential applications in photovoltaic devices [37].

An intriguing property of PFs is their ability to form liquid crystalline phases [38, 39]. Besides the different mesophases which are classified as nematic, poly(9,9-dioctyl)fluorene (PFO) shows several solid-state packing properties and polymorphs. The pioneering work of Grell et al. showed how solvent swelling and thermal cycling could be used to control the solid-state packing in PFO thin films [38, 40]. In addition, the different phases show remarkably diverse photophysical properties, which have tremendous implications for the material performances in optoelectronic devices. The so-called β phase displays a long-range order which is linked to a planarisation of the fluorene repeat units. This planarisation is visible in an increased electron coherence length observed in X-ray scattering and correlates directly with a decrease of the optical gap. This assignment to an increase in conjugation is consistent with the similarity in spectroscopic properties found in the ladder-type poly(para-phenylene) (LPPP), where the planarisation between phenylene rings is achieved by chemical bridging [41, 42]. The α and glassy phases are characterised by less ordered supramolecular structures, and their shorter conjugation length correlates with a more blue-shifted absorption and emission. Although many studies were performed on the nature of the β-phase formation and the involved thermodynamics, it is particularly interesting to address whether the β phase is purely intramolecular or also partly intermolecular in nature [29, 38, 43–46]. The first studies demonstrated the possibility to increase the β-phase spectral signatures while diluting the PFO chains in polystyrene [40], whereas recent neutron scattering experiments point to aggregation phenomena taking place in poor solvents [46]. We show how SMS can directly address these issues by providing straightforward experimental evidence for the interpretation of the electronic and optical properties in the PFO β phase. The pure β phase is observed in single chains of PFO, demonstrating that the β phase must be purely intramolecular—a fascinating example of one-dimensional (1D) crystallisation.

A device relevant problem is the blue colour degradation of PFs due to the formation of green luminescent defects. A broad green emission was first reported by Lemmer et al. and assigned to an aggregate state [42]. Efficient charge trapping was also reported in such degraded polymer chains, explaining the peculiar behaviour observed in electroluminescence [47]. Direct evidence for the presence of charge trapping came from time-resolved electroluminescence spectroscopy by Lupton et al. List et al. considered a possible explanation involving chemical defects for the green emission PFs by comparing to model systems containing fluorenone [48]. Indeed, it was shown that the oxidation of side groups along the chain can lead to the formation of a ketonic unit acting as a radiative trap for the excitons. The long-lived emission from such keto defects initially led to the conclusion that the photophysics of this state is of excimeric type, i.e. with two cofacial ketonic (fluorenone) units involved [49]. Although many experimental results support a monomolecular picture of the green fluorenone emission [48, 50], the topic remains under debate [51]. Here we will review the results on the SMS of PF–fluorenone copolymers [26]. The ability of SMS to distinguish between intramolecular and intermolecular processes is used to reveal the nature of such emission and to unequivocally assign it to isolated ketonic units (Fig. 1).

A long-term question which is the focus of many spectroscopic studies is the efficiency of intrachain and interchain energy transfer [21, 22, 52], which becomes particularly relevant when considering how emissive keto

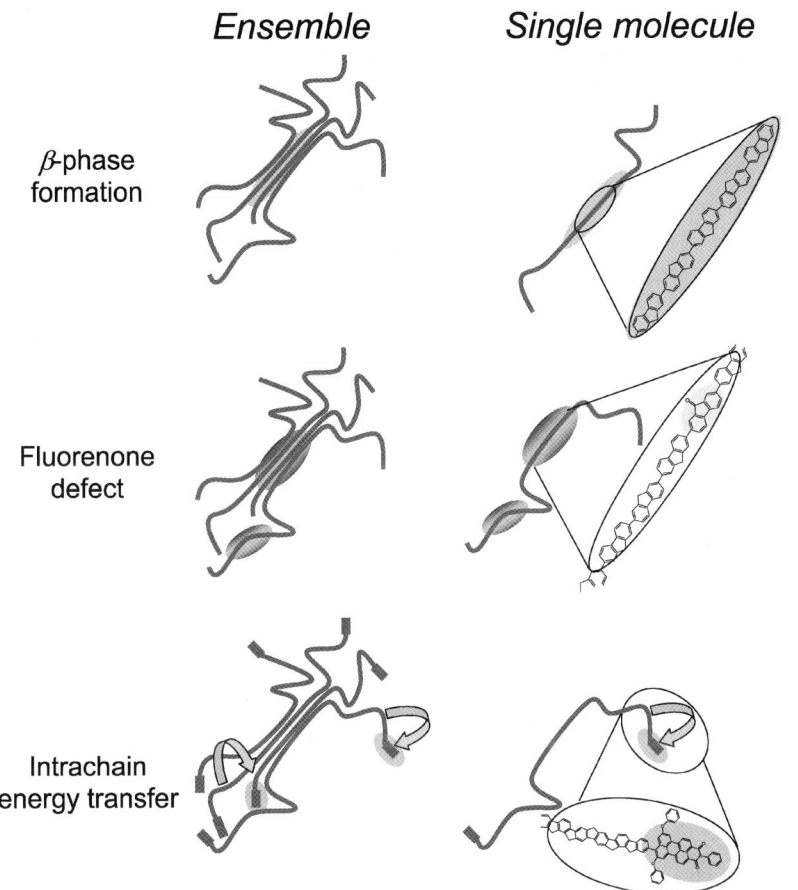

Fig. 1 Schematic illustration of the scientific issues relating to polyfluorene which can be addressed by using SMS. β-Phase formation, colour degradation by fluorenone defects and intrachain energy transfer are intriguing photophysical phenomena first observed in ensemble thin films, where chains aggregate forming disordered bundles (*left column*). SMS, observing single isolated chains (*right column*), can distinguish the intramolecular nature of the different light emission processes (*shaded areas*)

defects are excited. As a model system to investigate the efficiency and the dynamics of intramolecular energy transfer we have recently used a dye end-capped polyindenofluorene (Fig. 1). We refer the reader to our SMS studies of intrachain energy transfer in polyindenofluorene (donor)–perylene (acceptor) complexes, where the perylene moiety is covalently attached as an end group [53]. SMS reveals how an efficient energy transfer process takes place, even in the absence of appreciable spectral overlap between donor emission and acceptor absorption [25, 27]. Temperature impacts directly on the transition linewidth of the different chromophores and therefore influences energy funnelling along the polymer backbone. Figure 1 shows a cartoon summarising these issues, which can be addressed by performing single polymer chain spectroscopy on PFs. In the following, we will focus on the spectral properties of structural and chemical perturbations of the single PF molecule, the planarised β phase and the on-chain emissive keto defect. In all cases, excitation energy transfer to the emissive site is crucial.

2
Single Molecule Spectroscopy: Experimental Methods

SMS requires the experimenter to isolate single molecules and to study them free from external perturbations or contaminations. All the experiments reported in the following were performed by spin-coating thin films of the polymers studied dispersed at very high dilution in an inert polymer matrix such as Zeonex. Usually, thin films were deposited on quartz substrates with extremely flat surfaces so as to minimise light scattering. Film thicknesses of about 20 nm were used to ensure a large number of polymers with the backbone lying parallel to the substrate. Such a configuration favours efficient excitation and detection, with the transition dipole moment of the chromophores being aligned along the backbone. The quartz substrates were chosen to create a good thermal contact with the cold finger of the cryostat and to ensure very low background signals originating from the stray light of the exciting laser. Note that thermal conductivity as a function of temperature is a serious issue in cold finger cryostats, which necessitates the use of well-chosen substrates.

Once the molecules are dispersed in space, optical microscopy techniques are suitable to selectively study each molecule one by one. Wide-field fluorescence microscopy has become established as a facile tool to be combined with spectroscopy measurements [54]. Our setup is based on a home-built far-field microscope coupled to a monochromator and a CCD camera. Figure 2a shows a simplified scheme of the setup for SMS with the various components illustrated. The spatial photoluminescence (PL) images of single molecule samples are obtained by exciting with a 50-μm-diameter laser spot generated from the second harmonic of a tuneable Ti:sapphire oscillator (pulse length 100 fs, bandwidth 10 nm, 80 MHz repetition rate) and recording the image on a front-illuminated CCD camera coupled to a monochromator. The monochromator can be used both for imaging and spectroscopy. In the spectroscopy configuration two different gratings are available with resolutions of 0.4 and less than 0.1 nm. SMS can be performed using the monochromator slit as a pinhole for spatial filtering, once an image of the single molecules is obtained. The flexibility of our home-built setup allows for the implementation of several optical interference filters for high-sensitivity imaging detection in well-defined spectral windows. Additionally, light-polarising elem-

Single Molecule Spectroscopy of Polyfluorenes

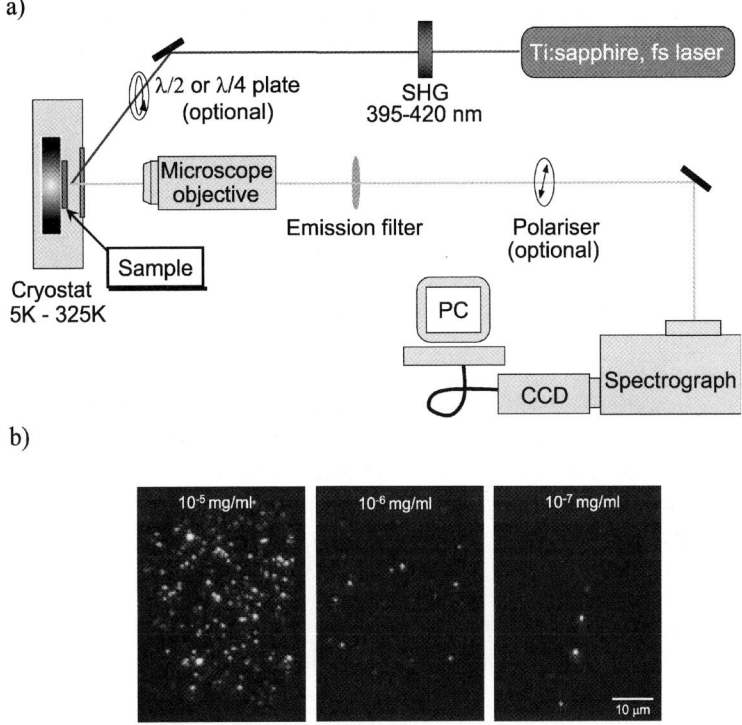

Fig. 2 Schematic description of the SMS setup. **a** Schematic diagram of the experimental setup for SMS. Laser pulses from a Ti:sapphire laser are frequency doubled in a second harmonic generation (SHG) crystal to generate the excitation beam in the wavelength range (395–420 nm). The PL of the SMS sample mounted in a liquid helium cold finger cryostat is imaged on a CCD camera by a microscope objective and optionally spectrally dispersed through a spectrograph. **b** Examples of wide-field PL images for three different single molecule samples obtained by spin-coating a conjugated polymer from solutions at different concentrations. The PL intensity is encoded in false colour in the online version. Adapted from [55]

ents, such as $\lambda/2$ or $\lambda/4$ plates and polarising foils, can be inserted in the excitation or emission path to study the anisotropy of optical transitions in excitation or emission, respectively.

Experimental indications of single molecule detection are provided by the wide-field images reported in Fig. 2b. In particular, the bright spots (Fig. 2b) show the typical blinking on a timescale of seconds as well as irreversible bleaching of the PL, which are both consistent with typical single molecule PL characteristics. As a further confirmation for the identification of bright fluorescence spots with single polymer chains, the density of spots per unit area scales with the initial polymer concentration in the spin-coated solution. Note that these spots appear much larger than the actual size of the polymer chain, due to the diffraction-limited resolution of the fluorescence microscope.

3
How Shape Controls Function: The Single Chain β Phase

3.1
Influence of Chain Conformation on β-Phase Formation

PFs differ by the side groups in position 9 of the fluorene moiety [29]. The PFO contains two octyl side groups and has attracted considerable attention because of its different phases [38]. Such a behaviour elects this system as a model conjugated polymer to study structure–property correlations, in the absence of chemical variations. Pioneering studies on PFO, using X-ray diffraction, have identified different distinct phases characterised by variations in the degree of inter- and intramolecular order [40]. In particular, the β phase is characterised by an extended planarisation of the fluorene repeat units. This planarisation clearly modifies the intersite coupling within the molecule and can potentially increase conjugation, while reducing excited-state relaxation. Due to the high sensitivity of optical spectra to variations in the conjugation length, the different phases are also characterised by a markedly different photophysical response. While PFO in the glassy or α phase is identified by a broad absorption band peaking at 390 nm, the β phase shows narrow features with the origin peaking at 440 nm. These spectral features are mirrored in emission with a very small Stokes shift for the β phase. Despite the observation of β-phase spectral signatures even in dilute solid solutions of PFO in polystyrene, the question surrounding the inter- or intramolecular nature of the β phase solicited many investigations involving several structural characterisation tools. For example, Knaapila et al. recently reported the formation of β-phase chains when PFO is dispersed in a poor solvent [46], suggesting a favourable role of physical interchain aggregation in driving intrachain planarisation.

Figure 3 reports two examples of low-temperature single molecule spectra for PFO. In panel (a) a spectrum with narrow (< 1 nm) features shows a zero phonon line at 440 nm, in remarkable agreement with the optical gap observed in ensemble studies [40]. The high dilution employed to perform SMS ensures the absence of interchain interactions (Fig. 1b), although we note that we cannot strictly exclude the possibility of self-collapsing of the PFO chain occurring at the present stage. Single PFO chains with a blue-shifted and broad spectrum are observed as well (Fig. 3b) and are reminiscent of a more disordered intramolecular structure where the fluorene units are twisted. We assign this spectrum to a glassy phase of PFO due to the similarity with the spectral features observed in the amorphous thin films of PFO [31, 40]. We note here that from luminescence experiments it is not possible to distinguish between glassy or α phases, both being characterised by a twisted conformation but different degrees of interchain order [40]. Interestingly, the overall number of single chains emitting as β phase is much lower than that of glassy

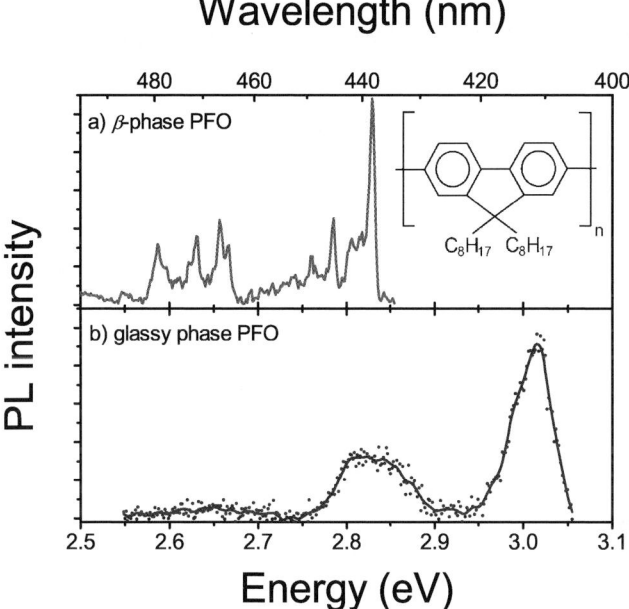

Fig. 3 Single molecule PL spectra of PFO at 5 K. **a** β Phase, **b** glassy phase. Adapted from [24]

phase ones [24], in agreement with previous studies which suggest the necessity of applying stress to induce β-phase formation [56].

Structure–property correlations are ubiquitous in soft semiconductors such as conjugated polymers [7, 22, 29, 57]. For PFO we expect that the markedly different spectroscopic behaviour is tightly entangled with the resulting photophysics. In order to probe the chain conformation of PFO in the two different phases, we performed polarisation anisotropy studies. Such kinds of experiments were first used by Hu et al. [6], giving remarkable results concerning the chain conformation of poly(phenylene vinylene) (PPV). Subsequently, Huser et al. [7] showed the influence of solvent quality on the resulting chain conformation. For good solvents extended conformations were observed with tremendous implications for the resulting photophysical properties. The experimental principle is based on the fact that the transition dipole moment of the low-lying singlet states is collinear with the polymer backbone [6, 9, 58]. By rotating the polarisation of the exciting laser beam and recording the PL intensity it is possible to probe the spatial organisation of the different chromophores making up the polymer chain (see Fig. 4b for a sketch of the experiment).

Figure 4a shows an example of an excitation anisotropy trace recorded for a single PFO chain in the β phase by collecting the fluorescence and rotating the plane of the exciting laser using a $\lambda/2$ plate. From this curve

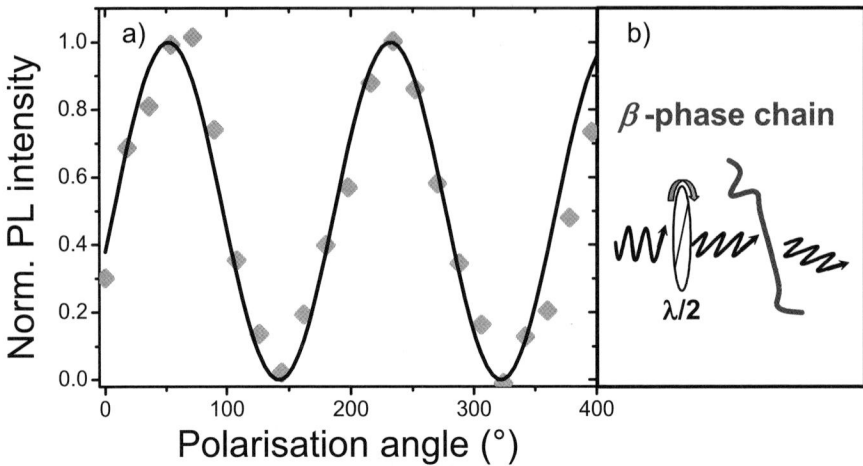

Fig. 4 **a** Polarisation anisotropy trace recorded by rotating the exciting laser polarisation and detecting the PL emission of a single chain in the β-phase conformation. The *green diamonds* are experimental data, whereas the *black line* is a fitting with a $\cos^2 \theta$ function. The extracted polarisation anisotropy is equal to 0.9, indicating an extended conformation. **b** The sketch illustrates the experiment where the λ/2 plate is rotated in the beam path of the exciting laser while the chain PL is monitored

a polarisation anisotropy parameter (P) can be extracted. P is defined as $P = (I_{max} - I_{min})/(I_{max} + I_{min})$, where the I_{max} and I_{min} correspond to the maximum and minimum of the anisotropy curve. Values of P tending towards 1 are symptomatic of a more extended structure of the polymer chains, assuming an equal probability for all of the chromophores in the chain to absorb light. $P = 0$ is characteristic of a system without any preferential alignment of the transition dipole moments and thus a coiled collection of chromophores, most probably held together by kinked saturated bonds [6]. The value extracted from Fig. 4 is 0.9, providing evidence for an extended conformation. We therefore exclude at this point that β-phase emission comes from a self-collapsed segment of the chain.

In order to obtain a detailed overview over the various PFO chain conformations possible in the two different phases, we analysed the anisotropy traces of more than 100 molecules in both phases, spin-coated from toluene solutions (Fig. 5). We note that the good solubility of PFO in toluene has been demonstrated recently [46], with experimental indications of a rodlike configuration observed even for long polymer chains. Remarkably, β-phase molecules show a large scatter in P, peaking at high anisotropy values (panel a), in contrast to the chains of the spectrally distinct glassy phase (panel b). These results demonstrate that the planarisation of fluorene repeat units not only has an impact on the emission colour and stability, but that it also characterises the degree of extension of the chain, resulting in the his-

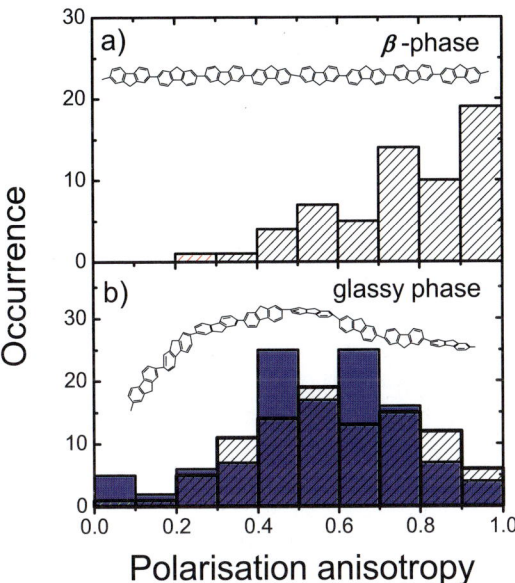

Fig. 5 Excitation polarisation anisotropy histograms for more than 200 single PFO chains. **a** Histogram for the β phase. **b** Histogram for chains in the glassy phase subjected (*shaded black bars*) or not subjected (*filled blue bars*) to the vapour swelling treatment. Adapted from [28]

togram of anisotropy values peaking at 0.9. A different behaviour is found in glassy-phase chains where the histogram shifts to lower values. While this is reminiscent of a more coiled structure (see the inset scheme in Fig. 5), the glassy phase and more generally PFO can still be considered as an extended polymer if compared with much more coiled systems, such as PPVs [6], where anisotropy histograms centred at $P = 0.2$ are usually observed.

The peculiar characteristic found in the β-phase chains can be connected to the chemical structure which is based on a planar and rather stiff repeat unit combined with the long-range planarisation between them [40]. In contrast, the torsional degrees of freedom between the adjacent repeat units are mostly responsible for the partial coiling found in most glassy-phase chains.

Similar results reporting high polarisation anisotropies were recently published by Link et al. [59]. In this case the effective extension of PPV single chains was achieved using a nematic liquid crystal as a solvent, where the chain segments can orient preferentially along the liquid crystal axis. We speculate that for the PFO β phase a similar process takes place where the long dioctyl side chains can act as a solvent in the immediate proximity of the polymer backbone. This hypothesis is supported by recent molecular dynamics simulations and Raman spectroscopy measurements, which suggest a 1D self-assembling of the side group around the β-phase chromophore [44, 45].

Long-range planarisation combined with the chain extension makes the β phase an example of crystallisation in 1D, a phenomenon which is rarely observed in nature [60]. Following the vapour swelling and annealing procedures reported in the literature [40], we found a dramatic increase in the number of PFO chains showing β-phase emission characteristics [28]. While this recipe turns out to be effective also at the single molecule level, the polymer chain *phase transition* involved still remains an open question. To clarify this point we measured the excitation polarisation anisotropy for the glassy phase after the swelling process (shaded black histogram in Fig. 5b). The swelling procedure only affects the frequency of planarisation of the chains, i.e. it does not actually induce stretching of the chains. As shown in the histogram the polarisation anisotropy of the glassy-phase single chains before (filled blue bars) and after (shaded black bars) swelling is found to remain virtually unchanged by the treatment.

3.2
Influence of Chain Planarity on the Photophysical Stability

According to the results of the previous section, the planarisation combined with the chain extension found in the β phase should result in a different photophysical behaviour with respect to glassy-phase chains. Figure 6 reports two spectral traces recorded as a function of time for a β phase (a) and a glassy phase (b) single chain. The colour code of the two-dimensional plot of photon energy versus time corresponds to intensity. Evidently, the planarisation results in an increased stability in the emission intensity (overlaid black curve) as well as in the spectral diffusion (note the temporal scale extending up to 2 h). In contrast to this unique behaviour, the glassy phase (panel b) shows a larger spectral diffusion accompanied by a short-timescale blinking. The photophysical stability of this system is also very different resulting in an irreversible photobleaching after 200 s. As mentioned before the two emission spectra differ with respect to the linewidth of the pure electronic transition; a few meV for the β phase and tens of meV for the glassy phase. While in the former case the origin of such a spectrally narrow emission is influenced by several factors (see next paragraph), it appears likely that the glassy-phase emission is mainly broadened by a fast spectral diffusion process taking place during the experimental integration time [24].

As far as the blinking behaviour is concerned, we recall here that extensive studies on small organic molecules have highlighted the major role played by the presence of the triplet states. After a certain number of excitation cycles triplet excited states can be populated, leaving the system in a dark state for several milliseconds [61], if not seconds, as in the case of PFO at low temperatures as witnessed by the phosphorescence lifetime [62]. The absence of blinking in the β phase suggests that triplet formation is indeed very inefficient under optical excitation. Generally, in the case of multichro-

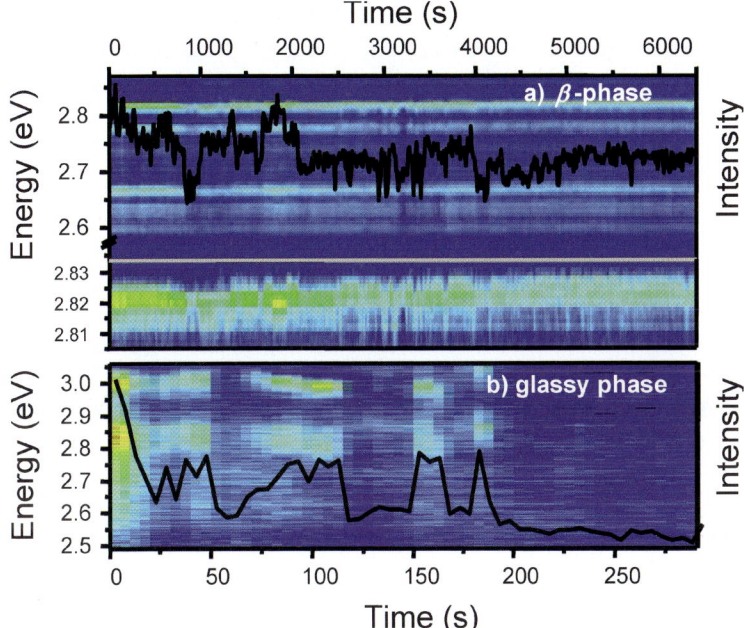

Fig. 6 Temporal evolution of the PL for a β phase (**a**) and a glassy phase (**b**) molecule at 5 K. The integrated intensity is overlaid as a *black curve*. A close-up of the zero phonon line of the β-phase PF is shown at the *bottom* of **a** which displays spectral diffusion. Reprinted from [24]

mophoric systems such as conjugated polymers the picture of blinking is more complicated [63]. In particular, the presence of different types of excitation (excitons, polaron pairs and polarons) extending on segments with a distribution of conjugation lengths can give rise to new phenomena. For example, polarons localised in one chromophore can quench the PL of excitons in the nearest-neighbour chain segments, resulting in blinking on a broad range of timescales [57, 64].

These effects add to the presence of triplet excitons which are more efficiently generated when conjugated systems are in a twisted conformation [65]. On the basis of these arguments it is not surprising that the glassy phase characterised by localised excitons and a twisted conformation shows the most unstable emission. While this result holds direct consequences for optoelectronic devices, where the photophysical stability is a crucial parameter, it also highlights the unique features of the β-phase PFO, which at the single molecule level is by far superior to any other solution processable conjugated polymer.

Similar properties were observed only in single polydiacetylene (PDA) quantum wires, embedded in their monomer crystal. Such a conjugated system has exhibited an almost undetectable spectral diffusion and blinking [66]. The unique electronic properties of these nanowires, where the exciton can be delocalised on a micrometre scale [12], are ascribed to the high degree of 1D order [67]. Although it is not possible to give an exact value for the conjugation length of the β-phase emitting chromophore, we note that the combined planarisation and chain extension characteristics are expected to result in an increased exciton delocalisation in 1D.

The implications for device applications are straightforward. β-Phase chromophores are much more resilient to optically induced and, most probably, also to electrically injected excitations. Therefore, OLEDs made out of PFO β phase should show extended lifetime operation with respect to those containing a large number of chains in the glassy phase. Optimisation of the quantity of β-phase chains by employing vapour swelling techniques or with the use of well-defined protocols involving different solvents will be crucial for device operation [56].

3.3
The β Phase: Strain-Induced Depolarisation in Extended π-Electron Systems

In this section we will discuss in detail how strain influences the emission properties of the PFO β phase. The remarkable photophysical stability (Fig. 6) elects this system as a model for single molecule studies of the intrinsic electronic properties of extended π-conjugated materials. Analysing in detail the histogram reported in Fig. 5 we note that while most of the β-phase chains are extended, there is a non-negligible number of chains with a low anisotropy. These might arise from slightly bent conformations of the single absorbing segment or, with an equal probability, the concomitant absorption by several segments with different orientations. The latter case is commonly observed in single molecule experiments on LPPP. Müller et al., observing the polarisation anisotropy in emission, reported evidence for several chromophores with perpendicular orientations emitting simultaneously at different energies [21, 68]. In the presently studied PFO β phase we found evidence of multichromophoric emission only in three single chains out of 400 studied at low temperature. While we do not exclude the presence of segments with a different degree of order, our statistics clearly suggest the presence of only one β-phase chromophore in most cases.

To further investigate the intriguing properties of such a system with a bent, planarised chain we studied the emission polarisation anisotropy (rather than the excitation anisotropy reported above). Such experiments were performed by rotating a polariser placed before the entrance slit of the spectrograph following unpolarised excitation of the single chain (Fig. 7b). As clearly observed in Fig. 7a, the PL intensity of the zero-phonon line (ZPL) at 2.806 eV is modulated with a 180° periodicity in polarisation. While the modulation does not reach the background intensity (dark blue in the intensity colour plot), the spectrum does not change either in energy or shape

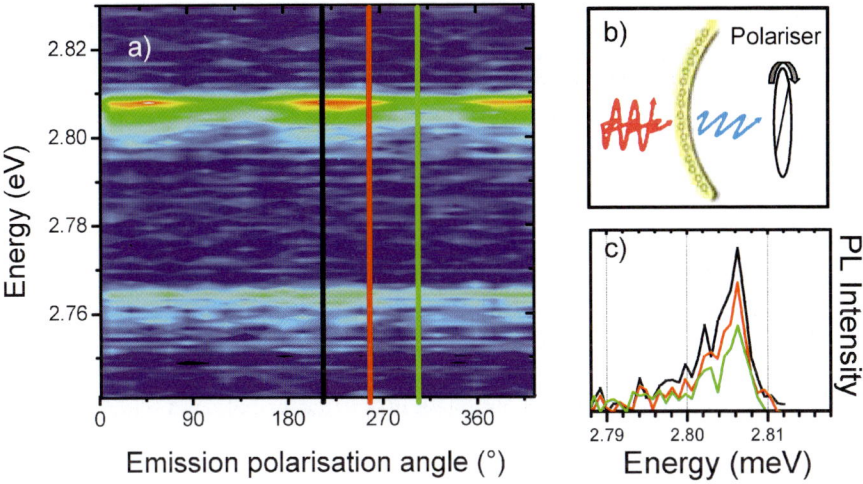

Fig. 7 a Temporal evolution of the PL spectrum of a single β-phase chromophore recorded during the rotation of a polariser placed in the emission light path as sketched in **b**. **c** PL spectra of the 0–0 transition taken at the polarisation angle marked in **a** by the respective *coloured lines*. Note that the PL intensity of the ZPL centred at 2.806 eV does not show a full modulation, as expected for emission from a linear dipole

during rotation (Fig. 7c). Therefore, we exclude the possibility that multiple chromophores are contributing to the observed emission, in contrast to the prior example of LPPP. We anticipate that a bending in the emitting chromophore could cause the observed partially polarised emission, since in the linear case (i.e. the situation of a spatially fixed dipole) full modulation must be observed.

To gain further insights into the nature of possibly bent chromophores, we correlated the spectroscopic information with polarisation anisotropy. Figure 8 reports the ZPL of two single chains rescaled at the respective peak energies together with their emission anisotropy traces. While the chain showing the narrowest spectrum (panel a) is characterised by a fully modulated emission with $P = 0.98$ (panel c), the broad one shows an anisotropy of only 0.33 (panels b, d). Since both chains are in the spectrally distinct β phase, such a behaviour implies a considerable amount of strain acting on the bent chromophore, which is characterised by an in-plane bending as suggested by the cartoon insets. The possibility of bending in polymer chains without the disruption of the π conjugation was recently predicted by quantum chemical modelling of distorted chains [69]. Before further discussing the result of Fig. 8 it is instructive to clarify the nature of optical excitations in organic conjugated systems.

Because of the strong Coulomb interactions inherent in such low dielectric molecular semiconductors, excitons are characterised by a very small radius (a few nanometres). At the same time the conjugation length can reach

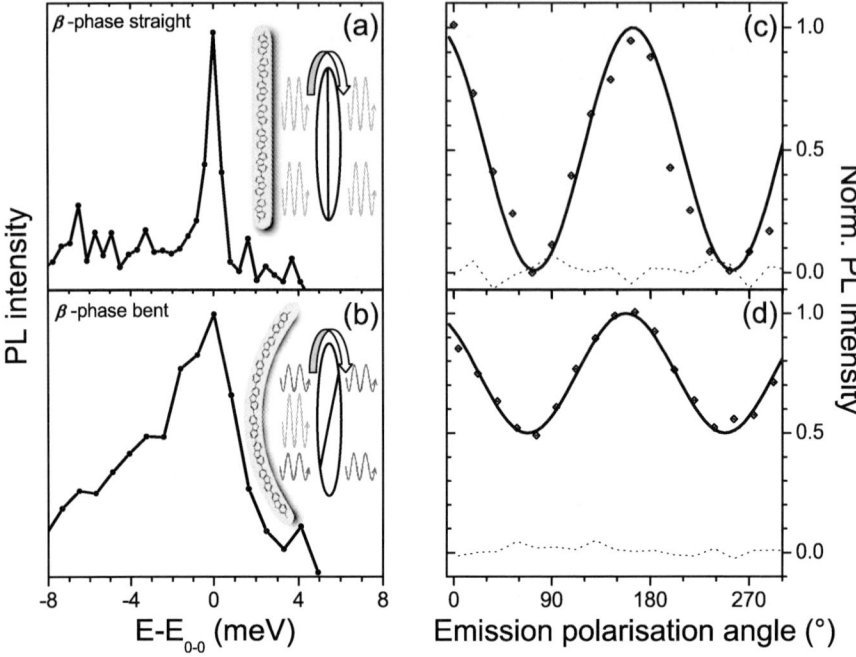

Fig. 8 a Narrow emission spectrum (linewidth 0.7 meV) shifted by the 0–0 transition energy, exhibiting linearly polarised emission ($P = 0.98$) as shown in (**c**). **b** The broad emission spectrum (linewidth 2.40 meV) of a different single chain is only weakly linearly polarised ($P = 0.33$), (**d**). The *dotted lines* in **c** and **d** indicate the fluorescence background. Reprinted from [28]

values of up to tens of nanometres or even micrometres, as in the case of PDAs [12]. As a consequence the exciton can, in principle, delocalise coherently along the segment where the π conjugation is preserved [70, 71], while maintaining a small electron–hole separation. The reasonable exclusion of multichromophoric emission in the results of Fig. 7d provides compelling evidence of emission from a bent segment where full conjugation is maintained, i.e. there is a similar probability for the exciton to recombine along the bent chromophore.

In order to establish a clear correlation between the two parameters studied in Fig. 8, namely linewidth and anisotropy, we plotted in Fig. 9a the emission anisotropy and ZPL for 80 single β-phase chains (circles). A clear correlation is observed with a statistical coefficient of –0.5, which translates into a probability of >99% to have these two variables correlated. The diamonds are obtained by averaging the linewidth of molecules within the same anisotropy interval of 0.1. Therefore, straight chromophores show linewidths which reach minimum values of 400 µeV, of the order of the resolution limit of the spectrometer used.

Recently, the origin of spectral broadening in the PL emission of conjugated polymers has received considerable attention [21, 66, 72, 73]. Electronic dephasing and spectral diffusion have been considered as the two main effects influencing the observed linewidths. Optical transitions are broadened by the intrinsic lifetime of the exciton. The linewidth (Γ) can therefore provide a lower limit for the intrinsic dephasing time of an electronic excitation ($T_2 \geq 2\hbar/\Gamma$), which for the completely extended chromophores in PFO can reach 3 ps. Figure 9b shows a scatter plot for the ZPL energy and the linewidth for the same molecules. Besides the lack of any correlation we also note a strong scattering in the values, which makes the possibility of having two chromophores merging into a single broad transition line very unlikely.

Very recently, the reports of Feist et al. on PPV have shown that the linewidth in emission can be orders of magnitude larger than what is observed in the excitation spectral profile [72]. This increased broadening has been ascribed to spectral diffusion, as also demonstrated by Hildner et al. on LPPP [73]. However, in both studies no correlation between linewidth and chromophore conformation was established.

In contrast to other polymers, the β-phase PFO, being characterised by a 1D crystalline structure where the repeat units all lie in the same plane,

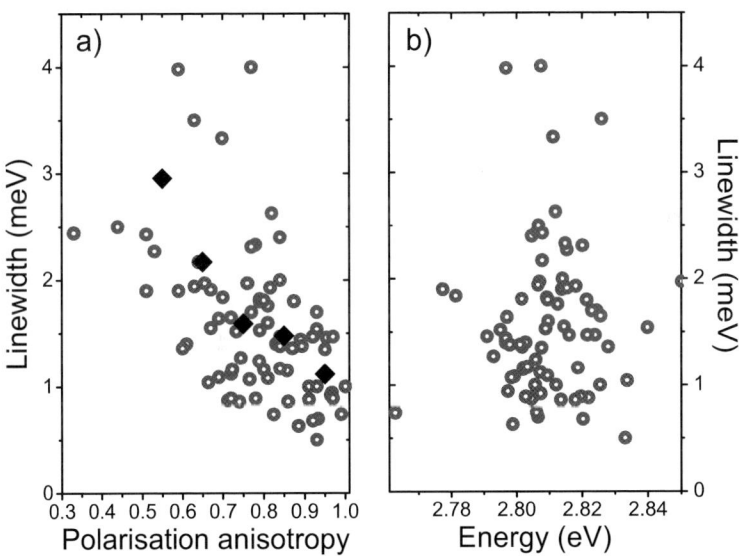

Fig. 9 a Correlation between transition linewidth and polarisation anisotropy in β-phase PFO for a total of 80 β-phase single molecules (*circles*), excited at 412 nm. The *diamonds* indicate an average over steps of 0.1 in polarisation anisotropy. The errors in the determination of linewidth and anisotropy are 0.25 meV and ∼0.03, respectively. **b** Scatter plot of linewidth and 0–0 transition energy, showing that the transition energy of the β phase does not correlate with the linewidth. Adapted from [28]

represents an ideal model system to study how the strain (i.e. the degree of bending in the plane) influences the electronic structure. Other systems, such as PPV or LPPP, show markedly different properties. For example, our recent results on PPV oligomers showed that the emission of this system is characterised by self-trapping [58], thus precluding the observation of any difference in the emission anisotropy from molecules of different conformations [74]. On the other hand the more rigid LPPP, where self-trapping is not expected to occur, has a minimum number of degrees of freedom, hindering the bending of the main chain and therefore making any accessible correlations with structure highly improbable. All LPPP chromophores are expected to have the same shape, and indeed only a small scatter in transition linewidths is typically observed for LPPP. Moreover, the LPPP chromophores are likely to be much more exposed to the surrounding matrix (phonons, extra charges) with respect to the PFO where the octyl side chains are expected to play the role of a *shell*.

While we cannot quantify the amount of spectral diffusion occurring in the individual chains of our experiment, we note that the reported linewidth in Fig. 7a is similar to the one recently observed in PDA. For this material a coherence time of 2 ps was extracted from the Lorentzian shaped ZPL [12, 66]. The most striking difference between these two systems is the fact that PDA is grown in its highly ordered monomer crystal, whereas PFO is dispersed in a supposedly amorphous matrix. Ballistic exciton transport has been demonstrated for this kind of chain leading to a macroscopic spatial coherence [12]. We propose here that ballistic transport could occur also in the β-phase chromophores. Signatures of such a behaviour are provided by the exciton delocalisation over bent segments, resulting in an equally probable emission from all the segments and resulting in an emission which is not fully linearly polarised. The direct conclusion is that exciton self-trapping does not occur in β-phase PFO. Effective exciton delocalisation over long distances might be an exclusive characteristic of systems with a low tendency to show exciton self-trapping. Indeed, both PDA and the β-phase PFO show a very small structural relaxation energy upon photoexcitation. Remarkably, in the case of PDA the observed Stokes' shift is less than 500 µeV [75].

We discuss now the possible reasons for the correlation reported in Fig. 9a. As mentioned above, the β phase can be considered as a 1D crystal where the long-range order is witnessed by the large diffraction coherence observed in X-ray scattering experiments [40]. Our observation of bent chromophores implies a structure under strain, where the planarisation between the repeat units must be preserved because of the typical β-phase emission characteristics, while a certain distortion within the plane can take place. The horizontal strain can reduce the intersite coupling impacting on the π conjugation, which is present all along the chromophore. We speculate here that this reduced intersite coupling may increase the scattering of the electronic wave function resulting in an accelerated dephasing. Therefore bent chromophores should present larger homogeneous linewidths.

At the same time, spectral diffusion takes place and can wash out the information extracted from the linewidth, precluding any definitive conclusion on the coherence time of the exciton. Spectral diffusion in organic molecules is ascribed to fluctuation in the surroundings of the chromophore due to the dissipation of the excess energy following photoexcitation. It has been demonstrated that for organic dyes this spectral jittering can substantially increase according to the number of degrees of freedom of the matrix [76, 77]. As pointed out above, there are several indications that the side groups play an important role in the photophysical response of the β phase. Most probably, they act as a crystalline environment in the immediate proximity of the chromophore. Such a phenomenon resembles the well-known Shpol'skii matrix effect, where organic molecules show remarkably narrow lines when crystallised as impurities in alkanes [78]. According to this picture, the bent chromophores should have a less ordered configuration of the side groups leading to a greater number of electronic degrees of freedom and therefore enhanced spectral diffusion.

Summarising, we have given an experimental demonstration of bending in a π-conjugated chromophore at the single molecule level. This is quite remarkable and should not be confused with prior demonstrations of the bending of the overall polymer chain [6, 7]. Moreover, we show how the strain resulting from the bending impacts on the exciton linewidth. The β-phase chromophore represents a model system to study the influence of shape on the resulting electronic properties of 1D semiconductors. The sensitivity of electronic material properties on the chain conformation demonstrated for PFO makes this system also suitable as a force nanosensor, where mechanical strain can be correlated with an optical response. A future goal is the control of the degree of planarisation and elongation in order to obtain samples with a large number of chains which display improved photophysical stability [24] and long-range exciton transport [36, 79]. In addition, the remarkable spectral distinction makes induced planarisation in PFO an interesting avenue to explore for reversible molecular switching.

4
Identification of Single Keto Defects on PF Chains

PFs have found applications in light-emitting diodes because of their high efficiency deep-blue ($\lambda < 450$ nm) emission [31, 32, 80]. Colour purity is an important characteristic for these materials, particularly when they are combined with green- and red-emitting materials to generate white-light-emitting devices or full-colour displays. For example, in applications such as displays or solid-state lighting, the long-term maintenance of colour purity according to the Commission Internationale de l'Éclairage (CIE) coordinates becomes crucial. A long-lasting issue for all kinds of PFs is colour degradation due to the rising of a broad green emission centred at 535 nm [42, 48–50, 81]. Several studies have tried to access the intrinsic nature of this lower energy emission, which can dominate once efficient trapping of the higher energy excitons occurs. Experiments based on time-resolved spectroscopy revealed the long-living (nanosecond) nature of the green transition [42, 49, 50]. Based on these initial photophysical characteristics and on the superlinear behaviour in the green emission intensity with increasing concentration [49], an excimer has been considered as a possible origin of colour degradation.

However, the work of List et al. [48], based on PL and IR spectroscopy, questioned this picture demonstrating a correlation between the green emission and the presence of keto defects [82] on the PF chain. These can be statistically introduced during the synthetic procedure or arise as the result of photo-oxidation.

Overwhelming evidence for the presence of an on-chain defect arising in pristine samples was reported simultaneously by Lupton et al. [50], who observed the long-lived green emission for short hexamer molecules in very dilute solutions (20 µg/ml). Spectroscopic investigations of the polymer in pristine thin films or poor solvent solutions inevitably leave the probability of chain entanglement or aggregation, potentially missing some crucial information on the physical processes taking place during the green emission.

Single molecule experiments on PF copolymers containing different amounts of fluorenone units (see Fig. 10a) represent the most convincing experimental way to prove the monomolecular origin of the green emission [26].

The low emission efficiency of the fluorenone moieties with respect to the PF chain blue emission requires the use of band-pass filters to obtain images of single polymer chains. Figure 10b shows wide-field images of the isolated fluorenone–PF copolymer obtained at 435–485 nm (blue, first column) and 500–600 nm (green, second column) spectral regions for PF–fluorenone copolymers: 0% (b) and 1% (c). In the third column the square root of the product of the intensity in the two detection channels shows the relative intensity between the two types of emission. Images in the two spectral regions are recorded without changing the spatial position of the sample, i.e. the same single chains are observed in the two images (black spots). While for the homopolymer the fluorescence signal is detected only in the blue image, in the 1% copolymer chains the PL signal is present in both. Remarkably, virtually all single chains emit in both channels. This result is interpreted by considering that the fluorenone units do not quench the overall blue emission coming from the PF segments, as this would require complete energy transfer from all polymer chromophores excited. The presence of molecules emitting in only one channel is reasonably expected from the different pho-

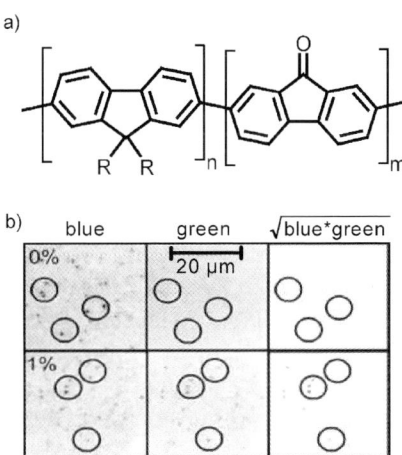

Fig. 10 PL from single PF molecules containing on-chain fluorenone defects. **a** Chemical structure of the dioctylfluorene–fluorenone copolymer investigated. **b** Room-temperature single molecule fluorescence wide-field images of the copolymer dispersed in a Zeonex matrix. Note the PL intensity encoded in a negative scale with respect to Fig. 2b. Excitation was performed at 400 nm in the tail of the backbone absorption. The PL was recorded in 5-s exposure windows. Spectral selection was performed by means of blue (*left column*) and green (*centre column*) band-pass filters centred at 460 and 550 nm, respectively. Adapted from [26]

tobleaching behaviour of the two units [26, 49]. Indeed, during the course of the measurements some PF chromophores can irreversibly photobleach leaving only green or blue emission. This observation also suggests the possibility of a direct excitation of the fluorenone units, in agreement with previous studies in solution [50]. Although the fluorenone molecule is characterised by a low quantum yield, our results suggest an increased emission efficiency when copolymerised in the PF backbone, as the oscillator strength of the delocalised π-electron system can be imparted on the localised fluorenone emissive centre.

While the blinking behaviour of the fluorenone is clearly a signature of single chain PL characterised by multichromophoric emission [21, 57, 71], self-aggregation could still play a role and potentially lead to the formation of excimeric fluorenone units. In principle, it is conceivable that some chains may collapse during the dispersion in Zeonex in the single molecule sample, although it appears extremely unlikely that two fluorenone units would approach each other to form an intramolecular excimer [49]. As the typical polymer chain contains, on average, 200–400 units, we expect less than two to four fluorenones per chain for the 1% copolymer. Without an extremely strong and selective attraction between fluorenone units, it is impossible for every chain to form an intramolecular aggregate of fluorenone units. Furthermore, PFs show a rather extended chain conformation, as observed for the case of PFO in the previous section. Such a folding behaviour is the opposite with respect to what is usually observed in PPVs where the large number of torsional degrees of freedom present in the polymer chain result in strongly collapsed and coiled conformations [6, 7]. An extended conformation again supports our interpretation, which considers a monomolecular origin of the green emission band in the absence of excimer formation.

In order to draw a clear picture of the single molecule data, we compare the intensity of the blue and green emissions for copolymers with an increasing percentage of fluorenone units. Figure 11 shows the correlation between the two intensities for a total of 1152 molecules together with the probability distributions. All of the copolymers studied show green emission with an increasing intensity as the number of copolymerised units is increased. This increase correlates directly with the blue emission that is strongly reduced for the 5% copolymer. We note here that the scattering in these single molecule data is due to the blinking characteristics of the single chains, which may vary

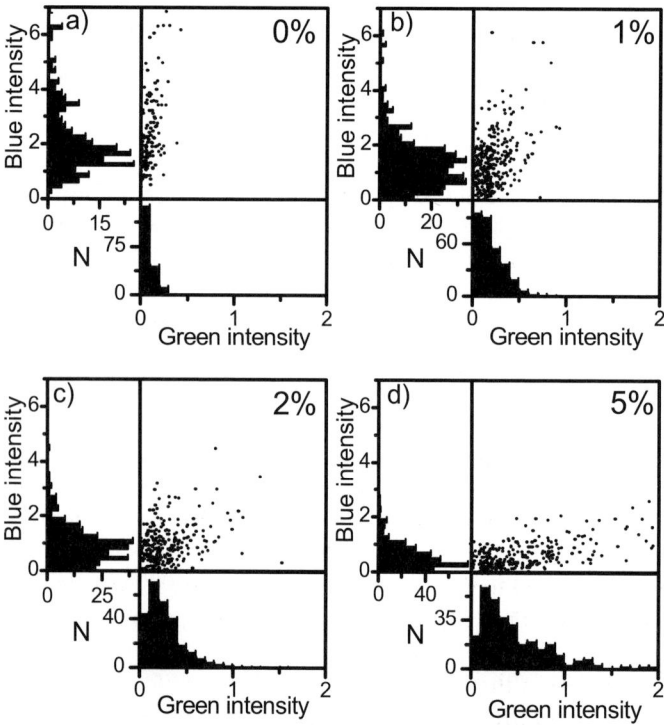

Fig. 11 Correlation of the blue emission and the green emission bands in a total of 1152 single PF molecules containing between 0 and 5% fluorenone units in the statistical copolymer (**a–d**). The side graphs are histograms for the blue (*vertical*) and green (*horizontal*) emission intensities. Reprinted from [26]

according to the slightly different environment or chain conformation of each single molecule [57].

Summarising our results on the green defect emission in PF, we stress the sensitivity of the conjugated polymer luminescence to a small amount of chemical defects (1%). The successful observation of green emission in *single chains* of the copolymerised fluorenone–PF gives a definitive answer to the question of the monomolecular origin of the green emission. We note here that in principle not all the oxidative defects need to result in green emission. In fact, the single chain green emission ultimately also vanishes, pointing to a more ultimate outcome of oxidation than the keto defect. Moreover, photo-oxidation can follow the same dynamics in films and isolated chains, although in the former case energy transfer to radiative defects is enhanced due to the close packing between the chains. The efficient energy transfer taking place in thin films of pure PF leads to a non-linear behaviour of the green band intensity with the defect concentration. SMS represents an ideal tool in observing single chain phenomena and distinguishing between interchain and intrachain photophysical processes. Further adaptations of this research include the study of end-capped conjugated polymers, such as polyindenofluorenes terminated with perylene dye acceptors. The covalent binding of the acceptor to the polymer backbone donor can significantly enhance energy transfer by enabling through-bond electronic coupling. Such model systems can provide further insight into the subtle role of intramolecular defects in large conjugated macromolecules. On the other hand, the fact that a small concentration of chemical impurities can so dramatically impact on the fluorescence properties can also be exploited in chemical sensing, by exploiting energy transfer to amplify a fluorescence-based sensor–analyte interaction.

5
Conclusions

PFs are a fascinating family of conjugated polymers. Their unique luminescence properties make them ideal model systems to study new physical phenomena in conjugated polymers at the single molecule level—effects which are usually masked in the ensemble photophysical studies, but are nevertheless absolutely crucial to determining the bulk material properties.

We showed that the PFO β phase can be likened to a quasi-1D crystal, a structure which is rarely encountered in disordered conjugated polymers. Moreover, because of this extended conformation, PFO has served as a model system to report on the first experimental evidence of chromophore bending detected by polarisation sensitive SMS. Besides the fundamental interest in such an observation, it can also result in new applications involving organic semiconducting wires which are able to sustain a certain degree of flexibility without losing the intrachain conjugation and electronic delocalisation.

We are of the opinion that our single molecule results on PFs copolymerised with fluorenone units represent compelling proof for the monomolecular nature of the green defect emission. The development of new synthetic strategies for reaching a blue stable emission in PF-based OLEDs should therefore concentrate on formulating protective groups, rather than on spacers to control intermolecular interactions, which inevitably lower the transport performances.

The virtually unlimited possibilities offered by organic chemistry in tailoring specific properties of PFs will give access to new polymeric systems which show remarkable and unexpected properties. As demonstrated in this chapter, SMS will be a valuable tool in answering scientific questions and will no doubt provide the experimental demonstration of many new phenomena. Ultimately, conjugated polymers may even prove versatile systems for sophisticated quantum optical experiments, due to the strong control over polarisation enabled by their anisotropic structure.

Acknowledgements We gratefully acknowledge our colleagues who contributed to this work. They are Jochen Feldmann, Frank Galbrecht, Dequing Gao, Andrew C. Grimsdale, Klaus Müllen, Benjamin S. Nehls, Sepas Satayesh, Ulrich Scherf and Florian Schindler. E.D. is grateful to the DFG for financial support through the project Polyspin.

References

1. van Oijen AM, Ketelaars M, Köhler J, Aartsma TJ, Schmidt J (1999) Science 285:400
2. Richter MF, Baier J, Southall J, Cogdell RJ, Oellerich S, Köhler J (2007) Proc Nat Acad Sci USA 104:20280
3. Dickson RM, Cubitt AB, Tsien RY, Moerner WE (1997) Nature 388:355
4. Tamarat P, Gaebel T, Rabeau JR, Khan M, Greentree AD, Wilson H, Hollenberg LCL, Prawer S, Hemmer P, Jelezko F, Wrachtrup J (2006) Phys Rev Lett 97:083002
5. Nirmal M, Dabbousi BO, Bawendi MG, Macklin JJ, Trautman JK, Harris TD, Brus LE (1996) Nature 383:802
6. Hu DH, Yu J, Wong K, Bagchi B, Rossky PJ, Barbara PF (2000) Nature 405:1030
7. Huser T, Yan M, Rothberg LJ (2000) Proc Nat Acad Sci USA 97:11187
8. Bässler H (1993) Phys Status Solidi B 175:15
9. Brédas JL, Beljonne D, Coropceanu V, Cornil J (2004) Chem Rev 104:4971
10. Tretiak S, Mukamel S (2002) Chem Rev 102:3171
11. Li ZQ, Podzorov V, Sai N, Martin MC, Gershenson ME, Di Ventra M, Basov DN (2007) Phys Rev Lett 99
12. Dubin F, Melet R, Barisien T, Grousson R, Legrand L, Schott M, Voliotis V (2006) Nat Phys 2:32
13. Basché T, Moerner WE, Orrit M, Talon H (1992) Phys Rev Lett 69:1516
14. Gerhardt I, Wrigge G, Bushev P, Zumofen G, Agio M, Pfab R, Sandoghdar V (2007) Phys Rev Lett 98:033601
15. Moerner WE, Fromm DP (2003) Rev Sci Instrum 74:3597
16. Kneipp K, Wang Y, Kneipp H, Perelman LT, Itzkan I, Dasari R, Feld MS (1997) Phys Rev Lett 78:1667

17. Sfeir MY, Wang F, Huang LM, Chuang CC, Hone J, O'Brien SP, Heinz TF, Brus LE (2004) Science 306:1540
18. Lagoudakis PG, de Souza MM, Schindler F, Lupton JM, Feldmann J, Wenus J, Lidzey DG (2004) Phys Rev Lett 93:257401
19. Walter MJ, Lupton JM, Becker K, Feldmann J, Gaefke G, Höger S (2007) Phys Rev Lett 98:137401
20. Hofkens J, Maus M, Gensch T, Vosch T, Cotlet M, Kohn F, Herrmann A, Müllen K, De Schryver F (2000) J Am Chem Soc 122:9278
21. Müller JG, Lemmer U, Raschke G, Anni M, Scherf U, Lupton JM, Feldmann J (2003) Phys Rev Lett 91:267403
22. Schwartz BJ (2003) Annu Rev Phys Chem 54:141
23. Pope M, Swenberg CE (1999) Electronic processes in organic crystals and polymers. Oxford University Press, Oxford
24. Becker K, Lupton JM (2005) J Am Chem Soc 127:7306
25. Becker K, Lupton JM (2006) J Am Chem Soc 128:6468
26. Becker K, Lupton JM, Feldmann J, Nehls BS, Galbrecht F, Gao DQ, Scherf U (2006) Adv Funct Mater 16:364
27. Becker K, Lupton JM, Feldmann J, Setayesh S, Grimsdale AC, Müllen K (2006) J Am Chem Soc 128:680
28. Da Como E, Becker K, Feldmann J, Lupton JM (2007) Nano Lett 7:2993
29. Scherf U, List EJW (2002) Adv Mater 14:477
30. Morteani AC, Dhoot AS, Kim JS, Silva C, Greenham NC, Murphy C, Moons E, Cina S, Burroughes JH, Friend RH (2003) Adv Mater 15:1708
31. Grell M, Knoll W, Lupo D, Meisel A, Miteva T, Neher D, Nothofer HG, Scherf U, Yasuda A (1999) Adv Mater 11:671
32. Grice AW, Bradley DDC, Bernius MT, Inbasekaran M, Wu WW, Woo EP (1998) Appl Phys Lett 73:629
33. Virgili T, Lidzey DG, Grell M, Walker S, Asimakis A, Bradley DDC (2001) Chem Phys Lett 341:219
34. Chua LL, Zaumseil J, Chang JF, Ou ECW, Ho PKH, Sirringhaus H, Friend RH (2005) Nature 434:194
35. Rothe C, Galbrecht F, Scherf U, Monkman A (2006) Adv Mater 18:2137
36. Virgili T, Marinotto D, Manzoni C, Cerullo G, Lanzani G (2005) Phys Rev Lett 94:117402
37. McNeill CR, Abrusci A, Zaumseil J, Wilson R, McKiernan MJ, Burroughes JH, Halls JJM, Greenham NC, Friend RH (2007) Appl Phys Lett 90:193506
38. Grell M, Bradley DDC, Inbasekaran M, Woo EP (1997) Adv Mater 9:798
39. Redecker M, Bradley DDC, Inbasekaran M, Woo EP (1999) Appl Phys Lett 74:1400
40. Grell M, Bradley DDC, Ungar G, Hill J, Whitehead KS (1999) Macromolecules 32:5810
41. Grüner J, Hamer PJ, Friend RH, Huber HJ, Scherf U, Holmes AB (1994) Adv Mater 6:748
42. Lemmer U, Heun S, Mahrt RF, Scherf U, Hopmeier M, Siegner U, Göbel EO, Müllen K, Bässler H (1995) Chem Phys Lett 240:373
43. Setayesh S, Grimsdale AC, Weil T, Enkelmann V, Müllen K, Meghdadi F, List EJW, Leising G (2001) J Am Chem Soc 123:946
44. Chunwaschirasiri W, Tanto B, Huber DL, Winokur MJ (2005) Phys Rev Lett 94:107402
45. Arif M, Volz C, Guha S (2006) Phys Rev Lett 96:025503
46. Knaapila M, Dias FB, Garamus VM, Almasy L, Torkkeli M, Leppanen K, Galbrecht F, Preis E, Burrows HD, Scherf U, Monkman AP (2007) Macromolecules 40:9398
47. Nikitenko VR, Lupton JM (2003) J Appl Phys 93:5973

48. List EJW, Guentner R, de Freitas PS, Scherf U (2002) Adv Mater 14:374
49. Sims M, Bradley DDC, Ariu M, Koeberg M, Asimakis A, Grell M, Lidzey DG (2004) Adv Funct Mater 14:765
50. Lupton JM, Craig MR, Meijer EW (2002) Appl Phys Lett 80:4489
51. Ferenczi TAM, Sims M, Bradley DDC (2008) J Phys Condens Matter 20:045220
52. Nguyen TQ, Wu JJ, Doan V, Schwartz BJ, Tolbert SH (2000) Science 288:652
53. Hennebicq E, Pourtois G, Scholes GD, Herz LM, Russell DM, Silva C, Setayesh S, Grimsdale AC, Müllen K, Brédas JL, Beljonne D (2005) J Am Chem Soc 127:4744
54. Müller JG, Anni M, Scherf U, Lupton JM, Feldmann J (2004) Phys Rev B 70:035205
55. Schindler F (2006) Ph.D. Thesis. Ludwig-Maximilians-Universität, Munich
56. Khan ALT, Sreearunothai P, Herz LM, Banach MJ, Kohler A (2004) Phys Rev B 69:085201
57. Schindler F, Lupton JM, Feldmann J, Scherf U (2004) Proc Nat Acad Sci USA 101:14695
58. Tretiak S, Saxena A, Martin RL, Bishop AR (2002) Phys Rev Lett 89:097402
59. Link S, Hu D, Chang WS, Scholes GD, Barbara PF (2005) Nano Lett 5:1757
60. Meyer RR, Sloan J, Dunin-Borkowski RE, Kirkland AI, Novotny MC, Bailey SR, Hutchison JL, Green MLH (2000) Science 289:1324
61. Basché T, Kummer S, Bräuchle C (1995) Nature 373:132
62. Hayer A, Khan ALT, Friend RH, Kohler A (2005) Phys Rev B 71:241302
63. VandenBout DA, Yip WT, Hu DH, Fu DK, Swager TM, Barbara PF (1997) Science 277:1074
64. Schindler F, Lupton JM, Müller J, Feldmann J, Scherf U (2006) Nat Mater 5:141
65. Beljonne D, Shuai Z, Pourtois G, Brédas JL (2001) J Phys Chem A 105:3899
66. Guillet T, Berrehar J, Grousson R, Kovensky J, Lapersonne-Meyer C, Schott M, Voliotis V (2001) Phys Rev Lett 87:087401
67. Bässler H (2006) Nat Phys 2:15
68. Müller JG, Lupton JM, Feldmann J, Lemmer U, Scherf U (2004) Appl Phys Lett 84:1183
69. Beenken WJD, Pullerits T (2004) J Phys Chem B 108:6164
70. Mukamel S, Tretiak S, Wagersreiter T, Chernyak V (1997) Science 277:781
71. Schindler F, Jacob J, Grimsdale AC, Scherf U, Müllen K, Lupton JM, Feldmann J (2005) Angew Chem Int Ed 44:1520
72. Feist FA, Tommaseo G, Basché T (2007) Phys Rev Lett 98:208301
73. Hildner R, Lemmer U, Scherf U, van Heel M, Köhler J (2007) Adv Mater 19:1978
74. Becker K, Da Como E, Feldmann J, Schelinga F, Thorn Csányi E, Lupton JM, Tretiak S (2008) J Phys Chem B, (in press)
75. Lecuiller R, Berrehar J, Ganiere JD, Lapersonne-Meyer C, Lavallard P, Schott M (2002) Phys Rev B 66:125205
76. Moerner WE, Orrit M (1999) Science 283:1670
77. Kiraz A, Ehrl M, Bräuchle C, Zumbusch A (2003) J Chem Phys 118:10821
78. Richards JL, Rice SA (1971) J Chem Phys 54:2014
79. Kim Y, Cook S, Tuladhar SM, Choulis SA, Nelson J, Durrant JR, Bradley DDC, Giles M, McCulloch I, Ha CS, Ree M (2006) Nat Mater 5:197
80. Gross M, Müller DC, Nothofer HG, Scherf U, Neher D, Bräuchle C, Meerholz K (2000) Nature 405:661
81. Lupton JM (2002) Chem Phys Lett 365:366
82. Ilharco LM, Garcia AR, daSilva JL, Lemos MJ, Ferreira LFV (1997) Langmuir 13:3787

Subject Index

Absorption   188
Alignment   251
Alkyl chain length   218
Alkylindenofluorene-based polymers   29
Amplified spontaneous emission (ASE)   172, 209, 219
Anthracene   154
Anthracene-dialkylfluorene copolymer   21
ASE   187

Bandgap chromophores   10
Beta "bulk" phase   216, 248, 293, 297, 300
–, single chain   300
Biphenylene units   26
Bis(cyclooctadiene)nickel(0)   8, 276
N,N-Bis(4-methylphenyl)-N-phenylamine   58
Bis(triphenylphosphino)dichloro-palladium(II)   152
Bisindenocarbazoles   99, 115
Blinking behaviour   313
Blue emission/emitters   7, 17, 55, 104, 162, 288, 314
Blue polymeric light-emitting devices   7, 273
2-Bromo-7-boronate   8

Cadogan ring closure   109
Calcium acetylacetonate [Ca(acac)$_2$]   79
Carbazoles   99, 101
–, 2,7-functionalized   102
2-Carbon bridges, ladder-type polymers   40
Cathode materials   78
Chain planarity, photophysical stability   304
Chain structures, adjustment   54
Charge injection   4, 20

Charge pairs   187
Charge-carrier transport   166
Chemical defects   273
Color control, copolymerisation   10
Color light emission   162
Comonomer units, incorporation   152
Conjugated polymers   85, 99, 227
π Conjugation   293
Crystallites   251
Crystallization vs. glass transition   157

Deep-blue emission   311
Defect emission   200
Defect sites   278
Defect trapping   207
Degradation   177
Dendrimers   17
Device performance   49
9,10-Dialkyl(diaryl)phenanthrenes   28
Dialkylfluorenes   17
9,9-Dialkylfluorenes   6
9,9-Diarylfluorenes   18
Dibenzophosphole oxide   86
Dibenzosilole   85, 86
Dibenzothiophene-S,S-dioxide   154
Dihexylfluorene   12
Diindolocarbazoles (DICs)   99, 115
Dioctylfluorene   27
Dioctylindolocarbazole   113
Diphenylfluorene polymer   18
Dithienosilole chromophore   66

Electroluminescence   1, 273
Electron-deficient moiety (TAZ)   63
Electron-injection/hole-blocking layer   78
Electron-transport moiety   49
Electronic devices, organic   93
Electronic states   51

Emission 189
Emission colors 65
Energy transfer 125
Europium(III) complexes 138
Excited electronic states 163
Exciton dynamics 187, 204
Exciton–exciton interactions 210
External quantum efficiency (EQE) 168

FFBFF (green emitter) 72
Field effect transistors, organic 173
Flourene oligomers, repeat units 156
Fluorene-based conjugated oligomers 145, 154, 176
Fluorene-S,S-dioxide-thiophene 172
Fluorene-ethynylene-pyrene teroligomers 162
Fluorene-fluorenone co-oligomers 176
Fluorenes, spiro-linked 26
Fluorenones 6, 16, 154
9-Fluorenyl anions, oxidation 16
Fluorescence lifetime 187, 192
Franck–Condon expression 232
FTBTF (red emitter) 72
Fullerene 174

Germanium 86
Glass transition 157
Green defect emission 286
–, thermal stress 284
Green emission, PF-based PLEDs 64, 314
Green emitters 29, 65

Hexafluorene 156
HLPPP 283
Hole-transport layer (HTL) 55
Hole-transport moiety 49
Hole-transporting layer 75
Hole-transporting triarylamines 22
HOMO 3, 51, 165

Indenofluorene-anthracene 31
Indium tin oxide (ITO) 4
Indolocarbazoles (ICs) 99, 108
Ionic states 165
Iridium(III) complexes 15, 71, 127, 135
ITO/PEDOT:PSS 56

Keto defects 187, 274, 293

Ladder polymers 1
Ladder-type poly(para-phenylene) (LPPP) 35, 283, 296
–, HLPPP 283
–, MLPPP 283
Laser dyes 171
Lasers, organic, solid-state 171
Lasing 1, 172
Layer self-organization 236
LEDs, organic 2, 49, 168
Light-emitting diodes 1, 85, 273
Light-emitting diodes, organic 168
Liquid crystallinity 159
LPPPs 35, 283, 296
LUMO 3, 51, 165

Material synthesis/characterization 145, 148
Metal complexes 125
MLPPP 283
Molecular design 145
Monoalkylfluorenes 17

Nanostructures, "bottom-up"/"top-down" 262, 264
Naphthalimide dyes 13
N-Octyl-2,7-dichlorocarbazole 103

OFETs 86, 104, 173
Oligofluorene-fullerene 174
Oligofluorenes 157, 250
–, synthetic approaches 148
Oligophenylenes, ladder-type 32, 36
Optoelectronic devices 125
Organic electronic devices 85, 93
Organic field effect transistors (OFET) 86, 104, 173
Organic light-emitting diodes (OLEDs) 86, 93, 168, 295
Organic photonics/electronics 145
Organic solar cells (OSC) 86, 174
Oxadiazole substituents 25

Palladium(II) acetate 152
PBS-PFP 239
PDAFs 6
–, defect emission 15
–, optical properties 9
PDMOF 235
PEDOT:PSS 52

Subject Index

Pentafluorene 156
Perylene dyes 10, 154
PF chains, single keto defects 311
PF oligomers 249
PF2/6, aligned films 253
–, branched side chain 240
PF8, linear side chain 246
PFO/PEO20/cathode 81
PFs/electrodes, alteration of interfaces 75
Phenanthrene 28
Phenylene polymers, bridged/unbridged 3
Phosphorescence 187, 194
Photoluminescence 1, 273
–, high solid-state 5
–, polarized 163
Photoluminescence quantum efficiencies (PLQEs) 51
Physical blending system 72
PIFs 29
–, defect emission 30
Platinum(II) complexes 135
Polaron 199
Poly(2-acyl-p-phenylene)s 28
Poly(bi-spirofluorene) (PSBF) 196
Poly[9,9-bis(2-ethylhexyl)fluorene] 8
Poly(2,7-carbazole)s 100, 103
Poly[9,9-di-n-(2-ethylhexyl)fluorene] (PF2/6) 188, 231
Poly(9,9-diarylfluorene)s 18
Poly(9,9-dihexyl-2,7-dibenzosiloles) 91, 288
Poly(9,9-dihexyl-2,7-fluorene) 6
Poly(9,10-dihydrophenanthren-2,7-diyl)s 26
Poly(9,9-dioctyl)fluorene (PFO) 8, 52, 192, 295
Poly(9,9-dioctyl-2,7-fluorene) (PF8) 91, 231
Poly(2,7-dioctyl-4,5,9,10-tetrahydropyrene) 27
Poly(diacylphenylene) 41
Poly(dialkylfluorene)s 6, 18
Poly(dibenzosilole)s 85
Poly(dibenzoylphenylene)s 41
Poly(2,7-fluorene-co-2,7-dibenzosilole) 88
Poly(ladder-type pentaphenylene)s 33
Poly(ladder-type tetraphenylene)s 32

Poly(2,7-phenanthrylene)s 28
Poly(p-phenylene) (PPP) 2
Poly(phenylene vinylene) (PPV) 301
Poly(tetraalkylindenofluorene)s 29
Poly[tetra(4-alkylphenyl)indenofluorene]s 32
Poly(tetraarylindenofluorene)s 31
Poly(tetrahydropyrene)s 27
Poly(vinyl oxadiazole) 21
Poly(vinyl triphenylamine) 21
Polydiacetylene (PDA) quantum wires 305
Polydialkylfluorenes (PDAFs) 3, 6
Polyfluorene beta-phase 187
Polyfluorene chains, macroscopic alignment 251
Polyfluorene devices, performance improvement 51
Polyfluorene-type polymers 273
Polyindenofluorene, double spirolinkages 32
Polyindenofluorenes (PIFs) 3, 29
–, emissive defect 30
Polyionene 7
Polyketal 23
Polymer light-emitting diodes (PLEDs) 49, 51
Polyparaphenylene derivatives, bridged 289
Polyphenylenes, ladder-type (LPPPs) 3, 34

Red emitters 65, 162
Red-light emitting Ir(III) complexes 133
Rhenium(I) complexes 140
Ruthenium(II) complexes 140

Self-organization 227
Shape controls function 300
Silicon 86
Silsesquioxanes 17
Single chain β phase 300
Single molecule spectroscopy 298
Single molecules 230, 293
Single-polymer approach 73
Singlet excited state absorption 198
Singlet migration 204
Singlet–singlet annihilation 210
Singlet–triplet annihilation 216
Sky-blue emitter 162

Solar cells, organic   86, 174
Solid state   240
Solid-state organic lasers   171
Solubility   157
Solutions   233
Spirobifluorene   25
Spiropolyfluorenes (SPFs)   54
Strain-induced depolarisation   306
Structural order   233
Supramolecules   227
Surface morphology   259
Suzuki coupling   151, 277

Ter(9,9-diaryl)fluorenes   162
Terfluorenes   162
–, spiro-configured   158
Thermal stability   157
Thiophene-S,S-dioxide   154

Time correlated single photon counting (TCSPC)   192
Transient absorption   197
Triarylamines, hole-transporting   22
Triplet diffusion   205
Triplet excited state absorption   198
Triplet–triplet annihilation, delayed fluorescence   213
Truxene core   154

White emitters   72
White emitting fluorene copolymer   14
White polymer light-emitting diodes (WPLEDs)   129

Yamamoto coupling   151, 277

Zero-phonon line (ZPL)   306

Printing: Krips bv, Meppel, The Netherlands
Binding: Stürtz, Würzburg, Germany

RETURN TO: CHEMISTRY LIBRARY
100 Hildebrand Hall • 510-642-3753

LOAN PERIOD	1	2	3
4		5	6

2 HOUR

ALL BOOKS MAY BE RECALLED AFTER 7 DAYS.

Renewals may be requested by phone or, using GLADIS, type **inv** followed by your patron ID number.

DUE AS STAMPED BELOW.

2/09/09

FORM NO. DD 10
3M 7-08

UNIVERSITY OF CALIFORNIA, BERKELEY
Berkeley, California 94720–6000